ORGANIC SYNTHESES

ORGANIC SYNTHESES

AN ANNUAL PUBLICATION OF SATISFACTORY METHODS FOR THE PREPARATION OF ORGANIC CHEMICALS

VOLUME 74

1997

JOHN WILEY & SONS, INC.

NEW YORK • CHICHESTER • WEINHEIM • BRISBANE • SINGAPORE • TORONTO

The procedures in this text are intended for use only by persons with prior training in the field of organic chemistry. In the checking and editing of these procedures, every effort has been made to identify potentially hazardous steps and to eliminate as much as possible the handling of potentially dangerous materials; safety precautions have been inserted where appropriate. If performed with the materials and equipment specified, in careful accordance with the instructions and methods in this text, the Editors believe the procedures to be very useful tools. However, these procedures must be conducted at one's own risk. Organic Syntheses, Inc., its Editors, who act as checkers, and its Board of Directors do not warrant or guarantee the safety of individuals using these procedures and hereby disclaim any liability for any injuries or damages claimed to have resulted from or related in any way to the procedures herein.

This text is printed on acid-free paper.

Published by John Wiley & Sons, Inc.

"John Wiley & Sons, Inc. is pleased to publish this volume of Organic Syntheses on behalf of Organic Syntheses, Inc. Although Organic Syntheses, Inc. has assured us that each preparation contained in this volume has been checked in an independent laboratory and that any hazards that were uncovered are clearly set forth in the write-up of each preparation. John Wiley & Sons, Inc. does not warrant the preparations against any safety hazards and assumes no liability with respect to the use of the preparations."

Library of Congress Catalog Card Number: 21-17747
ISBN 0-471-15656-6

Printed in the United States of America

10 9 8 7 6 5 4 3 2 1

ORGANIC SYNTHESES

VOLUME	EDITOR-IN-CHIEF	PAGES
I*	†ROGER ADAMS	84
II*	†JAMES BRYANT CONANT	100
III*	†HANS THACHER CLARKE	105
IV*	†OLIVER KAMM	89
V*	†CARL SHIPP MARVEL	110
VI*	†HENRY GILMAN	120
VII*	†FRANK C. WHITMORE	105
VIII*	†ROGER ADAMS	139
IX*	†JAMES BRYANT CONANT	108
Collective Vol. 1	A revised edition of Annual Volumes I–IX †HENRY GILMAN, *Editor-in-Chief* 2nd Edition revised by †A. H. BLATT	580
X*	†HANS THACHER CLARKE	119
XI*	†CARL SHIPP MARVEL	106
XII*	†FRANK C. WHITMORE	96
XIII*	†WALLACE H. CAROTHERS	119
XIV*	†WILLIAM W. HARTMAN	100
XV*	†CARL R. NOLLER	104
XVI*	†JOHN R. JOHNSON	104
XVII*	†L. F. FIESER	112
XVIII*	†REYNOLD C. FUSON	103
XIX*	†JOHN R. JOHNSON	105
Collective Vol. II	A revised edition of Volumes X–XIX †A. H. BLATT, *Editor-in-Chief*	654
20*	†CHARLES F. H. ALLEN	113
21*	†NATHAN L. DRAKE	120
22*	†LEE IRVIN SMITH	114
23*	†LEE IRVIN SMITH	124
24*	†NATHAN L. DRAKE	119
25*	†WERNER E. BACHMANN	120
26*	†HOMER ADKINS	124
27*	†R. L. SHRINER	121
28*	†H. R. SNYDER	121
29*	†CLIFF S. HAMILTON	119
Collective Vol. III	A revised edition of Annual Volumes 20–29 †E. C. HORNING, *Editor-in-Chief*	890
30*	†ARTHUR C. COPE	115
31*	†R. S. SCHREIBER	122

**Out of print.*
†*Deceased.*

VOLUME	EDITOR-IN-CHIEF	PAGES
32*	RICHARD T. ARNOLD	119
33*	CHARLES C. PRICE	115
34*	†WILLIAM S. JOHNSON	121
35*	†T. L. CAIRNS	122
36*	N. J. LEONARD	120
37*	JAMES CASON	109
38*	†JOHN C. SHEEHAN	120
39*	†MAX TISHLER	114
Collective Vol. IV	A revised edition of Annual Volumes 30–39 NORMAN RABJOHN, *Editor-in-Chief*	1036
40*	†MELVIN S. NEWMAN	114
41*	JOHN D. ROBERTS	118
42*	VIRGIL BOEKELHEIDE	118
43*	B. C. MCKUSICK	124
44*	†WILLIAM E. PARHAM	131
45*	WILLIAM G. DAUBEN	118
46*	E. J. COREY	146
47*	WILLIAM D. EMMONS	140
48*	†PETER YATES	164
49*	KENNETH B. WIBERG	124
Collective Vol. V	A revised edition of Annual Volumes 40–49 HENRY E. BAUMGARTEN, *Editor-in-Chief*	1234
Cumulative Indices to Collective Volumes, I, II, III, IV, V	†RALPH L. AND †RACHEL H. SHRINER, *Editors*	
50*	RONALD BRESLOW	136
51*	RICHARD E. BENSON	209
52*	HERBERT O. HOUSE	192
53*	ARNOLD BROSSI	193
54*	ROBERT E. IRELAND	155
55*	SATORU MASAMUNE	150
56*	GEORGE H. BÜCHI	144
57*	CARL R. JOHNSON	135
58*	†WILLIAM A. SHEPPARD	216
59*	ROBERT M. COATES	267
Collective Vol. VI	A revised edition of Annual Volumes 50–59 WAYLAND E. NOLAND, *Editor-in-Chief*	1208
60*	ORVILLE L. CHAPMAN	140
61*	†ROBERT V. STEVENS	165

**Out of print.*
†Deceased.

VOLUME	EDITOR-IN-CHIEF	PAGES
62	MARTIN F. SEMMELHACK	269
63	GABRIEL SAUCY	291
64	ANDREW S. KENDE	308
Collective Vol. VII	A revised edition of Annual Volumes 60–64 JEREMIAH P. FREEMAN, *Editor-in-Chief*	602
65	EDWIN VEDEJS	278
66	CLAYTON H. HEATHCOCK	265
67	BRUCE E. SMART	289
68	JAMES D. WHITE	318
69	LEO A. PAQUETTE	328
Reaction Guide to Collective Volumes I–VII and Annual Volumes 65–68	DENNIS C. LIOTTA AND MARK VOLMER, *Editors*	854
Collective Vol. VIII	A revised edition of Annual Volumes 65–69 JEREMIAH P. FREEMAN, *Editor-in-Chief*	696
70	ALBERT I. MEYERS	305
71	LARRY E. OVERMAN	285
72	DAVID L. COFFEN	333
73	ROBERT K. BOECKMAN, JR.	352
74	ICHIRO SHINKAI	341

Collective Volumes, Collective Indices to Collective Volumes I–VIII, Annual Volumes 70–74, and Reaction Guide are available from John Wiley & Sons, Inc.

**Out of print.*
†Deceased.

NOTICE

With Volume 62, the Editors of *Organic Syntheses* began a new presentation and distribution policy to shorten the time between submission and appearance of an accepted procedure. The soft cover edition of this volume is produced by a rapid and inexpensive process, and is sent at no charge to members of the Organic Divisions of the American and French Chemical Society, The Perkin Division of the Royal Society of Chemistry, and The Society of Synthetic Organic Chemistry, Japan. The soft cover edition is intended as the personal copy of the owner and is not for library use. A hard cover edition is published by John Wiley & Sons Inc. in the traditional format, and differs in content primarily in the inclusion of an index. The hard cover edition is intended primarily for library collections and is available for purchase through the publisher. Annual Volumes 65–69 have been incorporated into a new five-year version of the collective volumes of *Organic Syntheses* which have appeared as *Collective Volume Eight* in the traditional hard cover format. It is available for purchase from the publishers. The Editors hope that the new *Collective Volume* series, appearing twice as frequently as the previous decennial volumes, will provide a permanent and timely edition of the procedures for personal and institutional libraries. The Editors welcome comments and suggestions from users concerning the new editions.

NOMENCLATURE

Both common and systematic names of compounds are used throughout this volume, depending on which the Editor-in-Chief felt was more appropriate. The *Chemical Abstracts* indexing name for each title compound, if it differs from the title name, is given as a subtitle. Systematic *Chemical Abstracts* nomenclature, used in both the 9th and 10th Collective Indexes for the title compound and a selection of other compounds mentioned in the procedure, is provided in an appendix at the end of each preparation. Registry numbers, which are useful in computer searching and identification, are also provided in these appendixes. Whenever two names are concurrently in use and one name is the correct *Chemical Abstracts* name, that name is preferred.

SUBMISSION OF PREPARATIONS

Organic Syntheses welcomes and encourages submission of experimental procedures which lead to compounds of wide interest or which illustrate important new developments in methodology. The Editoral Board will consider proposals in outline format as shown below, and will request full experimental details for those proposals which are of sufficient interest. Submissions which are longer than three steps from commercial sources or from existing *Organic Syntheses* procedures will be accepted only in unusual circumstances.

Organic Syntheses Proposal Format

1) Authors
2) Title
3) Literature reference or enclose preprint if available
4) Proposed sequence
5) Best current alternative(s)
6) a. Proposed scale, final product:
   b. Overall yield:
   c. Method of isolation and purification:
   d. Purity of product (%):
   e. How determined?
7) Any unusual apparatus or experimental technique?

8) Any hazards?
9) Source of starting material?
10) Utility of method or usefulness of product

Submit to: Dr. Jeremiah P. Freeman, Secretary
Department of Chemistry
University of Notre Dame
Notre Dame, IN 46556

Proposals will be evaluated in outline form, again after submission of full experimental details and discussion, and, finally by checking experimental procedures. A form that details the preparation of a complete procedure (Notice to Submitters) may be obtained from the Secretary.

Additions, corrections, and improvements to the preparations previously published are welcomed; these should be directed to the Secretary. However, checking of such improvements will only be undertaken when new methodology is involved. Substantially improved procedures have been included in the Collective Volumes in place of a previously published procedure.

ACKNOWLEDGMENT

Organic Syntheses wishes to acknowledge the contributions of ArQule, Hoffmann-La Roche, Inc. and Merck & Co. to the success of this enterprise through their support, in the form of time and expenses, of members of the Boards of Directors and Editors.

HANDLING HAZARDOUS CHEMICALS
A Brief Introduction

General Reference: *Prudent Practices in the Laboratory,* National Academy Press, Washington, D.C. 1996.

Physical Hazards

Fire. Avoid open flames by use of electric heaters. Limit the quantity of flammable liquids stored in the laboratory. Motors should be of the nonsparking induction type.

Explosion. Use shielding when working with explosive classes such as acetylides, azides, ozonides, and peroxides. Peroxidizable substances such as ethers and alkenes, when stored for a long time, should be tested for peroxides before use. Only sparkless "flammable storage" refrigerators should be used in laboratories.

Electric Shock. Use 3-prong grounded electrical equipment if possible.

Chemical Hazards

Because all chemicals are toxic under some conditions, and relatively few have been thoroughly tested, it is good strategy to minimize exposure to all chemicals. In practice this means having a good, properly installed hood; checking its performance periodically; using it properly; carrying out most operations in the hood; protecting the eyes; and, since many chemicals can penetrate the skin, avoiding skin contact by use of gloves and other protective clothing.

a. Acute Effects. These effects occur soon after exposure. The effects include burn, inflammation, allergic responses, damage to the eyes, lungs, or nervous system (e.g., dizziness), and unconsciousness or death (as from overexposure to HCN). The effect and its cause are usually obvious and so are the methods to prevent it. They generally arise from inhalation or skin contact, so should not be a problem if one follows the admonition "work in a hood and keep chemicals off your hands." Ingestion is a rare route, being generally the result of eating in the laboratory or not washing hands before eating.

b. Chronic Effects. These effects occur after a long period of exposure or after a long latency period and may show up in any of numerous organs. Of the chronic effects of chemicals, cancer has received the most attention lately. Several dozen chemicals have been demonstrated to be carcinogenic in man and hundreds to be carcinogenic to animals. Although there is no simple correlation between carcinogenicity in animals and man, there is little doubt that a significant proportion of the chemicals used in laboratories have some potential for carcinogenicity in man. For this and other reasons, chemists should employ good practices.

The key to safe handling of chemicals is a good, properly installed hood, and the referenced book devotes many pages to hoods and ventilation. It recommends that in a laboratory where people spend much of their time working with chemicals there should be a hood for each two people, and each should have at least 2.5 linear feet (0.75 meter) of working space at it. Hoods are more than just devices to keep undesirable vapors from the laboratory atmosphere. When closed they provide a protective barrier between chemists and chemical operations, and they are a good containment device for spills. Portable shields can be a useful supplement to hoods, or can be an alternative for hazards of limited severity, e.g., for small-scale operations with oxidizing or explosive chemicals.

Specialized equipment can minimize exposure to the hazards of laboratory operations. Impact resistant safety glasses are basic equipment and should be worn at all times. They may be supplemented by face shields or goggles for particular operations, such as pouring corrosive liquids. Because skin contact with chemicals can lead to skin irritation or sensitization or, through absorption, to effects on internal organs, protective gloves are often needed.

Laboratories should have fire extinguishers and safety showers. Respirators should be available for emergencies. Emergency equipment should be kept in a central location and must be inspected periodically.

DISPOSAL OF CHEMICAL WASTE

General Reference: *Prudent Practices in the Laboratory*, National Academy Press, Washington, D.C. 1996

Effluents from synthetic organic chemistry fall into the following categories:

1. **Gases**

   1a. Gaseous materials either used or generated in an organic reaction.
   1b. Solvent vapors generated in reactions swept with an inert gas and during solvent stripping operations.
   1c. Vapors from volatile reagents, intermediates and products.

2. **Liquids**

   2a. Waste solvents and solvent solutions of organic solids (see item 3b).
   2b. Aqueous layers from reaction work-up containing volatile organic solvents.
   2c. Aqueous waste containing non-volatile organic materials.
   2d. Aqueous waste containing inorganic materials.

3. **Solids**

   3a. Metal salts and other inorganic materials.
   3b. Organic residues (tars) and other unwanted organic materials.
   3c. Used silica gel, charcoal, filter acids, spent catalysts and the like.

The operation of industrial scale synthetic organic chemistry in an environmentally acceptable manner* requires that all these effluent categories be dealt with properly. In small scale operations in a research or academic setting,

*An environmentally acceptable manner may be defined as being both in compliance with all relevant state and federal environmental regulations *and* in accord with the common sense and good judgment of an environmentally aware professional.

provision should be made for dealing with the more environmentally offensive categories.

1a. Gaseous materials that are toxic or noxious, e.g., halogens, hydrogen halides, hydrogen sulfide, ammonia, hydrogen cyanide, phosphine, nitrogen oxides, metal carbonyls, and the like.
1c. Vapors from noxious volatile organic compounds, e.g., mercaptans, sulfides, volatile amines, acrolein, acrylates, and the like.
2a. All waste solvents and solvent solutions of organic waste.
2c. Aqueous waste containing dissolved organic material known to be toxic.
2d. Aqueous waste containing dissolved inorganic material known to be toxic, particularly compounds of metals such as arsenic, beryllium, chromium, lead, manganese, mercury, nickel, and selenium.
3. All types of solid chemical waste.

Statutory procedures for waste and effluent management take precedence over any other methods. However, for operations in which compliance with statutory regulations is exempt or inapplicable because of scale or other circumstances, the following suggestions may be helpful.

Gases

Noxious gases and vapors from volatile compounds are best dealt with at the point of generation by "scrubbing" the effluent gas. The gas being swept from a reaction set-up is led through tubing to a (large!) trap to prevent suckback and on into a sintered glass gas dispersion tube immersed in the scrubbing fluid. A bleach container can be conveniently used as a vessel for the scrubbing fluid. The nature of the effluent determines which of four common fluids should be used: dilute sulfuric acid, dilute alkali or sodium carbonate solution, laundry bleach when an oxidizing scrubber is needed, and sodium thiosulfate solution or diluted alkaline sodium borohydride when a reducing scrubber is needed. Ice should be added if an exotherm is anticipated.

Larger scale operations may require the use of a pH meter or starch/iodide test paper to ensure that the scrubbing capacity is not being exceeded.

When the operation is complete, the contents of the scrubber can be poured down the laboratory sink with a large excess (10–100 volumes) of water. If the solution is a large volume of dilute acid or base, it should be neutralized before being poured down the sink.

Liquids

Every laboratory should be equipped with a waste solvent container in which *all* waste organic solvents and solutions are collected. The contents of these containers should be periodically transferred to properly labeled waste solvent drums and arrangements made for contracted disposal in a regulated and licensed incineration facility.**

Aqueous waste containing dissolved toxic organic material should be decomposed *in situ*, when feasible, by adding acid, base, oxidant, or reductant. Otherwise, the material should be concentrated to a minimum volume and added to the contents of a waste solvent drum.

Aqueous waste containing dissolved toxic inorganic material should be evaporated to dryness and the residue handled as a solid chemical waste.

Solids

Soluble organic solid waste can usually be transferred into a waste solvent drum, provided near-term incineration of the contents is assured.

Inorganic solid wastes, particularly those containing toxic metals and toxic metal compounds, used Raney nickel, manganese dioxide, etc. should be placed in glass bottles or lined fiber drums, sealed, properly labeled, and arrangements made for disposal in a secure landfill.** Used mercury is particularly pernicious and small amounts should first be amalgamated with zinc or combined with excess sulfur to solidify the material.

Other types of solid laboratory waste including used silica gel and charcoal should also be packed, labeled, and sent for disposal in a secure landfill.

Special Note

Since local ordinances may vary widely from one locale to another, one should always check with appropriate authorities. Also, professional disposal services differ in their requirements for segregating and packaging waste.

**If arrangements for incineration of waste solvent and disposal of solid chemical waste by licensed contract disposal services are not in place, a list of providers of such services should be available from a state or local office of environmental protection.

PREFACE

Most organic chemists agree that organic synthesis is not a mature art. We have come a long way, but we have to develop more reliable methods and procedures with improvement in synthetic efficiency. New procedures are needed for safe and economical large scale operation. Despite its great successes, organic synthesis still remains heavily dependent on empirical methods.

While we strive to attain these noble goals, *Organic Syntheses* continues to try to identify important current methods and preparations. Annual Volume 74 contains a series of twenty-eight checked and edited procedures that describe in detail the preparation of generally useful synthetic reagents, intermediates, and heterocycles. It also includes new synthetic methodology which is at the leading edge of organic chemistry.

This collection begins with a series of five procedures illustrating new methods for preparation of important chiral intermediates in enantiomerically pure form. Glyceraldehyde derivatives with one protected and one free hydroxyl function could offer new options. Both enantiomers of **2-O-BENZYLGLYCERALDEHYDE** can be prepared in three steps from commercially available diethyl D- and L-tartrate. The one-pot conversion of menthol to **1-MENTHOXY-1-BUTYNE** followed by the efficient stereoselective reduction to the corresponding Z- and E-enol ethers is described in the preparation of **(Z)- AND (E)-1-MENTHOXY-1-BUTENE.** Preparation of **(R)-(+)-2-(DIPHENYLHYDROXYMETHYL)PYRROLIDINE,** an important intermediate in the synthesis of chiral ligands, is illustrated. The chemistry reported here is based on the excellent utilization of an asymmetric lithiation/substitution sequence. The putative enantioenriched 2-lithiated pyrrolidine is obtained by the lithiation of Boc-pyrrolidine in the presence of (−)-sparteine. The B-methyloxazaborolidine-catalyzed asymmetric reduction of a cyclopentanone derivative is described in the procedure for **(R)-(−)-2,2-DIPHENYLCYCLOPENTANOL.** An alternative large scale preparation of (S)-1,1-diphenylprolinol via conventional (S)-proline-N-carboxyanhydride and the oxazaborole-borane complex is described in the procedure for the **(S)-TETRAHYDRO-1-METHYL-3,3-DIPHENYL-1H,3H-PYRROLO[1,2c]-[1,3,2]-OXAZABOROLE-BORANE COMPLEX.** This procedure provides a reliable process for oxazaborolidine preparation based on the reaction of trimethylboroxine with diphenylprolinol derivatives. The oxazaborolidine-borane complex is used for the enantioselective reduction of prochiral 1-indanone in high selectivity (ee = 97.8%).

The next six procedures illustrate the preparation of useful building blocks and intermediates. Large-scale preparation of **1,2,3-TRIPHENYLCYCLOPROPENIUM BROMIDE** by the addition of the phenylchloro carbenoid intermediate to acetylenes is illustrated. A sizeable amount of diphenylacetylene can be converted quantitatively to cyclopropenium bromides in a few hours. The reaction has wide generality and can be applied to other substituted acetylenes. Vinyl triflates are important intermediates in organic syntheses. New triflating reagents, **N-(PYRIDYL)TRIFLIMIDE AND N-(5-CHLORO-2-PYRIDYL)TRIFLIMIDE,** are presented. The preparation of **BIS(TRIMETHYLSILYL)PEROXIDE (BTMSPO)** provides a highly reliable and safe procedure for this widely used reagent. BTMSPO has been used in the preparation of stereochemically pure E- and Z-silyl enol ethers. BTMSPO can also be used as a mild aprotic oxidizing agent for the preparation of sulfoxides, phosphine oxides, and aldehydes. Another useful oxidation agent, **DIMETHYLDIOXIRANE,** and the preparation of **trans-STILBENE OXIDE** are presented. Dimethyldioxirane (DMD) is used widely for epoxidation. The reaction is usually carried out at room temperature in neutral solution and proceeds stereospecifically in essentially quantitative yields. The reaction is applicable to a variety of unsaturated systems. The next procedure describes the convenient preparation of the strong, non-nucleophilic base, **2-tert-BUTYL-1,1,3,3-TETRAMETHYLGUANIDINE.** This strong base provides an inexpensive alternative to the commonly used amidine bases such as DBN or DBU. The synthetic utility of this base is demonstrated in the preparation of **2,2,6-TRIMETHYLCYCLOHEXEN-1-YL IODIDE.** The preparation of **DIETHYL (DICHLOROMETHYL)PHOSPHONATE** is presented. The use of isopropylmagnesium chloride instead of butyllithium reduces the amount of by-products and simplifies the purification steps. The preparation of **(4-METHOXYPHENYL)ETHYNE** via generation and trapping of an unstable phosphorylated carbanion is illustrated. The methodology is well suited for the synthesis of a wide variety of terminal acetylenic compounds.

The next three procedures provide useful building blocks for general syntheses. **2,3-DIBROMO-1-(PHENYLSULFONYL)-1-PROPENE** (DBP) is utilized for the preparation of furans and cyclopentenones. The dibromopropene derivative can be obtained easily by the addition of bromine to a 1,2-propadiene. The resulting dibromide is a stable crystalline solid which may be viewed as a multielectrophilic reagent with a great potential as a nucleophilic acceptor for sequential additions. The reaction of 1,3-dicarbonyl compounds with DBP affords **2-METHYL-4-[(PHENYLSULFONYL)-METHYL]FURAN.** In contrast, anions derived from 1,3-dicarbonyl compounds substituted at the C-2 position are found to induce a complete reversal in the mode of ring closure to give **2-METHYL-3-[(PHENYLSULFONYL)METHYL]-2-CYCLOPENTEN-1-ONE.** The pendant sulfone group offers a convenient and versatile site for further elaborations. **PHENYL VINYL SULFIDE** shows a number of synthetically useful attributes as an electron-rich alkene in numerous cycloaddition reactions. The product obtained in high yield using common reagents and mild conditions via this procedure is stable at room temperature under a nitrogen atmosphere for months. **NITROACETALDEHYDE DIETHYL ACETAL** has been used to obtain various other acetals by transacetalization. Aliphatic nitro compounds are highly versatile building blocks in organic syntheses. For example, the nitroaldol addition leads to the formation of 1,2-nitro alcohols which are easily transformed into 1,2-aminoalcohols.

The next four procedures describe the regioselective preparation of bicyclic ring systems, specifically, condensed five-membered carbocyclic derivatives. A large number of methods have been developed for the construction of five membered carbocyclic systems. The method for the preparation of **(3aβ,9bβ)-1,2,3a,4,5,9b-HEXAHYDRO-9b-HYDROXY-3a-METHYL-3H-BENZ[e]INDEN-3-ONE** is based on the formal concept of employing cyclopropanone in a mixed aldol condensation with an enolate of another ketone. The aldolate thus obtained exhibits characteristics of a "homoenolate" and undergoes a subsequent annulation reaction. The chemistry of phenylsulfonyl-substituted butadienes is receiving increased attention due to their versatility in organic synthesis. Treatment of 2-butyne-1,4-diol with benzenesulfenyl chloride affords 2,3-bis(phenylsulfinyl)-1,3-butadiene as a result of a series of 2,3-sigmatropic rearrangements. The preparation of **2,3-BIS-(PHENYLSULFONYL)-1,3-BUTADIENE** by oxidation of the disulfoxide is demonstrated, and its [3+2] anionic cyclization to produce **trans-4,7,7-TRICARBOMETHOXY-2-PHENYLSULFONYLBICYCLO[3.3.0]-OCT-1-ENE** is detailed. This strategy can clearly be applied to more complex targets. The generation of an α,β-unsaturated ketene intermediate which undergoes intramolecular [2+2] cyclization to give **1,4-DIMETHYLBICYCLO-[3.2.0]HEPT-3-EN-6-ONE** is illustrated. This synthetic approach demonstrates the simplicity of the procedure and the selectivity by which the thermodynamically more stable isomer can be prepared in high purity and good yield. The last example in this class is the excellent regioselective preparation of **4,6a-DIMETHYL-4,5,6,6a-TETRAHYDRO-3a-HYDROXY-2,3-DIISOPROPOXY-4,6a-1(3aH)-PENTALENONE.** The principal pathway involves trans-1,2-addition of 2-lithiopropene to squarate esters to generate a cyclobutene dialkoxide which undergoes rapid ring opening to a doubly-charged 1,3,5,7-octatetraene. Following electrocyclization of the octatetraene intermediate to form a cyclooctenyl dienolate, the stage is set for intramolecular aldolization via transannular cyclization.

The next four procedures all involve the preparation of useful olefin derivatives. β,γ-Unsaturated carboxylic acids such as **(E)-4,8-DIMETHYL-3,7-NONADIENOIC ACID** are important intermediates for many natural products syntheses. Allylic barium reagents are prepared via reaction of in situ-generated **ACTIVE BARIUM** with various allylic chlorides, and react with excess carbon dioxide resulting in exclusive α-carboxylation, whereas γ-carboxylation occurs with the magnesium reagent. Allylic barium reagents show high regioselectivity for the α-position, and the double bond geometry of the allyl chloride precursor is completely retained. Geminal dimethylation at a carbon center is a useful method in organic synthesis. Although the Tebbe-like protocol is effective for converting a carbonyl group to a gem-dimethyl group, its application to an allylic carbonyl substrate is limited by poor regioselectivity. The next procedure describes a method based on the nickel-catalyzed cross coupling reactions of dithio acetals with Grignard reagents in the preparation of **(E)-1-PHENYL-3,3-DIMETHYL-1-BUTENE.** The conversion of ethyl propiolate to **(E)- or (Z)-1-IODOHEPT-1-EN-3-OL** and the transformation of a **(Z)-β-IODOACRYLATE** to **(Z)-β-IODOACROLEIN** is illustrated. Regio- and stereoselective conversion of ethyl propiolate with sodium iodide in acetic acid provides the Z-iodoacrylate. The thermal isomerization of the anion formed by Grignard addition provides either (Z)- or (E)-alkylated product. Organometallic derivatives of zirconium (IV) are readily obtained by hydrozirconation of alkenes and alkynes. Transmetalation of alkenylzirconocenes to the corresponding organozinc compounds occurs rapidly at low temperature. Subsequent addition of aldehydes provides an in situ protocol for the conversion of alkynes into allylic alcohols in good yields. Such an example is depicted in the preparation of **1-[(tert-BUTYLDIPHENYLSILYL)OXY]DEC-3-EN-5-OL.** Bromination of 1-alkynes with NBS in the presence of catalytic amounts of silver nitrate is successfully applied for the preparation of **METHYL and tert-BUTYL 3-BROMOPROPIOLATES.** The preparation of **MESITYLENESULFONYLHYDRAZINE** and its application for the preparation of racemic **2,6-DIMETHYLCYCLOHEXANECARBONITRILE** are illustrated. This procedure provides a simple one-pot process for conversion of a moderately hindered ketone to the next higher nitrile analog. Ketones with α,α'-alkyl substituents may be used as diastereomeric mixtures, since they equilibrate to one pair of enantiomers during hydrazone formation, and this stereochemistry is preserved during the cyanide ion reaction.

Volume 74 concludes with four procedures for heterocyclic compounds. Oxazoles bearing a 4-carboxy-derived group are of considerable importance in natural products. A more direct approach to oxazoles than that afforded by the Cornforth synthesis is desirable. Preparation of **4-CARBOMETHOXY-5-METHYL-2-PHENYL-1,3-OXAZOLE** based on rhodium-catalyzed hetero-cycloaddition is illustrated. Several examples of oxazole formation utilizing dimethyl diazomalonate and diazo formylacetate are included. The powerful method for the introduction of a masked hydrazine into an electron rich aromatic ring system is beautifully utilized for the preparation of **5,7-DIMETHOXY-3-METHYLINDAZOLE** in a two step procedure. This direct amination eliminates the traditional four step protocol. Lithiation at C-3 in indole has received little attention due to rearrangement to the more thermally stable 2-isomer at elevated temperature. The replacement of the traditional benzenesulfonyl nitrogen-protecting group with a trialkylsilyl group allows the preparation of 3-lithioindoles. Subsequent reaction with an electrophile is exemplified by the preparation of **3-ETHYLINDOLE** after the removal of the silyl group. The 3-lithioindole can be converted to 3-indolylzinc chloride which has been successfully employed in the heteroarylation of the indole 3-position by a palladium(0)-catalyzed cross-coupling reaction. The last procedure presents a very reliable preparation of 3-morpholinothioacrylic acid amides with a wide variety of 2-aryl substituents which can be utilized in the preparation of **5-(4-BROMOBENZOYL-2-(4-MORPHOLINO)-3-PHENYLTHIOPHENE.**

One of the reasons for the continuing success of Organic Syntheses is the dedicated teamwork of those organic chemists who act as Editorial Board members and their coworkers acting like a family. The scrupulous selection and follow up of these procedures is the key role played by Professor Jeremiah Freeman, Secretary to the Board. I would like to acknowledge Jerry's efforts and also those of Assistant Editor, Dr. Theordora W. Greene, whose invaluable editorial work made my task much easier. I am also indebted to my colleagues on the Editorial Board for their assistance and Merck Process Research chemists for the meticulous checking and numerous discussions of the procedures. I strongly believe these twenty-eight procedures cited in this volume will guide and provide a deeper prospect toward the "Act of Organic Synthesis" among the organic chemists at industrial and/or academic research institutes throughout the entire world.

ICHIRO SHINKAI

DBM, Parsippany, New Jersey
May 1996

WILLIAM S. JOHNSON
February 24, 1913–August 19, 1995

William S. Johnson was a highly respected leader among research chemists and educators while, at the same time, he was humble about his accomplishments. His career spanned an explosive period of rapid progress in science and he was at the cutting edge of many of the basic changes that have taken place. His deep respect and love for science led to a career that was characterized by creative, insightful and thorough research and resulted in a body of work on steroid synthesis that is unparalleled for its thorough comprehensive coverage.

Bill Johnson did his undergraduate studies at Amherst College, an institution that has spawned a number of chemical leaders. After finishing his doctoral work with Professor Louis Fieser at Harvard University in late 1939 and a brief postdoctoral stint at Harvard with Professor R. P. Linstead, Johnson began his independent academic career at the University of Wisconsin in 1940. The research program that Johnson initiated was directed at the development of methodology for the synthesis of steroids. While the approaches would change over the years, this theme would become the dominant direction for Bill Johnson's research effort over his entire academic career. The "Wisconsin era'' was devoted to a classical approach to the total synthesis of the steroid skeleton and resulted in the development of the benzylidene blocking group for the angular methylation of α-decalone type molecules, the use of the Stobbe reaction for the synthesis of the aromatic steroids equilenin and estrone, and the ''hydrochrysene approach'' to the total synthesis of nonaromatic steroids. These research efforts led to a remarkable collection of papers that described highly efficient classical syntheses of steroid hormones of increasing molecular complexity from equilenin to cortisone and aldosterone.

Bill Johnson rose through the academic ranks and became a full professor in 1946; in 1950 he was named Homer Adkins Professor of Chemistry. He was elected to the Board of Editors of *Organic Syntheses* in 1948 and edited the 34th volume of this important series. His election to the National Academy of Sciences in 1954 and receipt of the ACS Award for Creative Work in Synthetic Organic Chemistry in 1958 were testimony to the importance and creativity of his earlier synthetic achievements and were just the beginning of a long string of awards to come. In 1960 Johnson moved his research program to Stanford University and in addition to his scientific efforts took on the challenge of Executive Head of the Department of Chemistry. For the following nine years, he not only directed a vigorous research effort but also planned and developed the expansion of the Department that elevated it to one of the top-ranked schools in the world. This expansion added 13 new faculty members and included such stars as Carl Djerassi, Eugene van Tamelen, John Brauman, Paul Flory, Henry Taube and Harden McConnell.

His stint as Executive Head of the department ended when he resigned the post in 1969 and was appointed Jackson-Wood Professor of Chemistry. He held this endowed chair until 1978, when he became emeritus. As an emeritus professor, Johnson continued his research program in collaboration with postdoctoral students until the very end of his life and was particularly pleased and gratified that his research program continually earned the support of peer-reviewed funding agencies.

The Stanford era of Johnson' research program is characterized by his effort to mimic enzymatic syntheses of steroid molecules. Stimulated by the Stork-Eschenmoser proposals for the biosynthesis of sterols through the cationic olefin cyclization of squalene to form the polynuclear skeleton, Johnson first realized some success in the chemical cyclization of diene systems that formed the decalin ring system. By application of the characteristic thorough Johnson approach to research, he and his group of co-workers expanded these early beginnings with the development of methods for the synthesis and cyclization of polyene systems that were readily transformed into neutral steroid molecules. This effort not only led to the development of several different methods for the initiation of the polyene cyclization but also excellent methodology for the synthesis of the precursor polyene molecules themselves. Through the extensive development of acetylenic chemistry and the development and application of the Claisen rearrangement, Johnson's group was able to prepare large quantities of complex polyenes for cyclization experiments. Much of this methodology, of course, transcends the narrow purpose of polyene synthesis for which it was designed and is now in the general repertoire of synthetic organic chemistry and widely used for other objectives. Ultimately, Johnson was able to refine the polyene approach to steroid synthesis to the point that the process was nearly of commercial use. This beautiful body of work, like that of the more classical Wisconsin era, forms a lasting testament to the genius and insight of Bill Johnson. His work was marked by a passion for a thorough understanding of the science involved and the accuracy with which the experimental results were obtained and reported were hallmarks of the Johnson School of Synthesis. Many awards in recognition of this work were forthcoming. He received the Nichols Medal Award (1968), the Roussel Prize (1970), the Roger Adams Award (1977); the National Medal of Science (1987), the Arthur C. Cope Award (1989) and the Tetrahedron Prize (1991), several honorary degrees (Amherst College (1956), Long Island University (1968)), membership in the American Academy of Arts and Sciences (1963) and numerous endowed lectureships at universities throughout the world.

During a research career that spanned 53 years, Bill Johnson was as proud of his co-workers as their research. He enjoyed the job of teaching and watching young people light up and become turned on by science. He was the mentor of over 100 predoctoral and 200 postdoctoral students and he tried to follow all their careers as they developed in science. Many of his students made him very proud as their careers in academe and industry developed and he enjoyed his paternal part of their successes. In his teaching of research, he conveyed the respect that science requires and scrupulous attention to the accuracy of the observations made. Experiments worth doing were always to be well planned, executed with care and attention and recorded accurately and dispassionately. As a result Bill Johnson's research is a compilation of reliable and reproducible results that presents an understanding of steroid synthesis in a depth of unusual proportions.

He is survived by his wife and partner of 55 years, Barbara (nee Allen). In conclusion, I feel that as a Johnson graduate student, I am allowed a personal note of farewell to a mentor, a colleague and, most of all, a close and dearly loved friend. Were there a way to have it otherwise we would never have had to part.

ROBERT E. IRELAND

November 23, 1995

RALPH LLOYD SHRINER
October 9, 1899–June 7, 1994

Ralph Lloyd Shriner, Editor-in-Chief of Volume 27 of *Organic Syntheses*, published in 1947 and, with his wife, Rachel H., Coeditor of *Organic Syntheses, Cumulative Indices to Collective Volumes, I, II, III, IV, V*, published in 1976, died June 7, 1994 of pneumonia following surgery for a broken hip, in Lincolnshire, Illinois, at the age of 94. Following his term of service on the Board of Editors, during which Volume 27 was published, Dr. Shriner became a member of the Advisory Board, and was elected to the Board of Directors in 1951, and served as Vice President from 1974 to 1976, before retiring from the Board in 1977. Dr. Shriner was a dedicated and valuable member and senior statesman of the Board of Directors and of its Finance Committee, where his wise and conservative advice, particularly with respect to bonds, was much appreciated. Collective Volume VI of *Organic Syntheses* was dedicated to Ralph and Rachel Shriner.

Born in St. Louis, Missouri on October 9, 1899, Ralph Shriner was the oldest of the two children of George B. Shriner and Edith Barnett Shriner. His father worked as an agent for the railroad for some years. Ralph and sister, Ruth, attended public schools in St. Louis. During the summers, Ralph worked on his Uncle Al's farm in Cuba, Missouri, in the Ozarks, southwest of St. Louis. His uncle also owned the Ford Model T dealership, and one of Ralph's jobs was to go to the freight yard with his cousin Jim, assemble the cars, and drive them to the sales yard. Thus, Ralph learned to drive at the tender age of 12. Uncle Al also became the owner of the local telephone company, after it went bankrupt, and Ralph and his cousin Jim became the line repairboys, learning how to fix the phones, climb the poles, and maintain the wires. In his teens, Ralph built his own radio receiver and transmitter and became a ham radio operator. He was active in the Boy Scouts, became an Eagle Scout, and later served as a scoutmaster.

Ralph joined the U.S. Army and was finishing his basic training in 1918 when Germany surrendered, ending World War I. After receiving his honorable discharge from the army, Ralph entered Washington University in his hometown, where he majored in chemical engineering. He received a B.S. degree in 1921 and then stayed on for the year 1921–22 as Instructor in Chemistry. In 1922 he began graduate work in organic chemistry at the University of Illinois, receiving an M.S. degree in 1923 and a Ph.D. degree in 1925 under Roger Adams. From 1925 to 1927 he served as a Research Associate and Assistant Professor at the New York College of Agriculture Experiment Station in Geneva, New York, and also held a joint appointment at Cornell University. While in Geneva, Ralph met a young researcher, Rachel Haynes, and they were married on August 17, 1929 in Springfield, Massachusetts. Their daughter, Joan (now Palincsar), was born on July 25, 1932.

In 1927 Ralph returned to the University of Illinois as an Assistant Professor and worked his way through the ranks to become a full Professor. In 1941 he moved to Indiana University to become a Professor of Chemistry and Chairman of the Chemistry Department, which had not yet been accredited by the American Chemical Society. He led a revamping of the curriculum and, in 1942, the department was accredited. The library was improved, a trained chemistry librarian was appointed, and a shop was established. With the bombing of Pearl Harbor on December 7, 1941, and the subsequent entrance of the United States into World War II, there was much work to be accomplished, including the instruction of many soldiers. Ralph was also heavily involved in war research with the Committee on Medical Research (CMR), Office of Scientific Research and Development (OSRD), and the Army and Navy, particularly with respect to the synthesis of compounds having antimalarial properties. After the war ended in 1945, Ralph took a one-year sabbatical leave in 1946–47, during which he revised his textbook, "The Systematic Identification of Organic Compounds," and did some editing.

In 1947 Ralph moved to the University of Iowa (then called the State University of Iowa), in Iowa City, as Professor of Chemistry. He served as Chairman of the Chemistry Department from 1952 to 1963. Ralph and Rachel had a cabin on Lake McBride, and devoted quite a bit of work to it, particularly to the building of a dock, a rowboat from a kit, and a fireplace, where Ralph laid all the blocks and bricks and rolled in 150 wheelbarrow loads of cement. Ralph and Rachel also added bowling to their long-time expertise in bridge, and many other activities.

In 1963 Ralph "retired" and he and Rachel moved south to Southern Methodist University in Dallas, Texas, where his colleague, Harold Jeskey, became one of his closest friends. Ralph served as Visiting Professor of Chemistry from 1963 to 1978, when he retired again, but continued to live near the University, at 2709 Hanover St., in University Park, Dallas, until 1991. To escape the hot Texas summers, Ralph and Rachel would migrate in June like the birds in spring and fall to Deer River, in northern Minnesota. There, probably as a result of their University of Illinois connections, they had a little cabin on (Big) Deer Lake at Interlaken Camp, so named because it is located between Big and Little Deer Lakes (their cabin was near that of the late William E. Parham, Editor-in-Chief of Volume 44 of *Organic Syntheses*, and a member of the Board of Directors, Vice President, and Treasurer of *Organic Syntheses*).

While recovering from abdominal surgery, Rachel died unexpectedly on September 10, 1980, ending a happy partnership of 51 years. Ralph spent the next 11 years in Dallas, where he did all of his own housework and chores. He also continued to migrate to Deer River, Minnesota each summer through 1990. At the age of 90, he built a flight of stone steps up the cliff from Deer Lake. In 1991, Ralph moved to Skokie, Illinois to be closer to his daughter, Joan Palincsar (A.B. Mount Holyoke College; M.S. and Ph.D. Northwestern University) and her late husband Edward E., who, until recently, was a member of the faculty of the Biology Department at Loyola University in Chicago. Following Ed's death, Ralph and Joan purchased in December 1993 a ranch-style home in Lincolnshire, Illinois. In March of 1994, Ralph broke his hip and was recovering from hip surgery in a rehabilitation center, when he developed pneumonia in early June and died on June 7, 1994. He is survived by his daughter Joan Palincsar, his grandson Steven Palincsar (both of whom were very helpful in the preparation of this report), who also lives in Lincolnshire, and his granddaughters Suzanne Palincsar of Wheeling, Illinois, and Katherine Solk, of Chicago.

While at the University of Illinois, Ralph Shriner and his colleagues Carl S. Marvel and Reynold C. Fuson (Editor-in-Chief of *Organic Syntheses*, Volumes 5 and 11, and 18, respectively), established the senior research program in organic chemistry for undergraduate students. Following the system of identification of organic compounds originally published in the United States by Oliver Kamm (Editor-in-Chief of Volume 4 of *Organic Syntheses*), a former Illinois faculty member who had gone to Parke-Davis as Director of Research, R. L. Shriner and R. C. Fuson, and later D. Y. Curtin, published the classic laboratory text on qualitative analysis, "*The Systematic Identification of Organic Compounds*," which was first published in 1935 and has gone through six editions (the latest in 1980 with T. C. Morrill as an additional coauthor) and has been translated into Spanish, Japanese, and Russian. A seventh edition is scheduled to be published in 1996. Ralph Shriner was the author in 1942 of the first chapter (on the Reformatsky Reaction) of Volume I of *Organic Reactions*, which served as the editorial model for subsequent chapters in this valuable review series and sister publication of *Organic Syntheses.* He was the primary author of Chapter 4 on "Stereoisomerism" in Henry Gilman's classic "*Organic Chemistry, An Advanced Treatise*," John Wiley and Sons, Inc., New York, N.Y., 1st ed. (1938 pp. 150–405) and 2nd ed. (1943 pp. 214–488). The chapter was divided into 11 parts; Part II was coauthored by Roger Adams and Part XI was solely authored by Carl S. Marvel. Ralph Shriner was also a coauthor with Homer Adkins (Editor-in-Chief of Volume 26 of *Organic Syntheses*) of Chapter 9 on "Catalytic Hydrogenation and Hydrogenolysis" in the 2nd ed. (1943, pp. 779–834).

In 1950 Ralph Shriner was named Editor-in-Chief of *Chemical Reviews* and ably guided this important review journal in that capacity for 17 years (1950–1966). He was also a member of the Board of Editors of the *Journal of Organic Chemistry.* In 1956 Ralph Shriner was a coauthor with Walter T. Smith, Jr., of a short text, "*The Examination of New Organic Compounds, Macro and Semimicro Analytical Methods, a Laboratory Manual*," John Wiley and Sons, Inc., New York, N.Y., 136 pp.

Ralph Shriner became an American Chemical Society (A.C.S.) member as a student in 1919, and served as Secretary of the A.C.S. Division of Organic Chemistry from 1935 to 1940 and as its Chairman in 1944. He was an A.C.S. Councilor-at-Large from 1943 to 1946, and a Councilor from the Iowa Section in 1952 to 1953. He was a consultant to the Rohm and Haas Company of Philadelphia from 1938 until 1965. He was a member of the Chemistry panel on Cancer Chemotherapy at the National Service Center, National Cancer Institute, from 1959 to 1962 and served as its chairman from 1961 to 1962. Besides the A.C.S., Ralph Shriner was a member of the American Association for the Advancement of Science (A.A.A.S.), Phi Lambda Upsilon (honorary chemical fraternity), Tau Beta Pi (honorary engineering fraternity), Phi Kappa Phi, and Sigma Xi (the Research Society).

More than 100 students completed their Ph.D. degrees with Dr. Shriner. He and his students published over 150 papers in the areas of anthocyanins and flavylium salts, lignin model compounds, organic synthesis and structure, synthetic drugs, stereoisomerism, and organic chemical identification methods. Hundreds of students gained through Dr. Shriner a knowledge of organic chemistry, an inspiration to scholarship, and an introduction to that added spark, the zest for living a full life. In 1962 he received the James F. Norris Award of the Northeastern Section of the American Chemical Society (A.C.S.) in recognition of his excellence in teaching. In 1966, Ralph Shriner's former students and colleagues organized a special tribute to him at a luncheon at the New York Hilton Hotel, where many splendid tributes were given to this modest man. He received the Wilfred T. Doherty Award of the Dallas-Fort Worth Section of the A.C.S. in 1973 in recognition of his outstanding acnievements in teaching and research. In 1989 the Chemistry Department at the University of Iowa established a distinguished professorship to be entitled the "Ralph L. Shriner Professor of Synthetic Chemistry" in recognition of his contributions in teaching, research, and service as the department head.

Three "Shrinerisms" are:

(1) "A good research man leaves you with more research problems than when he came, and a poor research man can go through more research problems than you can think up."
(2) "Establish your credit when you don't need money so that if you ever do need it, the bank will know who you are when you want or need them."
(3) "If you do it right there is a way to get the job done."

We shall miss this giant of organic chemistry, this warm and understanding friend, adviser, and contributor to so much of the success of *Organic Syntheses*.

WAYLAND E. NOLAND

October 17, 1995

CONTENTS

B. Steuer, V. Wehner, A. Lieberknecht, and V. Jäger	1	(-)-2-O-BENZYL-L-GLYCERALDEHYDE AND ETHYL (R,E)-4-O-BENZYL-4,5-DIHYDROXY-2-PENTENOATE
Nina Kann, Vania Bernardes, and Andrew E. Greene	13	ACETYLENIC ETHERS FROM ALCOHOLS AND THEIR REDUCTION TO Z- AND E-ENOL ETHERS: PREPARATION OF 1-MENTHOXY-1-BUTYNE FROM MENTHOL AND CONVERSION TO (Z)- AND (E)-1-MENTHOXY-1-BUTENE
Nikola A. Nikolic and Peter Beak	23	(R)-(+)-2-(DIPHENYLHYDROXY-METHYL)PYRROLIDINE
Scott E. Denmark, Larry R. Marcin, Mark E. Schnute, and Atli Thorarensen	33	(R)-(-)-2,2-DIPHENYLCYCLO-PENTANOL
Lyndon C. Xavier, Julie J. Mohan, David J. Mathre, Andrew S. Thompson, James D. Carroll, Edward G. Corley, and Richard Desmond	50	(S)-TETRAHYDRO-1-METHYL-3,3-DIPHENYL-1H,3H-PYRROLO[1,2c]-[1,3,2]OXAZABOROLE-BORANE COMPLEX
Ruo Xu and Ronald Breslow	72	1,2,3-TRIPHENYLCYCLOPROPENIUM BROMIDE
Daniel L. Comins, Ali Dehghani, Christopher J. Foti, and Sajan P. Joseph	77	PYRIDINE-DERIVED TRIFLATING REAGENTS: N-(2-PYRIDYL)TRIFLIMIDE AND N-(5-CHLORO-2-PYRIDYL)-TRIFLIMIDE
P. Dembech, A. Ricci, G. Seconi, and M. Taddei	84	BIS(TRIMETHYLSILYL) PEROXIDE (BTMSPO)
Robert W. Murray and Megh Singh	91	SYNTHESIS OF EPOXIDES USING DIMETHYLDIOXIRANE: trans-STILBENE OXIDE
Derek H. R. Barton, Mi Chen, Joseph Cs Jászberényi, and Dennis K. Taylor	101	PREPARATION AND REACTIONS OF 2-tert-BUTYL-1,1,3,3-TETRAMETHYL-GUANIDINE: 2,2,6-TRIMETHYLCYCLO-HEXEN-1-YL IODIDE

Angela Marinetti and Philippe Savignac	108	DIETHYL (DICHLOROMETHYL) PHOSPHONATE. PREPARATION AND USE IN THE SYNTHESIS OF ALKYNES: (4-METHOXYPHENYL)ETHYNE
Scott H. Watterson, Zhijie Ni, Shaun S. Murphree, and Albert Padwa	115	2,3-DIBROMO-1-(PHENYLSULFONYL)-1-PROPENE AS A VERSATILE REAGENT FOR THE SYNTHESIS OF FURANS AND CYCLOPENTENONES: 2-METHYL-4-[(PHENYLSULFONYL)-METHYL]FURAN AND 2-METHYL-3-[(PHENYLSULFONYL)METHYL]-2-CYCLOPENTEN-1-ONE
Daniel S. Reno and Richard J. Pariza	124	PHENYL VINYL SULFIDE
V. Jäger and P. Poggendorf	130	NITROACETALDEHYDE DIETHYL ACETAL
Michael J. Bradlee and Paul Helquist	137	CYCLOPENTANONE ANNULATION VIA CYCLOPROPANONE DERIVATIVES: (3aβ,9bβ)-1,2,3a,4,5,9b-HEXAHYDRO-9b-HYDROXY-3a-METHYL-3H-BENZ[e]INDEN-3-ONE
Albert Padwa, Scott H. Watterson, and Zhijie Ni	147	[3+2]-ANIONIC ELECTROCYCLIZATION USING 2,3-BIS(PHENYLSULFONYL)-1,3-BUTADIENE: trans-4,7,7-TRICARBOMETHOXY-2-PHENYLSULFONYLBICYCLO[3.3.0]OCT-1-ENE
Goffredo Rosini, Giovanni Confalonieri, Emanuela Marotta, Franco Rama, and Paolo Righi	158	PREPARATION OF BICYCLO[3.2.0]-HEPT-3-EN-6-ONES: 1,4-DIMETHYLBICYCLO[3.2.0]HEPT-3-EN-6-ONE
Tina Morwick and Leo A. Paquette	169	PREPARATION OF POLYQUINANES BY DOUBLE ADDITION OF VINYL ANIONS TO SQUARATE ESTERS: 4,5,6,6a-TETRAHYDRO-3a-HYDROXY-2,3-DIISOPROPOXY-4,6a-DIMETHYL-1(3aH)-PENTALENONE
Akira Yanagisawa, Katsutaka Yasue, and Hisashi Yamamoto	178	REGIO- AND STEREOSELECTIVE CARBOXYLATION OF ALLYLIC BARIUM REAGENTS: (E)-4,8-DIMETHYL-3,7-NONADIENOIC ACID

Tien-Min Yuan and Tien-Yau Luh	187	NICKEL-CATALYZED, GEMINAL DIMETHYLATION OF ALLYLIC DITHIO ACETALS: (E)-1-PHENYL-3,3-DIMETHYL-1-BUTENE
Ilane Marek, Christophe Meyer, and Jean-F. Normant	194	A SIMPLE AND CONVENIENT METHOD FOR THE PREPARATION OF (Z)-β-IODO-ACROLEIN AND OF (Z)- OR (E)-γ-IODO ALLYLIC ALCOHOLS: (Z)- AND (E)-1-IODOHEPT-1-EN-3-OL
Peter Wipf and Wenjing Xu	205	ALLYLIC ALCOHOLS BY ALKENE TRANSFER FROM ZIRCONIUM TO ZINC: 1-[(tert-BUTYLDIPHENYLSILYL)-OXY]-DEC-3-EN-5-OL
J. Leroy	212	PREPARATION OF 3-BROMOPROPIOLIC ESTERS: METHYL AND tert-BUTYL 3-BROMOPROPIOLATES
Jack R. Reid, Richard F. Dufresne, and John J. Chapman	217	MESITYLENESULFONYLHYDRAZINE, AND (1α,2α,6β)-2,6-DIMETHYLCYCLO-HEXANECARBONITRILE AND (1α,2β,6α)-2,6-DIMETHYLCYCLO-HEXANECARBONITRILE AS A RACEMIC MIXTURE
Joshua S. Tullis and Paul Helquist	229	RHODIUM-CATALYZED HETERO-CYCLOADDITION OF A DIAZOMALONATE AND A NITRILE: 4-CARBOMETHOXY-5-METHOXY-2-PHENYL-1,3-OXAZOLE
Nicolas Boudreault and Yves Leblanc	241	SYNTHESIS OF 5,7-DIMETHOXY-3-METHYLINDAZOLE FROM 3,5-DIMETHOXYACETOPHENONE IN A TWO-STEP PROCEDURE
Mercedes Amat, Sabine Hadida, Swargam Sathyanarayana, and Joan Bosch	248	REGIOSELECTIVE SYNTHESIS OF 3-SUBSTITUTED INDOLES: 3-ETHYLINDOLE
Andreas Rolfs and Jürgen Liebscher	257	3-MORPHOLINO-2-PHENYLTHIO-ACRYLIC ACID MORPHOLIDE AND 5-(4-BROMOBENZOYL-2-(4-MORPHOLINO)-3-PHENYL-THIOPHENE
Unchecked Procedures	264	

ORGANIC SYNTHESES

(-)-2-O-BENZYL-L-GLYCERALDEHYDE AND ETHYL (R,E)-4-O-BENZYL-4,5-DIHYDROXY-2-PENTENOATE

(Propanal, 3-hydroxy-2-(phenylmethoxy)-, (R)- and 2-Pentenoic acid, 5-hydroxy-4-(phenylmethoxy)-, ethyl ester, [R-(E)]-)

A. $COOC_2H_5$, OH, HO, $COOC_2H_5$ + PhCHO (O, H) —p-TsOH→ C_2H_5OOC, C_2H_5OOC, O, O, Ph

B. C_2H_5OOC, C_2H_5OOC, O, O, Ph —$LiAlH_4$, $AlCl_3$→ OH, OH, BnO, OH

C. OH, OH, BnO, OH —$NaIO_4$→ BnO, =O, OH (liquid) ⇌ OBn, H, O, O, OH, BnO, OH (viscous wax)

D. OBn, HO, O —$(EtO)_2P(O)\text{-}CH_2COOEt$→ OBn, COOEt, HO

Submitted by B. Steuer, V. Wehner, A. Lieberknecht, and V. Jäger.[1]
Checked by Joseph P. Bullock and Louis S. Hegedus.

1. Procedure

Caution: Lithium aluminum hydride is sensitive to mechanical shock and very reactive towards moisture and other protic substances; its dust is very irritating to skin and mucous membranes. It should not be allowed to come into contact with metallic species or apparatus, including metal spatulas, because of the potential danger of metal ion-promoted detonation.

A. Diethyl (-)-2,3-O-benzylidene-L-tartrate. A 1-L, round-bottomed flask equipped with a Dean-Stark trap is charged with 103 g (500 mmol) of diethyl L-tartrate (Note 1), 53.1 g (500 mmol) of benzaldehyde (Note 2), 600 mL of cyclohexane and 2.80 g (14.7 mmol) of p-toluenesulfonic acid monohydrate (Note 3). The stirred mixture is heated with azeotropic removal of water (Note 4). The solution is allowed to cool to room temperature and then is concentrated by rotary evaporation (30 mm, 30°C). The residual yellow oil is dissolved in 400 mL of diethyl ether (Note 5), transferred to a 1-L separatory funnel, and washed with saturated aqueous potassium bicarbonate (200 mL) , and with water (2 x 200 mL). The ethereal layer is dried over magnesium sulfate and filtered. The solution is concentrated by rotary evaporation (30 mm, 30°C), followed by removal of solvent at 0.02 mm at room temperature. The solid crude product is collected by suction filtration through a sintered-glass funnel to give diethyl (-)-2,3-O-benzylidene-L-tartrate as yellow crystals. Trituration of the crude solid product with 150 mL of hexanes and collection of the purified product by suction filtration through a sintered-glass funnel gives 105 g (71%) of purified product, mp 47°C (Note 6).

B. (+)-2-O-Benzyl-L-threitol.[2a] A 2-L, three-necked, round-bottomed flask is oven-dried (140°C) and flushed with nitrogen. The flask is equipped with an efficient mechanical stirrer, a 250-mL pressure-equalizing addition funnel, and a reflux condenser equipped with a mineral oil bubbler that is connected to a nitrogen source.

A slight pressure of gas is maintained in the apparatus throughout the course of the reaction. The flask is charged with 16.5 g (434 mmol) of lithium aluminum hydride (Note 7), and cooled to -30°C using an isopropyl alcohol-dry ice bath. Then 167 mL of dry diethyl ether (Note 8) is added with vigorous stirring and a solution of 57.8 g (434 mmol) of aluminum chloride in 134 mL of dry diethyl ether (Notes 8, 9) is added dropwise during 40 min. Dry dichloromethane (134 mL) (Note 10) is placed in the addition funnel and added rapidly, while the temperature is allowed to rise to 0°C. A solution of 64.7 g (220 mmol) of diethyl (-)-2,3-O-benzylidene-L-tartrate in dry dichloromethane (134 mL) is added dropwise during 30 min (Note 11). The mixture is stirred for 1 hr at room temperature and heated to reflux for an additional 2 hr. The mixture is cooled to -20°C as above, and 14 mL of de-ionized water, followed by a solution of 31.5 g (561 mmol) of potassium hydroxide in 46 mL of de-ionized water, is added cautiously. The cooling bath is removed and the mixture is stirred at room temperature until the grey color (probably due to unreacted lithium aluminum hydride) has completely disappeared (Note 12). Efficient stirring is required throughout to ensure good yields. The mixture is filtered through a glass-sintered funnel containing a 2-cm pad of Celite, and the inorganic precipitate is extracted with 0.5 L of dichloromethane in a Soxhlet apparatus for 3 days. The combined extracts and filtrate are evaporated under reduced pressure (30 mm, 30°C). After drying over phosphorus pentoxide (P_4O_{10}) in an evacuated (1 mm) desiccator, 42.4 g (91%) of colorless crystals, mp 71-73°C, are obtained and used as such for step C. Recrystallization from dichloromethane (Note 10) gives 37.4 g (80%) of (+)-2-O-benzyl-L-threitol, mp 75-76°C (Note 13).

C. (-)-2-O-Benzyl-L-glyceraldehyde.[2b] A solution of 9.22 g (43.4 mmol) of (+)-2-O-benzyl-L-threitol (Note 14) in 100 mL of water is stirred vigorously, while 9.28 g (43.4 mmol) of sodium periodate (Note 15) is added in ca. 1-g portions over 40 min. The mixture is stirred for 2 hr at room temperature, then the pH is adjusted to 7.0 by addition of solid potassium carbonate (Note 16). The mixture is transferred to a 500-mL separatory funnel and extracted with dichloromethane (3 x 150 mL). The combined organic layers are dried over magnesium sulfate for 15 min and concentrated on a rotary evaporator (30 mm, 30°C). The remaining pale-yellow crude oil is transferred to a 25-mL round-bottomed flask and purified by short-path distillation in a preheated oil bath (0.025 mm, 160°C) (Note 17), to yield 6.32 g (80%) of (-)-2-O-benzyl-L-glyceraldehyde, a colorless oil that on standing turns more and more viscous, and after several weeks at room temperature forms a waxy solid (Note 18).

D. Ethyl (-)-[R-(E)]-4-O-benzyl-4,5-dihydroxy-2-pentenoate. To a 500-mL, round-bottomed, two-necked flask, fitted with a nitrogen inlet (Note 19) and a stopper, is added 3.32 g of a suspension of sodium hydride in paraffin, containing 65% sodium hydride. The suspension is washed three times with 40 mL and once with 20 mL of hexanes (Note 20) to remove the paraffin. The residue is freed from remaining hexanes under vacuum (0.01 mm) to give ca. 2.16 g (ca. 90 mmol) of sodium hydride. The flask is fitted with a magnetic stirring bar and the stopper is exchanged for a septum. Then 100 mL of tetrahydrofuran (Note 21) is added and the suspension is cooled to 0°C. Triethyl phosphonoacetate (22.87 g, 102 mmol) (Note 22) is added to the stirred sodium hydride/tetrahydrofuran suspension by means of a 50-mL syringe over a period of 20 min. The mixture is cooled to -78°C (acetone/dry ice) and a solution of 10.81 g (60 mmol) of (-)-2-O-benzyl-L-glyceraldehyde in 80 mL of tetrahydrofuran (Note 21) is added by means of a syringe over 20 min. (The checkers transferred this solution via a cannula.) The mixture is stirred for an additional 15 min at -78°C. The temperature is allowed to rise to 0°C within 30 min, and finally is kept at room temperature for additional 45 min. The reaction is quenched with 150 mL of saturated ammonium chloride solution and extracted three times with ether (500, 200, 200 mL). The combined organic layers are washed with a mixture of saturated sodium bicarbonate/brine (1:1, 160 mL). The aqueous layer is washed with ether (3 x 100 mL). The combined organic layers are dried over magnesium sulfate, filtered, and evaporated to dryness to give a pale yellow oil. The crude product is purified by column chromatography over silica gel (Note 23) with petroleum ether/ethyl acetate 1/1 as eluent (Note 24), to yield 11.7 g (78%) of analytically pure ethyl (-)-[R-(E)]-4-O-benzyl-4,5-dihydroxy-2-pentenoate (Note 25).

2. Notes

1. Diethyl L-tartrate (99%) was obtained from Janssen Chimica, Brüggen, Germany or Aldrich Chemical Company, Inc.

2. Benzaldehyde (99+%) from Aldrich Chemical Company, Inc., was used as received.

3. p-Toluenesulfonic acid monohydrate (99%) was obtained from Fluka Feinchemikalien GmbH, Neu-Ulm, Germany or from Aldrich Chemical Company, Inc.

4. The mixture becomes homogeneous at reflux temperature. The reaction usually takes about 16 hr at an 0.5 mole-scale as indicated by the amount of water separated.

5. Diethyl ether (technical grade) was distilled over potassium hydroxide.

6. The spectral properties of diethyl (-)-2,3-O-benzylidene-L-tartrate are as follows: ^1H NMR (250 MHz, $CDCl_3$) δ: 1.32, 1.35 (2 t, 6 H, J = 7.1, 2 $CH_2C\underline{H}_3$), 4.23, 4.28 (2 q, 4 H, J = 7.1, 2 $C\underline{H}_2CH_3$), 4.83, 4.95 (2 d, 2 H, J = 4.0, 2 CHO), 6.16 (s, 1 H, CHPh), 7.40, 7.58 (2 m, 5 H, C_6H_5),[3] $[\alpha]_D^{20}$ -30.7° ($CHCl_3$, *c* 2.20),[4] mp 45°C.[3,4]

7. Lithium aluminum hydride was obtained in 25-g samples (98%) from Merck-Schuchardt, Hohenbrunn, Germany or Aldrich Chemical Company, Inc.

8. Diethyl ether was dried first by distillation over potassium hydroxide, then by distillation from lithium aluminum hydride.

9. In order to dissolve aluminum chloride in dry diethyl ether, a flask is charged with the aluminum chloride and the diethyl ether is added in 10-mL portions with vigorous mechanical stirring to give a dark solution. The flask must be cooled in an ice bath.

10. Dichloromethane (technical grade) was distilled over phosphorus pentoxide.

11. If lithium aluminum hydride and aluminum chloride are not of high purity, an excess of 10% of each should be used. Otherwise a mixture with products of incomplete reduction is obtained.

12. The reaction mixture is heated to reflux with 200 mL of tetrahydrofuran; the precipitate then obtained is very easy to filter off.

13. The spectral properties of (+)-2-O-benzyl-L-threitol are as follows: ^{1}H NMR (250 MHz, $CDCl_3$) δ: 2.58, 2.76, 2.98 (bs, 3 H, 3 OH), 3.44-3.88 (m, 6 H, 2 CH_2OH, H-2, H-3), 4.56, 4.69 (AB, 2 H, J = 11.6, $CH_2C_6H_5$), 7.16-7.34 (m, 5 H, C_6H_5), $[\alpha]_D^{25}$ +17.5° (EtOH, *c* 1.14).[4]

14. In several experiments it was found that the yield of (-)-2-O-benzyl-L-glyceraldehyde is somewhat lower when the reaction is performed on a larger scale.

15. Sodium periodate (98%) was obtained from Fluka Feinchemikalien GmbH, Neu-Ulm, Germany or from Fisher Scientific Company.

16. The pH was controlled with Merck Universal-Indikatorpapier or pHydrion Vivid 1-11 Jumbo pH paper, Micro Essential Laboratory, Brooklyn, NY, USA.

17. A silicone-oil bath is preheated to 160°C (the checkers used a sand bath). The flask of the evacuated apparatus filled with the crude product is immersed totally in the bath until no more distillate is collected. For optimum results, the distillation should be performed within 10 to 15 min. Slower distillation leads to lower yields because of thermal decomposition.

18. It is not possible to give exact spectral properties of (-)-2-O-benzyl-L-glyceraldehyde because of rapid di- and/or oligomerization. In order to check the optical purity of the product, it is convenient to compare the equilibrium value of specific rotation, as obtained after 6 days in ethanol solution at room temperature; $[\alpha]_D^{22}$ -33.2° (EtOH, *c* 0.083). In succeeding reactions (see Discussion and Step D), it was determined from NMR shift experiments that these products [e.g., (i) the Z-selective Wittig enoate product (with ethoxycarbonylmethylene-triphenylphosphorane) and the penten-5-olide therefrom (after acid-catalyzed lactonization,[3,5]) and (ii) the E-enoate (Horner product)[3,5] contained enantiomers in ratios of >96:4.

19. Nitrogen was dried by means of a Sicapent® (E. Merck) drying tube. The checkers used argon.

20. The checkers used hexanes (technical grade) distilled over 3 Å molecular sieves. The submitters used pentane distilled from sodium.

21. Tetrahydrofuran was purified by distillation under nitrogen from a purple solution of sodium and benzophenone.

22. Triethyl phosphonoacetate was purified by distillation (bp 143°C, 9 mm). The checkers obtained this compound (99%) from Aldrich Chemical Company, Inc., and used it as received.

23. The submitters used 28 g of Kieselgel 60, E. MERCK, 0.040-0.063 mm (250-400 mesh); column: 28 cm x 2.5 cm. The checkers used 100 g of silica gel, 0.032-0.063 mm, Selecto Scientific, Norcross, GA, USA, catalog # 162824; column: 40 cm x 5.5 cm.

24. Ethyl acetate and petroleum ether (technical grade; boiling range 40-80°C) were purified by distillation. The checkers used ethyl acetate/hexanes 10/90 followed by ethyl acetate/hexanes 35/65 as eluent; ethyl acetate (HPLC grade) was obtained from Mallinckrodt Specialty Chemicals Company, Paris, KY, USA and hexanes (technical grade) were distilled over 3 Å molecular sieves.

25. The analytical data (after chromatography) were as follows: Calcd. for $C_{14}H_{18}O_4$ (250.29): C, 67.18; H, 7.25. Found: C, 67.00; H. 7.20. The E/Z ratio was found to be >97:3 (determined by HPLC). Rt_E = 3.72 min; Rt_Z = 4.28 min, eluent petroleum ether/ethyl acetate 6/4 (LiChrosorb Si 60 column, E. Merck]. TLC: R_f = 0.34 (petroleum ether/ethyl acetate 1/1). GLC analysis: Column PS086/.32 mm x 20 m glass capillary, 95:5 methyl/phenylsilicone. Program: T_1, 40°C/(1 min), rate 10°C/min, T_2, 300°C, 0.5 bar hydrogen pressure; Rt_Z = 16.47 min; Rt_E = 17.25 min. E/Z ratio was found to be >97:3 (determined by GLC). $[\alpha]_D^{22}$ -75.8° ($CHCl_3$, *c* 1.192, E/Z >97:3), bp 125-130°C (0.001 mm). ^{13}C NMR (63 MHz, $CDCl_3$) δ: 14.2 ($OCH_2\underline{C}H_3$), 60.7 ($O\underline{C}H_2CH_3$), 64.6 (C-5), 71.5 ($\underline{C}H_2Ph$), 79.0 (C-4), 123.9 (C-2), 127.9, 128.0, 128.3, 128.6, 137.6 (C_6H_5), 144.3 (C-3), 165.9 (C-1); ^1H NMR (250 MHz, $CDCl_3$), δ: 1.30 (t, 3 H, J = 7.1, CH_3), 2.32 (dd, 1 H, J = 5.2, 7.9, OH), 3.65 (m, 2 H, $C\underline{H_2}OH$), 4.14 (m, 1 H, 4-H), 4.22 (q, 2 H, J = 7.1, $OC\underline{H_2}CH_3$), 4.41, 4.45 (AB, 2 H, J = 11.6, $C\underline{H_2}Ph$), 6.11 (dd, 1 H, J = 1.3, 15.8, 2-H), 6.85 (dd, 1 H, J = 6.1, 15.8, 3-H), 7.34 (m, 5 H, C_6H_5).

Waste Disposal Information

All toxic materials were disposed of in accordance with "Prudent Practices in the Laboratory"; National Academic Press; Washington, DC, 1996 and "Neue Datenblätter für gefährliche Arbeitsstoffe nach der Gefahrstoffverordnung", Welzbacher, U. (Ed.); WEKA Fachverlage, Kissing, 1991.

3. Discussion

Optically active C_3-building blocks of type **1** are key starting compounds in organic synthesis.[6] The most important member of this class is 2,3-O-isopropylidene-glyceraldehyde **2**.[6,7]

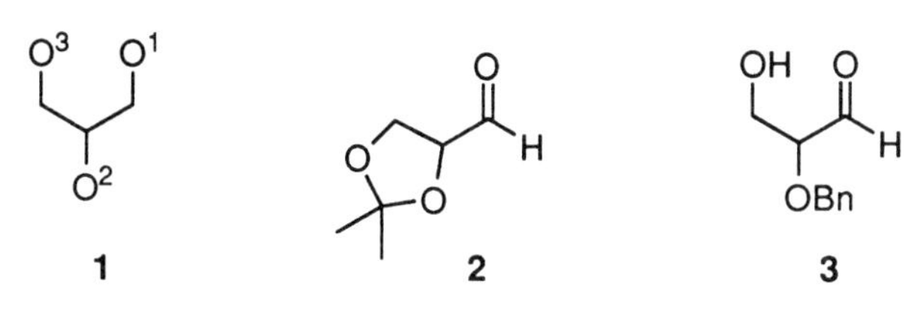

Both enantiomers of **2** are readily available from D-mannitol and from L-gulonolactone, respectively, and have been used in many reactions in a straightforward manner, because of the free aldehyde and the protected hydroxy functions. Unfortunately, **2** tends to trimerize and partial racemization has often been encountered on prolonged storage, thus preventing shipping/commercialization.[8] A recent paper on large-scale preparation of glyceraldehyde acetonide **2**, however, indicates suitable stability even at room temperature when certain precautions are met.[9]

Glyceraldehyde derivatives with one protected and one free hydroxy function could, in principle offer new options because 2O/3O are more strongly differentiated, and further, since the free hydroxy group does or may cause different regio- and stereoselectivity in the next or subsequent steps. 2-O-Benzylglyceraldehyde, **3**, because of its di- or oligomeric form, is configurationally stable at room temperature.[5] The D-form has previously been prepared from D-mannitol in four steps, with an overall yield of 5%,[10a] or in 9 steps with 4%,[10b] and from D-glucose in six steps with an overall yield of 50%.[10c]

As described here, both enantiomers of **3** can be prepared in three steps from commercially available diethyl D- and L-tartrate in up to 70% over-all yield.[2,3,5] Procedures to obtain the benzylidene acetal,[11,12] with the ensuing reduction step,[11,12] are based on previous literature reports. Both enantiomers of **3** have been used in highly stereoselective nitroaldol additions.[3,13] Imines, nitrones, oximes, and nitrile oxides derived therefrom were recently employed in a variety of additions/cycloadditions.[14,15] (-)-2-O-Benzyl-L-glyceraldehyde has further been used for the preparation of protected (2S,4R)-4-hydroxyornithine, via a Horner-Emmons reaction to the corresponding α,β-didehydroamino acid derivatives and subsequent diastereoselective hydrogenation.[16] Transformation of aldehyde **3** in Z-selective Wittig or E-selective Horner reactions[3,5,17] (see Step D[18]), formation of the dimethyl acetal[4,19] or of the corresponding nitro compound by oxidation of the oxime,[4,19] represent further recent uses of **3**.

1. Institut für Organische Chemie und Isotopenforschung der Universität Stuttgart, Pfaffenwaldring 55, D-70569 Stuttgart. Work carried out before 1993 was done at Universität Würzburg.
2. (a) (+)-2-O-Benzyl-L-threitol is available from Fluka Feinchemikalien GmbH at ca. DM 60/g; (b) 2-O-Benzylglyceraldehyde is available from Merck-Schuchardt, D- and L- at a price of ca. DM 140/g.
3. Wehner, V. Dissertation, Universität Würzburg, 1990.
4. Poggendorf, P. Dissertation, Universität Stuttgart, 1995.
5. Jäger, V.; Wehner, V. *Angew. Chem.* **1989**, *101*, 512; *Angew. Chem. Int. Ed. Engl.* **1989**, *28*, 469.
6. (a) McGarvey, G. J.; Kimura, M.; Oh, T.; Williams, J. M. *J. Carbohydr. Chem.* **1984**, *3*, 125; (b) Inch, T. D. *Tetrahedron* **1984**, *40*, 3161; (c) Jurczak, J.; Pikul, S.; Bauer, T. *Tetrahedron* **1986**, *42*, 447; (d) Mulzer, J. *Nachr. Chem., Techn. Lab.* **1984**, *32*, 146; (e) Altenbach, H. J. *Nachr. Chem., Techn. Lab.* **1988**, *36*, 33; (f) Takano, S. *Pure Appl. Chem.* **1987**, *59*, 353.
7. (a) Dumont, R.; Pfander, H. *Helv. Chim. Acta* **1983**, *66*, 814; (b) Hubschwerlen, C. *Synthesis* **1986**, 962; (c) Häfele, B.; Jäger, V. *Liebigs Ann. Chem.* **1987**, 85.
8. Personal communication, Merck-Schuchardt, Hohenbrunn, Germany.
9. (a) Schmid, C. R.; Bryant, J. D.; Dowlatzedah, M.; Phillips, J. L.; Prather, D. E.; Schantz, R. D.; Sear, N. L.; Vianco, C. S. *J. Org. Chem.* **1991**, *56*, 4056; (b)

Grauert, M.; Schöllkopf, U. *Liebigs Ann. Chem.* **1985**, 1817; (c) see also: Mulzer, J.; Brand, C. *Tetrahedron* **1986**, *42*, 5961.

10. (a) Ballou, C. E.; Fischer, H. O. L. *J. Am. Chem. Soc.* **1955**, *77,* 3329; (b) Suami, T.; Tadano, K.; Suga, A., Ueno, Y. *J. Carbohydr. Chem.* **1984**, *3*, 429; (c) Charon, D.; Szabó L. *J. Chem. Soc., Perkin Trans. 1*, **1980**, 1971.
11. Seebach, D.; Hungerbühler, E. *Mod. Synth. Methods, Conf. Pap. Int. Semin., 2nd* **1980**, as Ref. 2, 91-171, especially, 120, 152.
12. Takano, S.; Akiyama, M.; Sato, S., Ogasawara, K. *Chem. Lett.* **1983**, 1593.
13. Wehner, V.; Jäger, V. *Angew. Chem.* **1990**, *102*, 1180; *Angew. Chem. Int. Ed. Engl.* **1990**, *29*, 1169.
14. (a) Müller, R. Dissertation, Universität Würzburg, 1991; (b) Müller, R.; Leibold, T.; Pätzel, M.; Jäger, V. *Angew. Chem.* **1994**, *106*, 1305; *Angew. Chem. Int. Ed. Engl.* **1994**, *33*, 1295; (c) Jäger, V.; Müller, R.; Leibold, T.; Hein, M.; Schwarz, M.; Fengler, M.; Jaroskova, L.; Pätzel, M.; LeRoy, P.-Y. *Bull. Soc. Chim. Belg.* **1994**, *103*, 491.
15. (a) Franz, T. Dissertation, Universität Würzburg, 1993; (b) Franz, T.; Hein, M.; Veith, U.; Jäger, V.; Peters, E.-M.; Peters, K.; von Schnering, H. G. *Angew. Chem.* **1994**, *106*, 1308; *Angew. Chem. Int Ed. Engl.* **1994**, *33*, 1298; (c) Veith, U. Dissertation, Universität Stuttgart, 1995; (d) Veith, U.; Leurs, S.; Jäger, V. *J. Chem. Soc., Chem. Commun.* **1996**, 329; (e) Meunier, N.; Veith, U.; Jäger, V. *J. Chem. Soc., Chem. Commun.* **1996**, 331.
16. Schmidt, U.; Meyer, R.; Leitenberger, V.; Stäbler, F.; Lieberknecht, A. *Synthesis* **1991**, 409.
17. Review: Maryanoff, B. E.; Reitz, A. B. *Chem. Rev.* **1989**, *89*, 863-927.
18. We thank Mr. Timo Gräther for experimental assistance concerning Step D.
19. Poggendorf, P. Diplomarbeit, Universität Würzburg, 1990; see also Ref. 9a, b.

Appendix

Chemical Abstracts Nomenclature (Collective Index Number); (Registry Number)

(-)-2-O-Benzyl-L-glyceraldehyde: Propanal, 3-hydroxy-2-(phenylmethoxy)-, (R)- (10); (76227-09-3)

Diethyl (-)-2,3-O-benzylidene-L-tartrate: 1,3-Dioxolane-4,5-dicarboxylic acid, 2-phenyl-, diethyl ester, [4R-(2α,4α,5β)]- (9); (35572-31-7)

Diethyl L-tartrate: Tartaric acid, diethyl ester, L-(+)- (8); Butanedioic acid, 2,3-dihydroxy-, [R-(R*,R*)]-, diethyl ester (9); (87-91-2)

Benzaldehyde (8,9); (100-52-7)

p-Toluenesulfonic acid monohydrate (8); Benzenesulfonic acid, 4-methyl-, monohydrate (9); (6192-52-5)

(+)-2-(O)-Benzyl-L-threitol: 1,2,4-Butanetriol, 3-(phenylmethoxy)-, [S-(R*,S*)]- (12); (124909-02-0)

Sodium metaperiodate: Periodic acid, sodium salt (8,9); (7790-28-5)

Ethyl (-)-[R-(E)]-4-O-benzyl-4,5-dihydroxy-2-pentenoate: 2-Pentenoic acid, 5-hydroxy-4-(phenylmethoxy)-, ethyl ester, [R-(E)]- (12); (119770-84-2), [R-(Z)]- (12); (119770-88-6)

Sodium hydride (8,9); (7646-69-7)

Triethyl phosphonoacetate: Acetic acid, phosphono-, triethyl ester (8); Acetic acid, (diethoxyphosphinyl)-, ethyl ester (9); (867-13-0)

ACETYLENIC ETHERS FROM ALCOHOLS AND THEIR REDUCTION TO Z- AND E-ENOL ETHERS: PREPARATION OF 1-MENTHOXY-1-BUTYNE FROM MENTHOL AND CONVERSION TO (Z)- AND (E)-1-MENTHOXY-1-BUTENE

([Cyclohexane, 2-(1-butynyloxy)-4-methyl-1-(1-methylethyl)- [1S-(1α,2β,4β)]-], and [[[Cyclohexane, 2-(1-butenyloxy)-4-methyl-1-(1-methylethyl)-, [1S-[1α,2β(Z),4β]]- and [1S-[1α,2β(E),4β]]-)

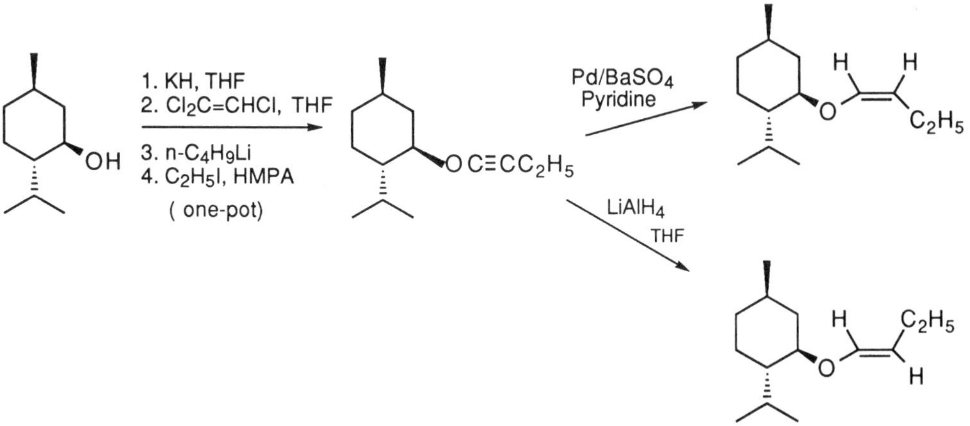

OH

1. KH, THF
2. Cl_2C=CHCl, THF
3. n-C_4H_9Li
4. C_2H_5I, HMPA
(one-pot)

O C≡CC_2H_5

Pd/$BaSO_4$
Pyridine

H H O C_2H_5

$LiAlH_4$
THF

H C_2H_5 O H

Submitted by Nina Kann, Vania Bernardes, and Andrew E. Greene.[1]

Checked by Rodolphe Tamion and Leon Ghosez.

1. Procedure

Caution! These transformations should be carried out in an efficient hood and only by persons familiar with the handling of air-sensitive and toxic materials.

A. 1-Menthoxy-1-butyne. A dry, 500-mL, one-necked, round-bottomed flask (Note 1), equipped with a Teflon-covered magnetic stirring bar, is flushed with argon and charged with 13.2 g (115 mmol) of a 35% suspension of potassium hydride in mineral oil (Note 2). The mineral oil is removed by washing with pentane (3 x 30 mL) and the potassium hydride is suspended in 100 mL of anhydrous tetrahydrofuran (Note 3). The flask is capped with a rubber septum and is connected to a Nujol-filled bubbler by means of a syringe needle. A solution of 12.0 g (76.8 mmol) of (-)-menthol (Note 2) in 120 mL of anhydrous tetrahydrofuran is then added dropwise by syringe over 20 min. The mixture is stirred until hydrogen evolution is complete (ca. 20 min) and cooled to -50°C by means of an acetone-liquid nitrogen bath. A solution of trichloroethylene (7.58 mL, 84.4 mmol) (Note 3) in 75 mL of anhydrous tetrahydrofuran is added dropwise by syringe over 15 min, after which the reaction mixture is allowed to warm to room temperature and is stirred for 1 hr. The resulting brown solution is then cooled to -70°C and treated dropwise with 67.6 mL (169 mmol) of 2.5 M butyllithium in hexanes (Note 2). After the reaction mixture is stirred for 30 min at -70°C, it is warmed to -40°C over 30 min (Note 4) and treated dropwise with a solution of ethyl iodide (18.5 mL, 230 mmol) (Note 5) in 25 mL of hexamethylphosphoramide (Notes 3, 6). The solution is stirred at room temperature for 3 hr, whereupon it is quenched by slow addition of 15 mL of methanol and poured into 300 mL of cold aqueous saturated ammonium chloride. The phases are separated and the aqueous phase is extracted with pentane (3 x 200 mL). The combined organic phases are washed with water (4 x 150 mL), dried over sodium sulfate, and filtered. Concentration by rotary evaporation affords 17.1 g of dark brown oil, which is purified by bulb-to-bulb distillation (0.02 mm, oven temperature 70-90°C) to yield 14.9 g (93%) of 1-menthoxy-1-butyne as a colorless oil (Note 7).

B. (Z)-1-Menthoxy-1-butene. A dry, 250-mL, one-necked, round-bottomed flask (Note 1), equipped with a Teflon-covered magnetic stirring bar, is charged with 14.0 g

(67.2 mmol) of 1-menthoxy-1-butyne, 0.7 g of 10% palladium on barium sulfate and 180 mL of pyridine (Note 8). The flask is capped with a rubber septum and by means of syringe needles is degassed and connected to a hydrogen-filled balloon. The mixture is stirred for 6 hr (Note 9), whereupon the hydrogen is replaced with argon, and the reaction mixture is filtered through a glass frit under water aspirator pressure. The filtrate is diluted with 300 mL of pentane and washed first with saturated copper sulfate solution (5 x 120 mL) and then with water (1 x 200 mL). The organic phase is dried over sodium sulfate and the solvent is removed by rotary evaporation to give 15.8 g of a yellow oil. Bulb-to-bulb distillation (0.02 mm, oven temperature 70-90°C) provides 12.7 g (90%) of (Z)-1-menthoxy-1-butene as a colorless oil (Note 10).

C. *(E)-1-Menthoxy-1-butene.* A dry, 500-mL, one-necked, round-bottomed flask (Note 1), equipped with a Teflon-covered magnetic stirring bar, is flushed with argon and charged with 7.65 g (202 mmol) of lithium aluminum hydride (Note 11) and 325 mL of anhydrous tetrahydrofuran (Note 3) and then capped with a rubber septum and connected by means of a syringe needle to an argon-filled balloon. To the stirred slurry is added dropwise by syringe neat 1-menthoxy-1-butyne (14.0 g, 67.2 mmol), followed by 2 mL of tetrahydrofuran wash. The septum is replaced with a reflux condenser connected to an argon-filled balloon and the mixture is refluxed for 4 hr. After the flask is allowed to cool to room temperature, it is placed in an ice bath and the reaction mixture is quenched by the slow addition (*Caution!* Note 12) of 12.5 mL of aqueous 10 % sodium hydroxide, followed by 15.5 mL of water. The resulting mixture is stirred for 30 min, after which anhydrous sodium sulfate (ca. 60 g) is added and the solid material is removed by filtration through a 1-cm pad of Celite, which is then rinsed with diethyl ether (4 x 100 mL) . Concentration of the combined filtrates yields 14.7 g of a pale yellow oil, which is purified by filtration through a 4-cm (80-g) pad of silica gel pretreated with triethylamine (2.5% v/v, Note 13) using hexane as the eluant.

Concentration gives 13.3 g (94%) of (E)-1-menthoxy-1-butene as a colorless oil (Note 14).

2. Notes

1. All glassware was flame-dried and allowed to cool in a desiccator before use.

2. Potassium hydride (35% in mineral oil) and (-)-menthol (99%) were purchased from the Aldrich Chemical Company, Inc., and butyllithium (2.5 M in hexanes) was obtained from Janssen Chimica.

3. Pentane and trichloroethylene were distilled from calcium chloride, tetrahydrofuran from the sodium ketyl of benzophenone, and hexamethylphosphoramide from calcium hydride under reduced pressure.

4. Quenching the reaction at this stage with methanol, followed by the work-up described below and bulb-to-bulb distillation (0.02 mm, oven temperature 60-80°C), gave 12.5 g (90%) of menthoxyacetylene as a colorless oil: $[\alpha]_D^{25}$ -74° (cyclohexane, *c* 0.76); IR (film) cm^{-1}: 3340, 2960, 2940, 2870, 2150, 1450, 1370, 1100, 940, 890, 840; ^1H NMR (200 MHz, $CDCl_3$) δ: 0.68-1.54 (m, 5 H; CH and CH_2), 0.81 (d, 3 H, J = 6.9, CH_3), 0.90 (d, 3 H, J = 7.0, CH_3), 0.94 (d, 3 H, J = 6.4, CH_3), 1.49 (s, 1 H, C≡CH), 1.57-1.74 (m, 2 H), 2.00-2.33 (m, 2 H), 3.83 (app td, 1 H, J = 10.9, 4.5, CH-O-); ^{13}C NMR (50.3 MHz, $CDCl_3$) δ: 16.3 (CH_3), 20.5 (CH_3), 22.0 (CH_3), 23.3 (CH_2), 25.9 (CH), 27.1 (C), 31.6 (CH), 33.9 (CH_2), 39.4 (CH_2), 46.7 (CH), 88.2 (CH), 89.7 (CH); mass spectrum (chemical ionization), m/e 181 (M^+ + 1, 5%), 156 (100%), 139 (58%); analytical TLC (pentane): R_f 0.54 (single spot). Proton NMR confirmed the complete absence (<1%) of menthol. Anal. Calcd for $C_{12}H_{20}O$: C, 79.94; H, 11.18. Found: 79.72; H, 10.98.

5. Ethyl iodide (99%) was purchased from Prolabo (Paris, France) and was washed with saturated aqueous sodium thiosulfate solution, dried over anhydrous sodium sulfate and then sodium metal, and distilled prior to use.

6. Dimethylpropyleneurea (75 mL) can be used in place of hexamethylphosphoramide; however, a small amount (5-10%) of menthoxyacetylene contaminates the final product.

7. Data for 1-menthoxy-1-butyne are as follows: $[\alpha]_D^{25}$ -78° (cyclohexane, *c* 0.69); IR (film) cm^{-1}: 2950, 2925, 2870, 2280, 1460, 1390, 1370, 1250, 1230, 1210, 980, 950, 910, 840 ; ^1H NMR (200 MHz, $CDCl_3$) δ: 0.75-1.20 (m, 3 H, CH and CH_2), 0.80 (d, 3 H, J = 6.9, CH_3), 0.89 (d, 3 H, J = 7.1, CH_3), 0.93 (d, 3 H, J = 6.4, CH_3), 1.08 (t, 3 H, J = 7.5, $CH_2C\underline{H_3}$), 1.27-1.51 (m, 2 H), 1.64 (br d, 2 H), 2.11 (q, 2 H, J = 7.5, $C\underline{H_2}CH_3$), 2.12-2.25 (m, 2 H), 3.69 (app td, 1 H, J = 10.9, 4.5, CHO); ^{13}C NMR (50.3 MHz, $CDCl_3$) δ: 11.0 (CH_2), 15.1 (CH_3), 16.2 (CH_3), 20.5 (CH_3), 21.9 (CH_3), 23.2 (CH_2), 25.7 (CH), 31.5 (CH), 34.0 (CH_2), 39.6 (2C: C and CH_2), 46.8 (CH), 86.9, (CH), 87.4 (C); mass spectrum (chemical ionization), m/e 209 (M^+ + 1, 4%), 156 (100%), 139 (26%); analytical TLC (pentane): R_f 0.54 (single spot). Proton NMR indicated a complete absence (<1%) of both menthol and menthoxyacetylene. Anal. Calcd for $C_{14}H_{24}O$: C, 80.71; H, 11.61. Found: C, 80.70; H, 11.55.

8. 10% Palladium on barium sulfate was purchased from Fluka Chemie AG and pyridine (99.5%) was obtained from Chimie-Plus Laboratories (St.-Priest, France) and dried over calcium hydride before use.

9. The reaction was followed by IR analysis of worked-up aliquots (disappearance of band at 2280 cm^{-1}).

10. Data for (Z)-1-menthoxy-1-butene are as follows: $[\alpha]_D^{25}$ -18° (cyclohexane, *c* 0.11); IR (film) cm^{-1}: 3030, 2970, 2920, 2870, 1660, 1460, 1380, 1350, 1250, 1140, 1110, 1090, 1070, 1050; ^1H NMR (200 MHz, $CDCl_3$) δ: 0.72-1.08 (m, 3 H, CH and CH_2), 0.80 (d, 3 H, J = 7.0, CH_3), 0.92 (d, 3 H, J = 7.1, CH_3), 0.93 (d, 3 H, J = 6.5, CH_3),

0.97 (t, 3 H, J = 7.5, $CH_2C\underline{H}_3$), 1.22-1.46 (m, 2 H), 1.56-1.70 (m, 2 H), 1.85-2.21 (m, 4 H), 3.35 (app td, 1 H, J = 10.7, 4.3, C$\underline{H}$OCH=CH), 4.32 (app q, 1 H, J = 7.1, OCH=C$\underline{H}$), 5.97 (d, 1 H, J = 6.2, -OC$\underline{H}$=CH); ^{13}C NMR (50.3 MHz, $CDCl_3$) δ: 14.6 (CH_3), 16.5 (CH_3), 17.5 (CH_2), 20.7 (CH_3), 22.2 (CH_3), 23.6 (CH_2), 25.8 (CH), 31.5 (CH), 34.5 (CH_2), 41.6 (CH_2), 47.9 (CH), 81.3 (CH), 108.2 (CH), 143.6 (CH); mass spectrum (electron impact) m/e 210 (M^+, 14%), 138 (32%), 83 (100%); analytical TLC (pentane): R_f 0.59 (major spot). Proton NMR indicated a complete absence (<1%) of menthol, menthoxybutyne, and menthoxybutane, but the presence of ca. 3% of the E -isomer. Anal. Calcd for $C_{14}H_{26}O$: C, 79.94; H, 12.46. Found: C, 79.84; H, 12.55.

11. Lithium aluminum hydride (95+%) was purchased from Janssen Chimica. Lesser amounts led to incomplete reaction under the stated conditions.

12. *Caution!* Quenching should be performed very carefully as a rapid, exothermic evolution of hydrogen occurs during the initial phase. Flushing with argon throughout the quenching is recommended. Toward completion of the sodium hydroxide addition, a thick slurry is produced, which makes stirring difficult. This difficulty is alleviated, however, upon the addition of water.

13. Silica gel (70-230 mesh) was treated with triethylamine (2.5% v/v) and then shaken to achieve homogeneity.

14. Data for (E)-1-menthoxy-1-butene are as follows: $[\alpha]_D^{25}$ -37° (cyclohexane, *c* 0.80); IR (neat) cm^{-1}: 2960, 2910, 2860, 1670, 1650, 1450, 1180, 1140, 920; ^1H NMR (200 MHz, $CDCl_3$) δ: 0.71-1.08 (m, 3 H, CH and CH_2), 0.75 (d, 3 H, J = 7.0, CH_3), 0.87 (d, 3 H, J = 7.1, CH_3), 0.89 (d, 3 H, J = 6.5, CH_3), 0.94 (t, 3 H, J = 7.4, $CH_2C\underline{H}_3$), 1.19-1.46 (m, 2 H), 1.54-1.70 (m, 2 H), 1.90 (app quint d, 2 H, J = 7.3, 1.3), 1.96-2.21 (m, 2 H), 3.36 (app td, 1 H, J = 10.6, 4.3, C$\underline{H}$OCH=CH), 4.88 (dt, 1 H, J = 12.3, 7.0, OCH=C$\underline{H}$), 6.06 (dt, 1 H, J = 12.3, 1.3, OC$\underline{H}$=CH); ^{13}C NMR (50.3 MHz, $CDCl_3$) δ: 15.2 (CH_3), 16.3 (CH_3), 20.8 (CH_3), 21.0 (CH_2), 22.2 (CH_3), 23.4 (CH_2), 25.7 (CH), 31.5 (CH), 34.4 (CH_2), 41.1 (CH_2), 47.8 (CH), 80.0 (CH), 107.5 (CH), 144.8 (CH); mass

spectrum (electron impact), m/e 210 (M^+, 4%), 83 (100%), 69 (40%); analytical TLC (pentane): R_f 0.53 (single spot). Proton NMR confirmed the complete absence (<1%) of menthol, menthoxybutyne, menthoxybutane, and the Z-isomer. Anal. Calcd for $C_{14}H_{26}O$: C, 79.94; H, 12.46. Found: C, 79.91; H, 12.41.

Waste Disposal Information

All toxic materials were disposed of in accordance with "Prudent Practices in the Laboratory"; National Academic Press; Washington, DC, 1996.

3. Discussion

The conversion of menthol to 1-menthoxy-1-butyne described in Part A illustrates an efficient, one-pot preparation of acetylenic ethers from alcohols that is relatively inexpensive and requires only common reagents and standard laboratory equipment. It is substantially higher-yielding and more direct and generally applicable than other available approaches to these versatile[2] compounds, which generally involve base treatment of haloalkenyl ethers, dihaloalkyl ethers, haloacetals, or related compounds.[2] The conversion in this procedure is an optimized example of a method that has previously been applied with success to a wide range of secondary alcohols.[3] Several primary alcohols have also been transformed to acetylenic ethers through this method.[4]

Parts B and C exemplify efficient procedures for the stereoselective reduction of acetylenic ethers to the corresponding Z- and E-enol ethers, synthetically useful intermediates.[5] These procedures, which are optimized versions of previously described methods,[3,6] also require only common reagents and standard laboratory equipment. Alternatively, acetylenic ethers can be converted stereoselectively to E-enol ethers under Birch conditions[3] and to Z- (or E-) enol ethers with Red-Al.[7]

Other known methods for preparing O-alkyl enol ethers include, most notably, alcohol elimination from acetals, double bond isomerization in allylic ethers, reduction of alkoxy enol phosphates, and phosphorane-based condensation approaches.[5] These methods, however, suffer from poor stereoselectivity, low yields, or lack of generality, if not a combination of these drawbacks.

1. LEDSS, Chimie Recherche, Université Joseph Fourier, BP 53X, 38041 Grenoble Cédex, France.
2. For reviews, see: Arens, J. F. In "Advances in Organic Chemistry"; Raphael, R. A.; Taylor, E. C.; Wynberg, H., Eds; Interscience Publishers: New York, 1960; Vol. II, pp 117-212; Brandsma, L.; Bos, H. J. T.; Arens, J. F. In "Chemistry of Acetylenes"; Viehe, H. G., Ed.; Marcel Dekker: New York, 1969; Chapter 11; Meerwein, H. In "Methoden der Organischen Chemie (Houben-Weyl)", 4th ed.; Georg Thieme Verlag: Stuttgart, 1965; Vol. 6/3, Chapter 1, pp 116-118; Ben-Efraim, D. A. In "The Chemistry of the Carbon-Carbon Triple Bond"; Patai, S., Ed.; Wiley-Interscience: New York, 1978; Vol. 2, Chapter 18; Radchenko, S. I.; Petrov, A. A. *Russ. Chem. Rev. (Engl. Transl.)* **1989**, *58*, 948-966.
3. Moyano, A.; Charbonnier, F.; Greene, A. E. *J. Org. Chem.* **1987**, *52*, 2919-2922.
4. Nopol, 2-cyclohexylethanol, and decanol have been converted to the corresponding acetylenic ethers in 64, 68, and 71% yield, respectively. Although 1-adamantanol can be transformed to adamantyloxyacetylene in 92% yield, other acetylenic ethers derived from tertiary alcohols have been found to be unstable to purification (unpublished results). See: Pericàs, M. A.; Serratosa, F.; Valentí, E. *Tetrahedron* **1987**, *43*, 2311-2316.

5. For reviews, see: Hesse, G. In "Methoden der Organischen Chemie (Houben-Weyl)", 4th ed.; Georg Thieme Verlag: Stuttgart, 1978; Vol. 6(1d), pp 1-216; Fischer, P. In "Chemistry of Ethers, Crown Ethers, Hydroxyl Groups and Their Sulphur Analogues Supplement E"; Patai, S., Ed.; Wiley: New York, 1980; Vol. 2, Chapter 17.
6. Johnson, F.; Paul, K. G.; Favara, D. *J. Org. Chem.* **1982**, *47,* 4254-4255.
7. Solà, L.; Castro, J.; Moyano, A.; Pericàs, M. A.; Riera, A. *Tetrahedron Lett.* **1992**, *33,* 2863-2866.

Appendix

Chemical Abstracts Nomenclature (Collective Index Number); (Registry Number)

1-Menthoxy-1-butyne: Cyclohexane, 2-(1-butynyloxy)-4-methyl-1-(1-methylethyl)-, [1S-(1α,2β,4β)]- (12); (108266-28-0)

(-)-Menthol: Menthol, (-)- (8); Cyclohexanol, 5-methyl-2-(1-methylethyl)-, [1R-(1α,2β,5α)]- (9), (2216-51-5)

(Z)-1-Menthoxy-1-butene: Cyclohexane, 2-(1-butenyloxy)-4-methyl-1-(1-methylethyl)-, [1S-[1α,2β(Z),4β]]- (12); (107941-62-8)

(E)-1-Menthoxy-1-butene: Cyclohexane, 2-(1-butenyloxy)-4-methyl-1-(1-methylethyl)-, [1S-[1α,2β(E),4β]]- (12); (107941-63-9)

Potassium hydride (8,9); (7693-26-7)

Trichloroethylene: Ethylene, trichloro- (8); Ethene, trichloro- (9); (79-01-6)

Butyllithium: Lithium, butyl- (8,9); (109-72-8)

Ethyl iodide: Ethane, iodo- (8,9); (75-03-6)

Hexamethylphosphoramide: HIGHLY TOXIC. CANCER SUSPECT AGENT: Phosphoric triamide, hexamethyl- (8,9); (680-31-9)

Pyridine (8,9); (110-86-1)

Copper(II) sulfate: Sulfuric acid copper(2+) salt (1:1) (9); (7758-98-7)

Lithium aluminum hydride: Aluminate (1-), tetrahydro-, lithium (8); Aluminate (1-), tetrahydro-, lithium, (T-4)- (9); (16853-83-3)

N,N'-Dimethylpropyleneurea: 1,3-Dimethyl-3,4,5,6-tetrahydro-2(1H)-pyrimidinone: DMPU: 2(1H)-Pyrimidinone, tetrahydro-1,3-dimethyl- (8,9); (7226-23-5)

(R)-(+)-2-(DIPHENYLHYDROXYMETHYL)PYRROLIDINE

(2-Pyrrolidinemethanol, α,α-diphenyl-, (R)-)

A. pyrrolidine + t-BuO(C=O)O(C=O)O-t-Bu $\xrightarrow{CH_2Cl_2}$ N-Boc-pyrrolidine

B. N-Boc-pyrrolidine $\xrightarrow[Et_2O,\ -78°C]{sec\text{-}BuLi,\ (-)\text{-}sparteine}$ [2-lithio-N-Boc-pyrrolidine (t-BuO, Li···O)] $\xrightarrow{PhC(O)Ph}$ N-Boc-2-(Ph₂C(OH))pyrrolidine

C. N-Boc-2-(diphenylhydroxymethyl)pyrrolidine $\xrightarrow[Reflux]{NaOH\ /\ EtOH}$ 2-(diphenylhydroxymethyl)pyrrolidine (N-H)

Submitted by Nikola A. Nikolic and Peter Beak.[1]

Checked by Michael R. Reeder and Robert K. Boeckman, Jr.

1. Procedure

A. N-(tert-Butoxycarbonyl)pyrrolidine. A 500-mL round-bottomed flask, equipped with magnetic stirring bar, is charged with dichloromethane (CH_2Cl_2) (120 mL) and pyrrolidine (11.3 mL, 133 mmol) (Note 1). The flask is fitted with a 60-mL, pressure-equalizing addition funnel vented through a mineral oil bubbler and charged with a solution of di-tert-butyl dicarbonate (24.6 g, 112 mmol) in CH_2Cl_2 (35 mL). After the pyrrolidine solution is cooled to 0°C in ice, the colorless dicarbonate solution is added dropwise over a period of 30 min, and the resulting solution is stirred at room temperature for 3 hr (Note 2). The solvents are then removed under reduced pressure, and two consecutive Kugelrohr distillations of the residual oil (oven temperature 80°C at 0.2 mm) afford 16.6 g (87%) of N-Boc-pyrrolidine as a colorless oil (Note 3).

B. (R)-(+)-2-(Diphenylhydroxymethyl)-N-(tert-butoxycarbonyl)pyrrolidine. An oven-dried, 2-L, three-necked flask, equipped with a magnetic stirring bar and a thermocouple (Note 4), is charged with (-)-sparteine (30.2 mL, 131 mmol) (Note 5), N-Boc-pyrrolidine (15.0 g, 87.6 mmol), and anhydrous ether (900 mL) (Note 6). The solution is cooled to ~-70°C (dry ice/acetone bath) (Note 4). To this solution is added sec-butyllithium (96 mL, 1.16 M in cyclohexane, 111 mmol) (Notes 7 and 8) dropwise over a period of 35 min (Note 9). The reaction is then stirred at ~-70°C for 5.5 hr (Note 10).

After this interval, a solution of benzophenone (25.5 g, 140 mmol) (Note 11) in anhydrous ether (200 mL) is added dropwise over a period of 1.25 hr (Note 9). The dark green to greenish-yellow suspension is maintained at -70°C for 2.0 hr, and the reaction is then quenched by dropwise addition of glacial acetic acid (8.5 mL, 150 mmol) over a period of 15 min. The resulting lemon-yellow suspension is allowed to warm slowly to room temperature over a period of 12 hr, during which time the mixture becomes cream colored.

After the solution is warmed to 25°C, 5% phosphoric acid (H_3PO_4) (150 mL) is added to the reaction mixture, and the resulting biphasic mixture is stirred for 20 min. The layers are partitioned and the organic phase is washed with additional 5% H_3PO_4 (3 x 150 mL). Combined aqueous phases are extracted with ether (3 x 200 mL). The original organic phase and the ethereal extracts are combined, washed with brine (200 mL), dried over magnesium sulfate ($MgSO_4$), filtered, and the solvents are removed under reduced pressure to afford crude product as an off-white solid. The crude (R)-(+)-2-(diphenylhydroxymethyl)-N-(tert-butoxycarbonyl)pyrrolidine is purified by recrystallization from a mixture of hexanes-ethyl acetate (~675 mL, 20 : 1, v/v) affording in two crops 20.9 - 22.0 g (73 - 74%) of analytically pure product as a white solid (Note 12) having greater than 99.5% ee (Note 13).

Sparteine is recovered by making the aqueous phases basic with aqueous 20% sodium hydroxide (NaOH) (160 mL) (Note 14). The aqueous phase is extracted with Et_2O (4 x 150 mL), and the combined organic phases are dried over potassium carbonate (K_2CO_3), filtered, and the solvents removed under reduced pressure to afford 30.3 g (98%) of crude, recovered sparteine as a pale yellow oil (Note 15). Fractional distillation of the residual oil from calcium hydride (CaH_2) (Note 5) affords 27.0 g of sparteine (88%) suitable for reuse.

C. (R)-(+)-2-(Diphenylhydroxymethyl)pyrrolidine. A 1-L, round-bottomed flask, equipped with a magnetic stirring bar, is charged with 325 mL of absolute ethanol and NaOH (27.0 g, 675 mmol). The NaOH is dissolved with vigorous stirring, and (R)-(+)-2-(diphenylhydroxymethyl)-N-(tert-butoxycarbonyl)pyrrolidine (22.0 g, 62.3 mmol) is added (Note 16). The flask is fitted with a reflux condenser, and the resulting milky white suspension is heated to reflux for 2.5 hr. The suspension is cooled to room temperature, and the solvents are removed under reduced pressure. To the residual off-white solids are added ether (800 mL) and deionized water (400 mL). The suspension is stirred until the solids are dissolved, and the resulting biphasic mixture is transferred to a 2-L separatory funnel. The layers are partitioned, and the aqueous phase is extracted with ether (4 x 200 mL). The organic phases are combined, dried over K_2CO_3 (ca. 100 g), filtered, and the solvents are removed under reduced pressure (Note 17) to afford 14.5 - 15.8 g (92 - 100%) of the pure title compound as a white solid (Note 18).

2. Notes

1. Dichloromethane was obtained from Mallinckrodt Inc., and was used without further purification. Pyrrolidine and di-tert-butyl dicarbonate were obtained from Aldrich Chemical Company, Inc., and used as received.

2. Toward the end of the addition, gas evolution (CO_2) occurs. Care should be taken to provide adequate venting to avoid pressure buildup.

3. The product, N-Boc-pyrrolidine, has the following spectral characteristics: ^{1}H NMR ($CDCl_3$, 300 MHz) δ: 1.43 (s, 9 H), 1.81 [s (br), 4 H], 3.27 (m, 4 H); ^{13}C NMR ($CDCl_3$, 75 MHz) δ: 24.78, 25.54, 28.32, 45.38, 45.71, 78.57, 154.41; IR (film) cm^{-1}: 2974, 2875, 1698, 1403, 1168, 877, 772 .

4. The internal temperature was monitored throughout the reaction with an Omega D730 or equivalent thermocouple.

5. Sparteine is liberated from the commercially available sulfate salt (Aldrich Chemical Company, Inc.) as follows: Sparteine sulfate pentahydrate (100 g, 240 mmol) is dissolved in deionized water (125 mL), and to this solution is slowly added aqueous 20% NaOH (100 mL). The resulting milky-white, oily mixture is then extracted with ether (4 x 150 mL). The combined ethereal extracts are dried over anhydrous K_2CO_3, filtered, and the solvent is removed under reduced pressure. Vacuum distillation of the residual oil from CaH_2 affords 52 g (92%) of sparteine as a clear, colorless to slightly yellow, viscous oil (bp 115-120°C/0.3 mm). The sparteine free base readily absorbs atmospheric carbon dioxide (CO_2) and should be stored under argon at -20°C in a freezer.

6. Anhydrous ethyl ether was obtained by distillation under nitrogen from sodium benzophenone ketyl.

7. sec-Butyllithium (1.3 M in cyclohexane) was titrated in toluene immediately before use using a standard solution (1 M) of sec-butyl alcohol in o-xylene with 0.2% 2,2'-biquinoline in toluene as the indicator according to Watson and Eastham.[2]

8. The checkers found that the yields obtained in this procedure are critically dependent on the quality of the sec-butyllithium employed. Best results are obtained with fresh (< 3 months shelf life) commercial samples (FMC Lithium Division, and Aldrich Chemical Company, Inc.) that are colorless to deep yellow, largely free of precipitated salts, and that have been kept refrigerated and have not been exposed to traces of moisture or oxygen by extensive previous sampling. Samples of sec-butyllithium that contain alkoxide or hydroxide undergo alkoxide/hydroxide-catalyzed decomposition to butene and lithium hydride (LiH), particularly when stored at room temperature; the latter cannot be readily removed. In the hands of the checkers, such aged sec-butyllithium samples provide the N-Boc amino alcohol of comparable enantiomeric purity in ~5-15% lower yield.

9. During the addition, the internal temperature of the reaction did not exceed -68°C.

10. The reaction mixture became milky white during this interval.

11. Commercially available benzophenone (Aldrich Chemical Company, Inc., 99+ %) was used without further purification.

12. The product has the following characteristics: ^{1}H NMR ($CDCl_3$, 300 MHz) δ: 0.65-0.80 (m, 1 H), 1.40-1.60 (s, 11 H), 1.82-1.95 (m, 1 H), 1.98-2.14 (m, 1 H), 2.75-2.90 (m, 1 H), 3.20-3.45 (m, 1 H), 4.86 (dd, 1 H, J_1 = 8.9, J_2 = 3.8), 7.20-7.45 (m, 10 H, Ar-H); ^{13}C NMR ($CDCl_3$, 75 MHz) δ: 22.84, 28.29, 29.67, 47.77, 65.50, 80.54, 81.62, 126.96, 127.00, 127.27, 127.61, 127.77, 128.14, 143.72, 146.41, 159.00; IR (film) cm^{-1}: 3370, 2979, 1659, 1415, 1164, 763, 70 . Anal. Calcd for $C_{22}H_{27}NO_3$: C, 74.76; H, 7.70; N, 3.96. Found: C, 74.62; H, 7.72; N, 4.13. $[\alpha]_D^{25}$ +150° ($CHCl_3$, *c* 3.62); mp 150.5-152°C. The checkers found $[\alpha]_D^{25}$ +144° ($CHCl_3$, *c* 3.89).

13. The enantiomeric excess was determined by alkaline ethanolysis of the Boc group followed by conversion of the amine to the 3,5-dinitrobenzamide. HPLC analysis of a saturated solution of the benzamide in 5% 2-propanol in hexane using a Pirkle Covalent S-N1N-Naphthylleucine Column (Regis Chemical Company) with 5% 2-propanol in hexane as the eluent, and a flow rate of 1.5 mL/min indicates a single peak with retention time of 35 min. HPLC analysis of the corresponding racemic benzamide affords two peaks at 27 and 32 min corresponding to the (S)- and (R)-enantiomers respectively. The differences in retention times arise from the size of sample that was injected; the ^{1}H NMR of the racemic and enantio-enriched benzamides were identical.

14. The aqueous solution was brought to ca. pH 11 as tested by Hydrion Paper (Micro Essential Laboratories, Brooklyn, NY).

15. The ^{1}H NMR of the recovered (-)-sparteine was identical to that of pure (-)-sparteine.

16. The concentration of the reactants is ~0.2 M. If more dilute solutions of base are employed (0.01 M), the checkers found that the reaction required at least 24 hr to completion and that impure product was obtained. At higher dilution, formation of significant amounts of the cyclic urethane was observed, and this by-product required removal by chromatography.

17. The residual, colorless, viscous oil solidified slowly upon exposure to high vacuum.

18. The title compound has the following characteristics: ^{1}H NMR ($CDCl_3$, 300 MHz) δ: 1.40-1.95 (m, 5 H), 2.89-3.05 (m, 2 H), 4.23 (t, 1 H, J = 7.4), 4.55 [(s (br), 1 H], 7.10-7.35 (m, 4 H), 7.45-7.60 (m, 6 H); ^{13}C NMR ($CDCl_3$, 75MHz) δ: 25.47, 26.23, 46.71, 64.42, 76.65, 125.47, 125.80, 126.29, 126.40, 127.91, 128.17, 145.36, 148.13; IR (film) cm^{-1}: 3352, 2968, 1598, 1492, 1448, 1173, 748, 701. Anal. Calcd for $C_{17}H_{19}NO$: C, 80.60; H, 7.56; N, 5.53. Found: C, 80.56; H, 7.60; N, 5.68. $[\alpha]_D^{25}$

+73.8° ($CHCl_3$, *c* 3.37); mp 76-77°C; R_f = 0.13 (CH_2Cl_2 : MeOH, 95 : 5). The checkers found $[\alpha]_D^{25}$ +67.9° ($CHCl_3$, *c* 3.37)

Waste Disposal Information

All toxic materials were disposed of in accordance with "Prudent Practices in the Laboratory"; National Academic Press; Washington, DC, 1996.

3. Discussion

α,α-Diaryl-2-pyrrolidinemethanols represent an important class of ligands for asymmetric synthesis.[3] For example, reaction of these amino alcohols with boranes affords oxazaborolidinones that are effective enantioselective reagents for asymmetric reduction of ketones.[3,4,5] The chiral non-racemic amino alcohols have been prepared through addition of organometallic agents to enantio-enriched proline,[3,5,6] or by resolution of racemic pyrrolidinemethanols.[5e] The procedure reported here describes a new approach to the synthesis of the title compound based on an asymmetric lithiation/substitution sequence.[7]

Treatment of Boc-pyrrolidine with sec-butyllithium in the presence of (-)-sparteine affords the putative enantio-enriched organolithium reagent. This organolithium reagent can be quenched with electrophiles to afford 2-substituted-Boc-pyrrolidines.[7] This approach offers several advantages over existing methodologies. First, the use of the (-)-sparteine ligand affords 2-substituted pyrrolidines with high enantioselectivity. Second, the ligand can be recovered and purified in high yields; [in this example, (-)-sparteine was recovered and purified in 88% yield]. Third, this approach obviates the preparation of enolizable proline derivatives that have been shown to racemize.[3] Finally, this two-step approach affords the (R)-α,αdiphenylpyrrolidine enantiomer that has previously been obtained from relatively expensive "unnatural" D-proline.

The approach reported here should facilitate the preparation of α,α-disubstituted-pyrrolidinemethanol analogs. By using this methodology, a single enantio-enriched organolithium intermediate can be treated with a variety of electrophiles (e.g., diaryl ketones) to afford aryl-substituted analogs of the title compound. Previously reported syntheses involve a variety of nucleophilic organometallic reagents that must be prepared and treated with proline derivatives.

An interesting feature of this study is the enantiomeric purity analysis of the products. By converting the amine functionality of the pyrrolidine to a 3,5-dinitrobenzamide, the substrate can be analyzed by chiral HPLC. To date, the 3,5-dinitrobenzamide of 2-substituted pyrrolidines that the submitters have prepared have been baseline-resolved by the Pirkle S-N1N-Naphthylleucine column. This approach obviates the need for MPTA-derivativization which has been previously employed in enantiomeric purity determinations.[3,5]

1. Department of Chemistry, University of Illinois, Urbana, IL 61801.
2. Watson, S. C.; Eastham, J. F. *J. Organomet. Chem.* **1967**, *9*, 165.
3. See: Mathre, D. J.; Jones, T. K.; Xavier, L. C.; Blacklock, T. J.; Reamer, R. A.; Mohan, J. J.; Jones, E. T. T.; Hoogsteen, K.; Baum, M. W.; Grabowski, E. J. J. *J. Org. Chem.* **1991**, *56*, 751-762, and references therein.
4. For recent reviews of oxazaborolidinone-catalyzed reductions, see: (a) Deloux, L.; Srebnik, M. *Chem. Rev.* **1993**, *93*, 763-784; (b) Wallbaum, S.; Martens, J. *Tetrahedron: Asymmetry* **1992**, *3*, 1475-1504; (c) Singh, V. K. *Synthesis* **1992**, 605-617.

5. For diaryl amino alcohols as enantio-enriched oxazaborolidine precursors, see: (a) Itsuno, S.; Ito, K.; Hirao, A.; Nakahama, S. *J. Chem. Soc., Chem. Commun.* **1983**, 469-470; (b) Itsuno, S.; Ito, K.; Hirao, A.; Nakahama, S. *J. Org. Chem.* **1984**, *49*, 555-557; (c) Itsuno, S.; Sakurai, Y.; Ito, K.; Hirao, A.; Nakahama, S. *Bull. Chem. Soc. Jpn.* **1987**, *60*, 395-396. For proline-derived diaryl amino alcohols as enantio-enriched oxazaborolidine precursors, see: (d) Corey, E. J.; Bakshi, R. K.; Shibata, S. *J. Am. Chem. Soc.* **1987**, *109*, 5551-5553; (e) Corey, E. J.; Bakshi, R. K.; Shibata, S. Chen, C.-P.; Singh, V. K. *J. Am. Chem. Soc.* **1987**, *109*, 7925-7926; (f) Corey, E. J.; Shibata, S.; Bakshi, R. K. *J. Org. Chem.* **1988**, *53*, 2861-2863.
6. Jones, T. K.; Mohan, J. J.; Xavier, L. C.; Blacklock, T. J.; Mathre, D. J.; Shohar, P.; Jones, E. T. T.; Reamer, R. A.; Roberts, F. E.; M. W.; Grabowski, E. J. J. *J. Org. Chem.* **1991**, *56*, 763-769.
7. (a) Kerrick, S. T.; Beak, P. *J. Am. Chem. Soc.* **1991**, *113*, 9708-9709; (b) Beak, P.; Kerrick, S. T., Wu, S.; Chu, J. *J. Am. Chem. Soc.* **1994**, *116* , 3231-3239.

Appendix

Chemical Abstracts Nomenclature; [Registry Number]

(R)-(+)-2-Diphenylhydroxymethyl)pyrrolidine: 2-Pyrrolidinemethanol, α,α-diphenyl-, (+)- (8); 2-Pyrrolidinemethanol, α,α-diphenyl-, (R)- (9); (22348-32-9)

N-(tert-Butoxycarbonyl)pyrrolidine: 1-Pyrrolidinecarboxylic acid, 1,1-dimethylethyl ester (11); (86953-79-9)

Pyrrolidine (8,9); (123-75-1)

Di-tert-butyl dicarbonate: Formic acid, oxydi-, di-tert-butyl ester (8); Dicarbonic acid, bis(1,1-dimethylethyl) ester; (9); (24424-99-5)

(R)-(+)-2-(Diphenylhydroxymethyl)-N-(tert-butoxycarbonyl)pyrrolidine: 1-Pyrrolidinecarboxylic acid, 2-(hydroxydiphenylmethyl)-, 1,1-dimethylethyl ester, (R)- (12); (137496-68-5)

sec-Butyllithium: Lithium, sec-butyl- (8); Lithium, (1-methylpropyl)- (9); (598-30-1)

(-)-Sparteine: Sparteine (8); 7,14-Methano-2H,6H-dipyrido[1,2-a:1',2'-e]-[1,5]diazocine, dodecahydro-, [7S-(7α,7aα,14α,14aβ)]- (9); (90-39-1)

(-)-Sparteine sulfate pentahydrate: Spateine, sulfate (1:1), pentahydrate (9); (6160-12-9)

Benzophenone (8); Methanone, diphenyl- (9); (119-61-9)

Phosphoric acid (8,9); (7664-38-2)

(-)-Sparteine sulfate pentahydrate: Sparteine, sulfate (1:1), pentahydrate; (6160-12-9)

Glacial acetic acid: Acetic acid; (64-19-7)

(R)-(-)-2,2-DIPHENYLCYCLOPENTANOL

(Cyclopentanol, 2,2-diphenyl-, (R)-)

A. Ph_2CHCN → (LDA, THF; Br(CH₂)₃CN; KOt-Bu, t-BuOH, THF) → 2-amino-3,3-diphenyl-1-cyclopentene-1-carbonitrile

B. → (HCl, H_2O, 100°C) → 2,2-diphenylcyclopentanone

C. 2,2-diphenylcyclopentanone + oxazaborolidine (B-CH₃) → ($BH_3 \cdot SMe_2$, THF, 40°C) → **1**

Submitted by Scott E. Denmark, Lawrence R. Marcin, Mark E. Schnute, and Atli Thorarensen.[1]

Checked by David J. Mathre, Khateeta M. Emerson, and Ichiro Shinkai.

1. Procedure

A. 2-Amino-3,3-diphenyl-1-cyclopentene-1-carbonitrile. To a 2-L, three-necked, round-bottomed flask equipped with a 250-mL, pressure-equalizing addition funnel, magnetic stirrer, nitrogen/vacuum adapter, and a thermometer is added 39.9 mL (0.29 mol) of diisopropylamine and 200 mL of tetrahydrofuran (THF) (Notes 1, 2). The solution is cooled to 0°C, and 101.6 mL of butyllithium (2.55 M in hexane, 0.26

mol) is added slowly (Note 3). After 10 min, a solution of 50.0 g (0.26 mol) of diphenylacetonitrile (Note 4) in 200 mL of THF is added over 30 min forming a deep yellow solution. A solution of 28.3 mL (0.29 mol) of 4-bromobutyronitrile (Notes 5, 6) in 200 mL of THF is then added over 20 min. The resulting bright-yellow solution is allowed to warm slowly to room temperature overnight (10 hr) (Note 7). The reaction mixture is quenched by the slow addition of water (25 mL) and then is diluted with 400 mL of tert-butyl methyl ether (MTBE), and washed with water (2 x 100 mL) and brine (100 mL). The aqueous layers are back-extracted with MTBE (100 mL). The combined organic layers are dried with sodium sulfate (Na_2SO_4), concentrated on a rotary evaporator, and the resulting crude dinitrile is placed under high vacuum (0.2 mm) for 1 hr. The dinitrile is transferred to a 2-L, three-necked, round-bottomed flask equipped with a reflux condenser, magnetic stirrer, nitrogen/vacuum adapter, and a thermometer and is dissolved in a mixture of tert-butyl alcohol (400 mL) and THF (200 mL). To the solution is added 23.25 g (0.21 mol) of potassium tert-butoxide, and the suspension is heated at 60°C (internal temperature) for 2 hr. After the reaction mixture is cooled to room temperature, it is quenched with water (25 mL), diluted with MTBE (500 mL), and washed with water (100 mL) and brine (3 x 100 mL). The aqueous layers are back-extracted with MTBE (100 mL), and the combined organic layers are dried (Na_2SO_4) and concentrated on a rotary evaporator to afford an off-white granular solid. The crude product is suspended in MTBE (75 mL), cooled (0°C), filtered, and recrystallized from absolute ethanol (400 mL). The mother liquor is concentrated on a rotary evaporator, purified by column chromatography on silica gel (330 g) (Note 8) eluting with hexane/EtOAc (4/1), and recrystallized from absolute ethanol to afford a combined yield of 57.7 g (86%) of the enaminonitrile as a white solid (Note 9,10).

B. 2,2-Diphenylcyclopentanone. To a 3-L, three-necked, round-bottomed flask equipped with a mechanical stirrer, thermometer, and wide inner-spiral reflux condenser (Note 11) is added 57.5 g (0.22 mol) of the enaminonitrile from step A and

800 mL of concd hydrochloric acid (HCl, Note 12). The mixture is stirred for 5 min, and 800 mL of water is added. The reaction mixture is heated to reflux (heating mantle, 110°C internal temperature) with vigorous stirring for 4 days (Notes 13, 14, 15). The reaction mixture is cooled to room temperature, and extracted with dichloromethane (CH_2Cl_2, 5 x 200 mL). The organic layers are washed with saturated aqueous sodium bicarbonate ($NaHCO_3$, 100 mL) and brine (100 mL), and the aqueous layers are back-extracted with dichloromethane (100 mL). The combined organic layers are dried ($MgSO_4$) and concentrated on a rotary evaporator. The crude product is recrystallized from MTBE (300 mL). The mother liquor is concentrated on a rotary evaporator, and purified by column chromatography on silica gel (250 g) eluting with hexane/EtOAc (4/1), decolorized with carbon, and recrystallized from MTBE to afford a combined yield of 48.2 g (92%, Note 16) of the ketone as a white solid (Notes 17, 18).

C. (R)-(-)-2,2-Diphenylcyclopentanol **1**. In a 500-mL, three-necked, round-bottomed flask, equipped with a 125-mL graduated, pressure-equalizing addition funnel, 30-mm, egg-shaped magnetic stir bar, nitrogen/vacuum adapter, and an internal temperature probe is placed 1.76 g (6.34 mmol) of the B-methyloxazaborolidine catalyst (Notes 1, 19, 20). The apparatus is evacuated, flushed with nitrogen, charged with 86 mL of dry THF and 6.34 mL (63.4 mmol) of borane-methyl sulfide complex, and then warmed to 40°C (internal temperature) (Note 21). In a 250-mL, three-necked, round-bottomed flask, equipped with a nitrogen/vacuum adapter and magnetic stirrer is placed 15 g (63.4 mmol) of 2,2-diphenylcyclopentanone. The flask is evacuated, flushed with nitrogen, and charged with 111 mL of dry THF. After dissolution, the ketone solution is transferred to the addition funnel via cannula and added dropwise over 8 hr to the stirred catalyst solution maintained at 40°C (Note 22). After complete addition, the funnel is rinsed into the reaction vessel with 10 mL of dry THF, and the resulting reaction mixture is stirred at 40°C for an additional 30 min. Finally, the reaction mixture is cooled to 0-5°C and carefully quenched by the dropwise addition of 100 mL of methanol (*Caution: considerable hydrogen evolution occurs after a short induction period*) (Note 23). The cold bath is removed and the reaction is stirred until gas evolution ceases (Note 24). The resulting solution is poured into a 1-L, round-bottomed flask and the reaction vessel is rinsed with 50 mL of methanol. A simple distillation head is attached to the 1-L flask and 100 mL of solvent is distilled (*Caution: the distillate contains malodorous methyl sulfide*). An additional 100 mL of fresh methanol is added and 100 mL of solvent is again distilled. The residue is cooled to room temperature and concentrated on a rotary evaporator to afford a slightly yellow oil. The oil is dissolved in MTBE (250 mL), washed with aqueous 0.1 N aqueous hydrochloric acid (3 x 100 mL), and the combined acidic, aqueous phases are back-extracted with MTBE (100 mL). The combined organic phases are washed with water (100 mL) and brine (100 mL), dried (Na_2SO_4), filtered, and concentrated on a rotary evaporator to afford 15.1 g of an off-white solid. The solid is purified by bulb-to-bulb distillation (bp 180°C/0.2 mm) to afford 14.7 g (97%) of (R)-(-)-2,2-diphenylcyclopentanol (92% ee) as an analytically-pure, white solid (Notes 25, 26). Multiple recrystallizations of the product from hexane afford 11.3 g (75%) of (R)-(-)-2,2-diphenylcyclopentanol (>97% ee) (Notes 26, 27, 28). To recover the catalyst precursor, (S)-α,α–diphenyl-2-pyrrolidinemethanol, the acidic, aqueous phase is made basic (blue to litmus) by addition of 25 mL of aqueous 25% sodium hydroxide solution. The aqueous phase is extracted with dichloromethane (3 x 100 mL), and the combined organic phases are washed with brine (100 mL), dried (Na_2SO_4), filtered, and concentrated on a rotary evaporator to afford a clear oil that crystallizes under high vacuum (0.1 mm, several hours). The solid is recrystallized from hexane to afford 1.5 g (93% recovery) of (S)-α,α-diphenyl-2-pyrrolidinemethanol as a white crystalline solid (Note 29).

2. Notes

1. All glassware was dried in an oven (140°C) and after assembly was allowed to cool under an atmosphere of dry nitrogen.

2. THF was freshly distilled from sodium/benzophenone. tert-Butyl alcohol was purchased from Aldrich Chemical Company, Inc., and was used without further purification. Solvents for extraction and chromatography were technical grade and distilled from the indicated drying agents: hexane ($CaCl_2$); dichloromethane ($CaCl_2$); tert-butyl methyl ether (MTBE) ($CaSO_4/FeSO_4$); ethyl acetate (K_2CO_3).

3. Butyllithium was freshly titrated by the method of Gilman.[2] Excess strong base, either butyllithium or lithium diisopropylamide (LDA), (exceeding 1 equiv per diphenylacetonitrile) should be avoided since the resulting Thorpe-Ziegler cyclization product is susceptible to fragmentation under the reaction conditions to afford 1,1-dicyano-4,4-diphenylbutane.

4. Diphenylacetonitrile was purchased from Aldrich Chemical Company, Inc., and recrystallized from hexane (mp 73-75°C).

5. 4-Bromobutyronitrile was purchased from Aldrich Chemical Company, Inc., and was freshly distilled (bp 95-98°C, 15 mm).

6. 4-Iodobutyronitrile may also be used as a less expensive alternative available from 4-chlorobutyronitrile[3] by a modification of the above procedure. The increased reactivity of the iodide, however, requires a more tedious procedure, but is provided as follows:

A. 4-Iodobutyronitrile. To a 1-L, three-necked, round-bottomed flask equipped with a mechanical stirrer and a reflux condenser is placed a solution of 70 mL (0.77 mol) of 4-chlorobutyronitrile in 420 mL of acetone. To the solution is added 123.4 g (0.82 mol) of sodium iodide, and the resulting clear solution is heated to reflux for 23 hr. Over time, the formation of large amounts of a white precipitate is observed. The resulting suspension is cooled to room temperature, filtered, and the filter cake is washed with dichloromethane (200 mL). The combined organic layers are concentrated on a rotary evaporator. The residue is redissolved into dichloromethane (200 mL), washed with saturated aqueous sodium thiosulfate ($Na_2S_2O_3$, 50 mL) and brine (50 mL), dried (Na_2SO_4), and concentrated on a rotary evaporator. The resulting oil is distilled (bp 80-92°C/0.8 mm) to afford 141.6 g (92%) of 4-iodobutyronitrile as a clear colorless oil.

B. 2-Amino-3,3-diphenyl-1-cyclopentene-1-carbonitrile. In a 1-L, three-necked, round-bottomed flask, a solution of 53.8 mL (0.38 mol) of diisopropylamine in 250 mL of dry THF is cooled to -70°C. To the solution is slowly added 118 mL of butyllithium (2.95 M in hexane, 0.34 mol) at such a rate that the internal temperature never exceeds -50°C. The resulting solution is cooled to -70°C, and a solution of 67.5 g (0.34 mol) of diphenylacetonitrile in 250 mL of dry THF is added over 30 min forming a black solution that is stirred an additional 20 min. In a 2-L, three-necked, round-bottomed flask, a solution of 4-iodobutyronitrile (74.9 g, 0.38 mol) in 250 mL of dry THF is cooled to -77°C. Using a Teflon cannula (0.5-cm diameter) the anion of diphenylacetonitrile is added very rapidly to the 4-iodobutyronitrile solution (internal temperature rose by only 3°C). The resulting light-yellow solution is stirred for 80 min at -78°C, warmed to 0°C for 60 min, and allowed to stir at room temperature for 30 min. The reaction mixture is quenched with water (34 mL), diluted with MTBE (500 mL), and washed with water (2 x 250 mL) and brine (250 mL). The aqueous layers are back-extracted with MTBE (200 mL), and the combined organic layers are dried (Na_2SO_4) and concentrated on a rotary evaporator. The resulting crude dinitrile is dissolved in a mixture of tert-butyl alcohol (540 mL) and THF (270 mL). To the solution is added 31.0 g (0.28 mol) of potassium tert-butoxide, and the suspension is heated at 60°C (internal temperature) for 2 hr. After the reaction mixture is cooled to room temperature, it is diluted with MTBE (600 mL), and washed with water (150 mL) and brine (3 x 150 mL).

The aqueous layers are back-extracted with MTBE (150 mL), and the combined organic layers are dried ($MgSO_4$) and concentrated on a rotary evaporator. The crude product is suspended in MTBE (75 mL), cooled, filtered, and recrystallized from absolute ethanol (600 mL) to afford 67.5 g of pure ketone. The mother liquor is concentrated on a rotary evaporator, purified by column chromatography [hexane/EtOAc (8/1, 4/1)], and crystallized from absolute ethanol to afford 8.6 g (9.5%) of additional material for a combined yield of 76.1 g (84%) of analytically pure enaminonitrile as a white solid giving identical spectral data to that reported above (Anal. Calcd for $C_{18}N_{16}N_2$: C, 83.04; H, 6.20; N, 10.76. Found: C, 83.06; H, 6.20; N, 10.74).

7. The following reverse-phase HPLC assay was developed to monitor steps A-C. Column: YMC J'Sphere H80 (4.6 x 250 mm); eluent: 45:55 H_2O (20 mM H_3PO_4)/MeCN; flow rate: 1.0 mL/min; column temp.: 45°C; detection: UV (210 nm). Retention times: "diphenylprolinol" (1.85 min, with solvent front); 4-bromobutyronitrile (4.0 min); "enaminoamide" (6.1 min); "dinitrile" (11.4 min); diphenylacetonitrile (12.0 min); "cyanoketone" (12.5 min); diphenylcyclopentanol (13.1 min); "enaminonitrile" (14.4 min); "diphenylcyclopentanone" (19.5 min).

8. Kieselgel 60 (230-400 mesh) was purchased from EM Science.

9. Rather than purifying the mother liquors by column chromatography, the checkers obtained a second crop of crystals for a combined yield of 57.3 to 60.3 g (85-89%). The checkers also note that the crude product can be used "as is" in the next step after being suspended and washed with cold MTBE.

10. The physical properties are as follows: mp 145-148°C; ^1H NMR (400 MHz, $CDCl_3$) δ: 2.46 (dd, 2 H, J = 6.8, 5.6), 2.63 (dd, 2 H, J = 7.6, 6.3), 4.38 (br, 2 H, NH_2), 7.23-7.37 (m, 10 H); ^{13}C NMR (100 MHz, $CDCl_3$) δ: 27.94, 41.48, 62.76, 76.19, 118.77, 127.08, 128.19, 128.40, 142.82, 164.78; IR (CCl_4) 3490 (w), 3395 (w), 3063 (w), 2954 (w), 2863 (w), 2197 (m), 1643 (s), 1595 (m); MS (EI, 70 eV) 260 (M^+, 100),

259 (31), 183 (49), 182 (36); TLC R_f = 0.38 (hexane/EtOAc, 4/1). Anal. Calcd for $C_{18}H_{16}N_2$: C, 83.05; H, 6.19; N, 10.76. Found: C, 83.34; H, 6.07; N, 10.84.

11. The product may sublime into the condenser causing it to become clogged. This can be prevented by periodically washing down the solids with 6 N HCl.

12. To ensure complete consumption of the intermediate cyano ketone, the ratio of enaminonitrile to solvent volume cannot be altered.

13. Efficient stirring and heating to a vigorous reflux is crucial for complete consumption of the intermediate cyano ketone.

14. The reaction progress can be monitored by ^1H NMR integration of the signals of the ketone (dt, 1.95 ppm, 2 H) and the cyano ketone intermediate (t, 3.43 ppm, 1 H) determined from a crude reaction sample following the same work-up as reported above. Spectral data for 2-cyano-5,5-diphenylcyclopentanone are as follows: ^1H NMR (400 MHz, $CDCl_3$) δ: 2.22 (m, 1 H), 2.47 (m, 1 H), 2.70 (ddd, 1 H, J = 12.5, 9.8, 6.1), 2.90 (dt, 1 H, J_d = 12.5, J_t = 6.1), 3.43 (t, 1 H, J = 9.0), 7.15-7.40 (m, 10 H). Alternatively, high pressure liquid chromatography (HPLC) analysis may be used employing a Supelco LC-Si (5μ, 250 x 4.5 mm) column (hexane/EtOAc, 9/1, 1.5 mL/min, detector λ = 254 nm); R_t: "ketone" (4.6 min, response factor = 1.50), R_t: "cyano ketone" (16.8 min, response factor = 0.76).

15. The progress of the reaction was monitored by HPLC. The checkers found the reaction to take from 4 to 7 days to reach > 99% completion.

16. Yields ranged from 84% to 92%.

17. Rather than purifying the mother liquors by column chromatography, the checkers obtained a second crop of crystals for a combined yield of 87-95%.

18. The physical properties are as follows: mp 85-88°C; ^1H NMR (400 MHz, $CDCl_3$) δ: 1.95 (dt, 2 H, J_d = 13.4, J_t = 7.3) 2.46 (t, 2 H, J = 7.7), 2.73 (t, 2 H, J = 6.6), 7.21-7.32 (m, 10 H); ^{13}C NMR (100 MHz, $CDCl_3$) δ: 18.79, 38.07, 38.16, 62.44, 126.69, 127.96, 128.31, 142.02, 217.77; IR (CCl_4) cm^{-1}: 3061 (m), 3033 (m), 2969 (m),

2886 (m), 1744 (s), 1494 (s), 1446 (m), 1406 (m), 1143 (m), 1104 (m); MS (EI, 70 eV) 236 (M^+, 47), 208 (11), 180 (100), 179 (37), 178 (25), 165 (43); TLC R_f = 0.48 (hexane/EtOAc, 8/1). Anal. Calcd for $C_{17}H_{16}O$: C, 86.41; H, 6.82. Found: C, 86.57; H, 6.75.

19. A Baxter Diagnostics Inc. Type K Thermo-Couple Thermometer was used to monitor the internal temperature of the reaction solution.

20. (S)-Tetrahydro-1-methyl-3,3-diphenyl-1H,3H-pyrrolo[1,2-c][1,3,2]oxazaborole was prepared from (S)-proline in two steps according to the literature procedure[4] and purified by bulb-to-bulb distillation (170°C, 0.2 mm). The enantiomeric purity of the intermediate, (S)-α,α-diphenyl-2-pyrrolidinemethanol, was determined to be >99% ee by chiral HPLC analysis of its corresponding N-p-toluenesulfonamide derivative (DIACEL Chiralcel OD column; hexane/ethanol, 92/8; 1.0 mL/min; R_t(S) 8.6 min; R_t(R) 12.8 min). The checkers used the crystalline β-methyloxazaborolidine-borane complex (1.84 g, 6.34 mmol) as the catalyst.

21. Borane-methyl sulfide complex was purchased from Aldrich Chemical Company, Inc., and was used without purification.

22. Addition of the ketone solution over a 6-hr period affords almost identical results. However, variation of the reaction temperature can have a dramatic effect on product ee.[5]

23. Anhydrous, reagent-grade methyl alcohol was purchased from Mallinckrodt Inc. and used without purification.

24. Hydrogen evolution stops after 2-3 hr; however, for convenience the reaction can be allowed to stand at room temperature overnight with no deleterious effect on yield or enantioselectivity.

25. Anal. Calcd for $C_{17}H_{18}O$: C, 85.67; H, 7.61. Found: C, 85.65; H, 7.63.

26. Enantiomeric excess is determined by chiral HPLC analysis (DIACEL Chiralcel OJ column; hexane/ethanol, 70/30; 1.0 mL/min; R_t: S-isomer (8.8 min); R_t: R-isomer (17.9 min).

27. The checkers determined enantiomeric purity by supercritical fluid chromatography (SFC) using a Chiralpak AD (4.6 x 250 mm) column. Eluent: carbon dioxide (300 Bar); modifier: methanol (24%); flow rate: 1.5 mL/min; detection: UV (210 nm). Retention times were as follows: "diphenylcyclopentanone" (3.9 min); (R)-diphenylcyclopentanol (5.9 min); (S)-diphenylcyclopentanol (10.4 min).

28. The product is recrystallized two times by dissolution in boiling hexane (60 mL and 50 mL) and cooling to room temperature to provide 9.8 g of material with greater than 97% ee. The mother liquors are then combined, concentrated and recrystallized four times from hexane (20 mL, 15 mL, 10 mL and 10 mL) to provide 1.5 g of additional (R)-(-)-2,2-diphenylcyclopentanol with greater than 97% ee. The physical properties are as follows: mp 76-77°C; ^1H NMR ($CDCl_3$, 400 MHz) δ: 1.28 (dd, 1 H, J = 4.9, 0.7), 1.55-1.75 (m, 2 H), 1.93 (m, 1 H), 2.10 (m, 1 H), 2.32 (ddd, 1 H, J = 12.9, 8.7, 3.3), 2.66 (dt, 1 H, J_t = 12.9, J_d = 8.9), 4.88 (dd, 1 H, J = 9.7, 4.8), 7.14-7.33 (m, 10 H); ^{13}C NMR (100 MHz, $CDCl_3$) δ: 19.95, 31.67, 34.60, 59.71, 77.57, 125.89, 126.33, 126.92, 128.17, 128.44, 128.53, 144.26, 146.87; MS (EI, 70 eV) 239 (8), 238 (M^+, 44), 78 (12), 167 (100), 115 (16), 91 (11); IR (CCl_4) cm^{-1}: 3585 (m), 3061 (m), 3025 (w), 2967 (s), 2916 (w), 1495 (s), 1446 (s), 1288 (w), 1094 (m), 1074 (m), 1034 (m), 1015 (m); $[\alpha]_D^{26}$ -114.8° (EtOH, *c* 1.17). Anal. Calcd for $C_{17}H_{18}O$: C, 85.67; H, 7.61. Found: C, 85.65; H, 7.59.

29. The solid is recrystallized by dissolution in boiling hexane (20 mL) and cooling to 0°C to afford 1.37 g of (S)-α,α-diphenyl-2-pyrrolidinemethanol. The mother liquor is then concentrated and recrystallized from hexane (10 mL) to afford an additional 0.13 g of material. The physical properties are as follows: mp 75-76°C; ^1H

NMR (400 MHz, $CDCl_3$) 1.58-1.74 (m, 4 H), 2.96 (m, 1 H), 3.02 (m, 1 H), 4.26 (t, 1 H, J = 7.6), 7.14-7.59 (m, 10 H).

Waste Disposal Information

All toxic materials were disposed of in accordance with "Prudent Practices in the Laboratory"; National Academic Press; Washington, DC, 1996. The malodorous, methanol distillate resulting from the borane reduction was first treated with commercial bleach before disposal.

3. Discussion

(R)-(-)-2,2-Diphenylcyclopentanol (**1**) is a highly effective chiral auxiliary in asymmetric synthesis. Hydrogenation of chiral β-acetamidocrotonates derived from this alcohol has afforded the corresponding β-amido esters with high diastereoselectivity (96% de).[6] In addition, (R)-**1** has been used as a chiral auxiliary in Mn(III)-based oxidative free-radical cyclizations to provide diastereomerically enriched cycloalkanones (60% de).[7] Our interest in (R)-(-)-2,2-diphenylcyclopentanol is its utility as a chiral auxiliary in Lewis acid-promoted, asymmetric nitroalkene [4+2] cycloadditions. The 2-(acetoxy)vinyl ether derived from alcohol (R)-**1** is useful for the asymmetric synthesis of 3-hydroxy-4-substituted pyrrolidines from nitroalkenes (96% ee).[8] In a similar fashion, a number of enantiomerically enriched (71-97% ee) N-protected, 3-substituted pyrrolidines have been prepared in two steps from 2-substituted 1-nitroalkenes and (R)-2,2-diphenyl-1-ethenoxycyclopentane (**2**) (see Table).[9]

TABLE

SYNTHESIS OF OPTICALLY ACTIVE 3-SUBSTITUTED PYRROLIDINES

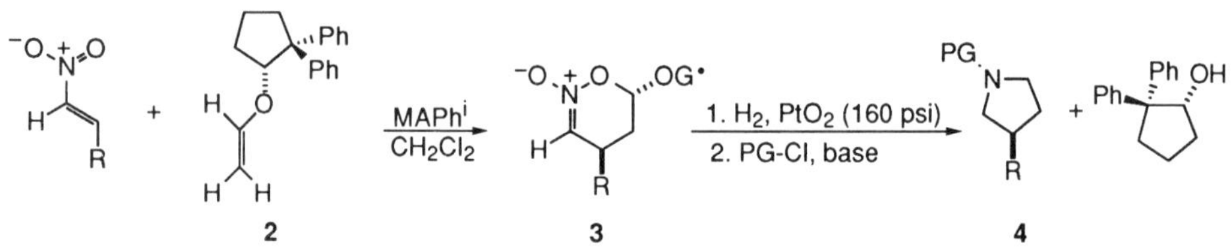

R	Yield **3**, %	PG	Base	Yield **4**, %	ee, %
phenyl	89	Ts	Et_3N	80	91
veratryl	91	Troc	pyridine	76	71
pentyl	88	Troc	pyridine	72	93
cyclohexyl	84	Troc	pyridine	79	91
tert-butyl	91	Troc	pyridine	80	77
$-(CH_2)_4CO_2$-tert-Bu	92	Troc	pyridine	69	97

i) Methylaluminum bis-2,6-diphenylphenoxide.

2,2-Diphenylcyclopentanone previously has been synthesized several times employing three general approaches.[10-12] The most common approach involves alkylation of diphenylacetonitrile with a 4-halobutyronitrile followed by Thorpe-Ziegler cyclization and acid hydrolysis to afford the ketone.[10] One- and two-step preparations of the enaminonitrile have been reported. The best results have been obtained by using sodium amide in liquid ammonia (70% yield).[10b] Acid hydrolysis (H_2SO_4) of the enaminonitrile is reported to afford the desired ketone in only moderate yields employing tedious procedures. Alternatively, the ketone is available by allylation of diphenylacetic acid followed by Lewis acid promoted Friedel Crafts acylation to afford the 5,5-diphenyl-2-cyclopentenone.[11] Subsequent reduction provided the ketone, however, only in 20% overall yield.[11b] A more recent and more conceptually attractive approach involves the one step diphenylation of cyclopentanone trimethylsilyl enol ether with diphenyliodonium fluoride (DIF) giving the desired ketone in 51% yield.[12] The disadvantage of this approach lies in the preparation of the DIF reagent, available in one step from very expensive diphenyliodonium iodide or in five steps from inexpensive, available starting materials.

The procedure described here represents a reliable modification of the alkylation/Thorpe-Ziegler approach to 2,2-diphenylcyclopentanone applicable to large-scale preparations. The use of LDA as the alkylation base, in exact stoichiometry, has been found to avoid over-reaction and the undesired fragmentation pathway of the enaminonitrile observed with sodium amide. The Thorpe-Ziegler cyclization proceeds smoothly employing modified literature conditions[10a] with potassium tert-butoxide as the base. Hydrolysis of the enaminonitrile with hydrochloric acid,[13] was found to be superior to previously reported methods, which led to incomplete consumption of the intermediate cyano ketone in our hands. The ratio of solvent volume to enaminonitrile is very important and more concentrated reaction mixtures result in incomplete conversion of the cyano ketone even after prolonged heating.

Asymmetric reduction of the ketone on a 1.0-g (4.0-mmol) scale to provide (R)-(-)-2,2-diphenylcyclopentanol (96% ee) has been reported employing (+)-β-chlorodiisopinocampheylborane; however, the reaction is extremely slow and inefficient [70% yield, 5 days, 2.6 equiv of (+)-β-chlorodiisopinocampheylborane].[6] Other efforts to obtain enantiomerically pure **1** by means of enzymatic hydrolysis of the corresponding racemic acetates using horse liver acetone powder (HLAP) and pig liver acetone powder (PLAP) have been only moderately successful and are of limited utility [4.0-mmol scale, 28% yield, 96.5% ee (R)].[14]

The oxazaborolidine-catalyzed borane reduction of 2,2-diphenylcyclopentanone provides an efficient and useful alternative for the asymmetric synthesis of (R)-**1** on a preparative scale. Variation of several reaction parameters such as catalyst loading, solvent, temperature, and addition order, have led to the development of an optimized procedure for this reduction. To achieve a selectivity of >90% ee, the reaction requires the use of 10 mol% of the oxazaborolidine catalyst, which is easily prepared in two steps from natural proline[4] or in one step from commercially available (S)-α,α-diphenylpyrrolidinemethanol. When using a smaller catalyst loading a significant decrease in selectivity is observed [5 mol% cat. provides 87% ee (R)-**1**]. The oxazaborolidine catalyst used in these experiments was purified by bulb-to-bulb distillation prior to use and quickly weighed in the open atmosphere. Examination of the purified catalyst by ^{1}H NMR confirmed the presence of B-methyloxazaborolidine as well as varying amounts of the hydrated adduct (approximately 7-20%).[4] Unfortunately, it is not clear whether the hydrated adduct was the result of trace amounts of water in the NMR solvent, exposure to atmospheric moisture, or simply insufficient purification. Regardless, the use of different batches of the catalyst provided reproducible results that are within experimental error [(R)-**1** with 91-94% ee]. Several solvents such as toluene, dichloromethane and THF have been reported to be useful in oxazaborolidine reductions;[15] however, for the reduction of 2,2-diphenylcyclopentanone the use of THF was found to provide the highest enantioselectivity. An extremely important parameter in this reaction is temperature. The reaction displays an inverse temperature effect with respect to enantioselectivity, where decreased selectivity is observed at lower temperatures. This interesting phenomenon in oxazaborolidine-catalyzed reductions has precedents,[5] and can be attributed to the slow breakdown of the catalyst-product complex at low temperatures.

The catalyst-product complex is a highly active but less selective catalyst for the reduction of the starting ketone.[16] Accumulation of this undesired intermediate can be avoided by running the reaction at higher temperatures (40°C) as well as using a slow inverse addition of the ketone to the catalyst-borane mixture.

The oxazaborolidine-catalyzed borane reduction to prepare (R)-**1** is an improvement over existing methods such as the β-chlorodiisopinocampheylborane reduction,[6] and enzymatic resolution[14] for several reasons. First, the reaction uses an easily obtained *catalytic* reducing agent that provides the chiral alcohol in 92% ee. Secondly, the reaction proceeds at a reasonable rate (6-8 hr) and affords the chiral alcohol (92% ee) in nearly quantitative yield (97%). Finally, the work-up, isolation and purification of the product is straightforward and requires no column chromatography, only bulb-to-bulb distillation and recrystallization, affording (R)-**1** in 75% yield with 97% ee. In addition, the catalyst precursor, (S)-α,α-diphenylpyrrolidinemethanol, can be easily recovered in excellent yield.

1. Department of Chemistry, University of Illinois at Urbana-Champaign, Urbana, IL 61801. We are grateful to Dr. David J. Mathre (Merck) for valuable discussions and the generous gift of B-methyloxazaborolidine catalyst.
2. Gilman, H.; Cartledge, F. K. *J. Organometal. Chem.* **1964**, *2*, 447.
3. Ruzicka, L.; Koolhaas, D. R.; Wind, A. H. *Helv. Chim. Acta* **1931**, *14*, 1151.
4. Mathre, D. J.; Jones, T. K.; Xavier, L. C.; Blacklock, T. J.; Reamer, R. A.; Mohan, J. J.; Jones, E. T. T.; Hoogsteen, K.; Baum, M. W.; Grabowski, E. J. J. *J. Org. Chem.* **1991**, *56*, 751; Xavier, L. C.; Mohan, J. J.; Mathre, D. J.; Thompson, A. S.; Carroll, J. D.; Corley, E. G.; Desmond, R. *Org. Synth.* **1996**, *74*, 50.
5. Stone, G. B. *Tetrahedron: Asymmetry* **1994**, *5*, 465.
6. Potin, D.; Dumas, F.; d'Angelo, J. *J. Am. Chem. Soc.* **1990**, *112*, 3483.

7. Zhang, Q.; Mohan, R. M.; Cook, L.; Kazanis, S.; Peisach, D.; Foxman, B. M.; Snider, B. B. *J. Org. Chem.* **1993**, *58*, 7640.
8. Denmark, S. E.; Schnute, M. E. *J. Org. Chem.* **1994**, *59*, 4576.
9. Denmark, S. E.; Marcin, L. R. *J. Org. Chem.* **1995**, *60*, 3221.
10. (a) Easton, N. R.; Nelson, S. J. *J. Am. Chem. Soc.* **1953**, *75*, 640; (b) Sneen, R. A.; Jenkins, R. W., Jr.; Riddle, F. L., Jr. *J. Am. Chem. Soc.* **1962**, *84*, 1598; (c) Salmon-Legagneur, F.; Neveu, C. *Bull. soc. chim. France* **1956**, 929; (d) Salmon-Legagneur, F.; Rabadeux, J. *Bull. soc. chim. France* **1967**, 1310.
11. (a) Craig, P. N.; Witt, I. H. *J. Am. Chem. Soc.* **1950**, *72*, 4925; (b) Halterman, R. L.; McEvoy, M. A. *J. Am. Chem. Soc.* **1990**, *112*, 6690.
12. Chen. K.; Koser, G. F. *J. Org. Chem.* **1991**, *56*, 5764.
13. Snider, T. E.; Morris, D. L.; Srivastava, K. C.; Berlin, K. D. *Org. Synth. Coll. Vol. 6*, **1988**, 932.
14. Randrianasolo-Rakotozafy, L. R.; Azerad, R.; Dumas, F.; Potin, D.; d'Angelo, J. *Tetrahedron: Asymmetry* **1993**, *4*, 761.
15. Mathre, D. J.; Thompson, A. S.; Douglas, A. W.; Hoogsteen, K.; Carroll, J. D.; Corley, E. G.; Grabowski, E. J. J. *J. Org. Chem.* **1993**, *58*, 2880.
16. Personal communication with Dr. David J. Mathre of Merck Research Laboratories.

Appendix

Chemical Abstracts Nomenclature (Collective Index Number); (Registry Number)

(R)-(-)-2,2-Diphenylcyclopentanol: Cyclopentanol, 2,2-diphenyl-, (R)- (12); (126421-67-8)

2-Amino-3,3-diphenyl-1-cyclopentene-1-carbonitrile: 1-Cyclopentene-1-carbonitrile, 2-amino-3,3-diphenyl- (8,9); (3597-67-9)

Diisopropylamine (8); 2-Propanamine, N-(1-methylethyl)- (9); (108-18-9)

Butyllithium: Lithium, butyl (8,9); (109-72-8)

Diphenylacetonitrile: Acetonitrile, diphenyl- (8); Benzeneacetonitrile, α-phenyl- (9); (86-29-3)

4-Bromobutyronitrile: Butyronitrile, 4-bromo- (8); Butanenitrile, 4-bromo- (9); (5332-06-9)

tert-Butyl methyl ether: Ether, tert-butyl methyl (8); Propane, 2-methoxy-2-methyl- (9); (1634-04-4)

2,2-Diphenylcyclopentanone: Cyclopentanone, 2,2-diphenyl- (8,9); (15324-42-2)

(S)-Tetrahydro-1-methyl-3,3-diphenyl-1H,3H-pyrrolo[1,2-*c*][1,3,2]oxazaborole: 1H,3H-Pyrrolo[1,2-*c*][1,3,2]oxazaborole, tetrahydro-1-methyl-3,3-diphenyl-, (S)- (12); (112022-81-8)

(S)-Tetrahydro-1-methyl-3,3-diphenyl-1H,3H-pyrrolo[1,2-*c*][1,3,2]oxazaborole-boran complex: Boron, trihydro(tetrahydro-1-methyl-3,3-diphenyl-1H,3H-pyrrolo[1,2-*c*][1,3,2]oxazaborole-N^7)-, [l-4-(3aS-cis)]- (12); (112022-90-9)

Borane-methyl sulfide complex: Methyl sulfide, compd. with borane (1:1) (8); Borane, compd. with thiobis[methane] (1:1) (9); (13292-87-0)

(S)-α,α-Diphenyl-2-pyrrolidinemethanol: 2-Pyrrolidinemethanol, α,α-diphenyl-, (S)- (12); (112068-01-6)

4-Iodobutyronitrile: Butyronitrile, 4-iodo- (8); Butanenitrile, 4-iodo- (9); (6727-73-7)

4-Chlorobutyronitrile: Butyronitrile, 4-chloro- (8); Butanenitrile, 4-chloro- (9); (628-20-6)

(S)-(-)-Proline: L-Proline (8,9); (147-85-3)

(S)-TETRAHYDRO-1-METHYL-3,3-DIPHENYL-1H,3H-PYRROLO-[1,2-*c*][1,3,2]OXAZABOROLE-BORANE COMPLEX

(Boron, trihydro(tetrahydro-1-methyl-3,3-diphenyl-1H,3H-pyrrolo-[1,2-*c*][1,3,2]oxazaborole-N^7)-, [I-4-(3aS-cis)]-)

A.
N H O OH
$COCl_2$
N O OH O Cl
Et_3N
H O N O O

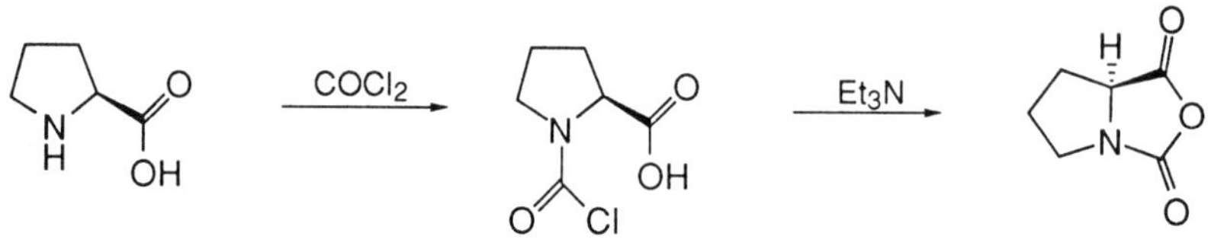

B.
H O N O O
1. PhMgCl
2. H_2SO_4
N H Ph Ph OH
· 1/2 H_2SO_4
NaOH
N H Ph Ph OH

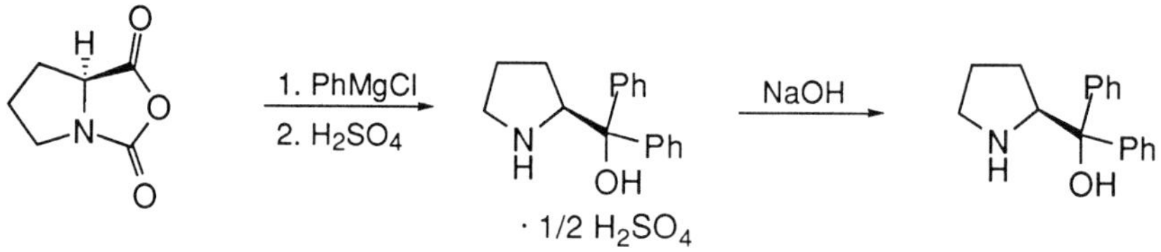

C.
N H Ph Ph OH
2/3 $(CH_3BO)_3$
Ph Ph O N B H CH_3 O H O B CH_3
xylene
Δ, $-MeB(OH)_2$
Ph Ph O N B CH_3

Ph Ph O N B CH_3
$Me_2S\text{-}BH_3$,
xylene/hexane
crystallize, $-Me_2S$
Ph Ph O N B BH_3 CH_3

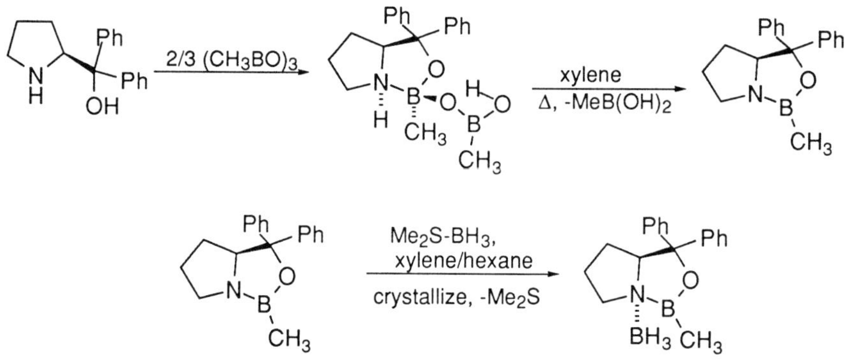

D.
O
Ph Ph O N B BH_3 CH_3
CH_2Cl_2, $Me_2S\text{-}BH_3$, -20°C
OH

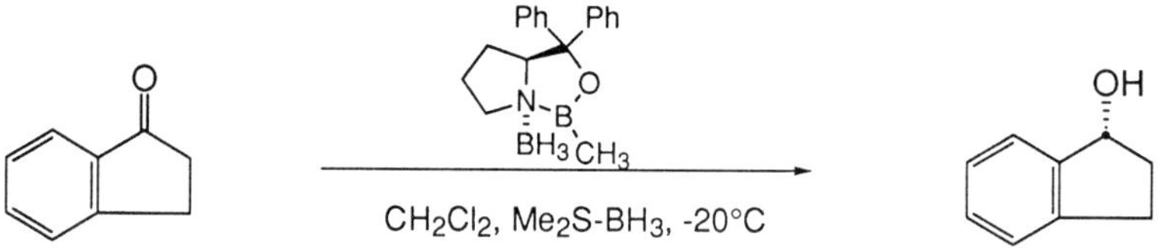

Submitted by Lyndon C. Xavier, Julie J. Mohan, David J. Mathre, Andrew S. Thompson, James D. Carroll, Edward G. Corley, and Richard Desmond.[1]
Checked by Peter G. Dormer and Amos B. Smith, III.

1. Procedure

Caution! This reaction should be carried out in a fume hood since phosgene is an insidious poison.

A. (S)-Proline-N-carboxyanhydride. A dry, 3-L, three-necked flask, fitted with a mechanical stirrer, nitrogen inlet tube, 500-mL, pressure-equalized addition funnel and a Teflon-coated thermocouple probe, containing dry tetrahydrofuran (THF) (600 mL) is charged with (S)-proline (60.0 g, 0.521 mol) (Note 1). To the cooled (15-20°C) suspension is added phosgene (324 mL, 0.625 mol, 1.93 M in toluene), via a pressure-equalized addition funnel, over a 0.5-1.0 hr period maintaining a temperature of 15-20°C (Note 2). The reaction mixture is then aged for 0.5-0.75 hr at 30-40°C. The reaction mixture should become homogeneous as the proline reacts with the phosgene to afford the intermediate N-carbamoyl chloride. Once homogeneous, the reaction mixture is aged an additional 0.5 hr, then concentrated under reduced pressure (15-20°C, 1000 down to 50 mBar) to a volume of 80 mL (Note 3). Dry tetrahydrofuran (600 mL, KF < 50 µg/mL) is added to the mixture and it is cooled to 0-5°C. Dry triethylamine (72.6 mL, 0.521 mol, KF < 50 µg/mL) is added over 0.25 hr (Note 4). The mixture is aged for 0.5 hr at 0-5°C, then filtered through an enclosed, 2-L, medium-frit Schlenk funnel under an atmosphere of nitrogen (with careful exclusion of moisture). The triethylammonium hydrochloride cake is washed with dry tetrahydrofuran (3 x 105 mL, KF < 50 µg/mL). The filtrate and washes are combined and used as is in the next reaction (Notes 5 and 6).

Caution! Benzene is generated in this reaction. Benzene is a known carcinogen.

B. (S)-1,1-Diphenylprolinol. A 3-L, three-necked flask, fitted with a mechanical stirrer, nitrogen inlet tube, 1-L addition funnel and Teflon coated thermocouple probe, is charged with phenylmagnesium chloride (0.80 L, 1.6 mol, 2.0 M in tetrahydrofuran) (Note 7) and cooled to -10°C (Note 8). To the suspension is added, with stirring, the tetrahydrofuran solution of (S)-proline-N-carboxyanhydride over a 1-hr period maintaining the temperature at -15° to -10°C (Note 9). After the addition is complete, the reaction mixture is aged for 3 hr at -15°C, then 1 hr at 0°C (Note 10). The reaction is quenched into a 5-L, mechanically-stirred flask containing 2 M aqueous sulfuric acid (1.05 L, 2.1 mol) (0°C) over a 0.5-1.0 hr period (Note 11). The reaction mixture is aged for 1 hr at 0-5°C, filtered, and the magnesium sulfate cake washed with tetrahydrofuran (3 x 500 mL) (Note 12). The filtrate and tetrahydrofuran washes are combined and concentrated (1 atm) to a volume of 1.0 L (Note 13). The mixture is cooled to 0°C, aged 1 hr, and then filtered. The cake is washed with water (5°C, 2 x 100 mL) and ethyl acetate (3 x 180 mL) (Note 14). The cake is dried under reduced pressure (40°C, 50 mBar) to yield 79-88 g (50-56% based on proline) of a free-flowing white crystalline solid, mp 275-290°C (Note 15).

A 3-L, three-necked flask, fitted with a mechanical stirrer, nitrogen inlet tube, and Teflon coated thermocouple, is charged with (S)-diphenylprolinol sulfate (81.0 g, 0.268 mol), tetrahydrofuran (268 mL), and 2 M aqueous sodium hydroxide (268 mL). The mixture is stirred at 20-25°C until all the solid dissolves (0.5 hr) (Note 16). Toluene (1.1 L) is added and the mixture is aged 0.5 hr. The two-phase mixture is filtered through a medium frit sintered glass funnel and partitioned (Note 17). The upper layer is washed with water (130 mL) and the toluene is removed under reduced pressure to afford a light-tan colored oil that crystallizes upon standing at ambient temperature (Note 18). Recrystallization of the resulting light-tan crystals from heptane (ca. 2 mL of heptane per gram of prolinol) affords (S)-1,1-diphenylprolinol; [yield 64-68 g (94-99% yield based upon diphenylprolinol sulfate)] as white crystals, mp 76-78°C, >98% ee (Notes 19-20).

Caution! Neat (10 M) borane-methyl sulfide is an air- and moisture-sensitive flammable liquid. Dimethyl sulfide has a noxious odor. All reactions should be conducted in an efficient fume hood.

C. (S)-Tetrahydro-1-methyl-3,3-diphenyl-1H,3H-pyrrolo[1,2-c][1,3,2]oxaza-borole-borane complex. A 500-mL, four-necked, round-bottomed flask (Note 21) is equipped with a Teflon-coated thermocouple probe, 60-mL pressure-equalizing addition funnel, nitrogen inlet, mechanical stirrer, and distillation head connected to a wide-bore, air-cooled condenser. The flask is charged with 200 mL of xylenes (Note 22) and (S)-diphenylprolinol (50.0 g, 198 mmol). To the vigorously stirred solution at 20°C is added a 50% (w/w) solution of trimethylboroxine (TMB) in tetrahydrofuran (38 g, containing 19 g, 152 mmol of TMB) (Note 23) dropwise over a 15-min period. The addition funnel is rinsed with dry xylenes (10 mL). During the addition an exothermic reaction occurs (warming the mixture to ca. 35°C) resulting in the formation of the bismethylboronic acid adduct as a white precipitate (Note 24). After the addition, the mixture is stirred for 0.5 hr, then heated and concentrated by distillation (1 atm) (Note 25) to a volume of ca. 100 mL (Note 26). The mixture is allowed to cool to ca. 60°C, and is charged with additional dry xylenes (200 mL) (Note 22). The distillation and nitrogen sweep are continued, this time until essentially all the solvent is removed. The mixture is then heated so that the internal temperature is maintained between 160-180°C for a 0.5-hr period. The pale yellow oil (free oxazaborolidine) is cooled to 20°C (Notes 26 and 27), and diluted with dry xylenes (Note 22) to a volume of 100 mL. The distillation head is replaced with a 500-mL, pressure-equalizing addition funnel. At this point, the nitrogen sweep is discontinued; however, the vessel is maintained under a static atmosphere of nitrogen for the remainder of the reaction. To the well stirred solution is added neat (10 M) borane-methyl sulfide (21 mL, 210 mmol) (Note

23) over a 5-min period via syringe. The mixture is stirred for 0.5 hr at 20°C, then slowly diluted with dry hexane to a volume of 450 mL over a 1-2 hr period (Note 28). After the addition is complete, the mixture is aged for 1 hr at 20°C, cooled to -10°C, and then aged for 2 hr at -10°C. The mixture is filtered in an enclosed Schlenk filter (Note 29). The cake is washed with dry hexane (3 x 50 mL) (Note 30). The product is dried under reduced pressure (100 mBar, nitrogen sweep, 20°C) for 1 hr to afford 51-53 g of the title compound as a free-flowing white crystalline solid (Notes 31-32).

Caution! Neat (10 M) borane-methyl sulfide is an air- and moisture-sensitive flammable liquid. Dimethyl sulfide has a noxious odor. All reactions should be conducted in an efficient fume hood.

D. (R)-2,3-Dihydro-1H-inden-1-ol. A 250-mL, three-necked, round-bottomed flask (Note 33) is equipped with a Teflon-coated thermocouple probe, a 125-mL pressure-equalizing addition funnel, nitrogen inlet, and Teflon-coated magnetic stirring bar. The flask is charged with the (S)-oxazaborolidine-borane complex (2.91 g, 10.0 mmol) (Note 34), purged with nitrogen, and charged with dry dichloromethane (10 mL) (Note 35). To the stirred solution is added borane-methyl sulfide (10 M, 20 mL, 200 mmol) (Note 36). The solution is cooled to -20°C (Note 37) and the addition funnel charged with a solution of 1-indanone (26.4 g, 200 mmol) (Note 38) in dry dichloromethane (75 mL) (Note 35). The 1-indanone solution is then added dropwise over a 4-6 hr period while maintaining the internal temperature of the reaction mixture at -20±5°C. After the addition is complete, the reaction mixture is stirred for 2 hr at -20°C (Note 39). After the reaction is complete, the mixture is cautiously poured into a 1-L flask containing magnetically-stirred, pre-cooled (-20°C) methanol (300 mL) (Note 40). The cooling bath is removed, and the stirred mixture allowed to warm to room temperature (20-25°C). After evolution of hydrogen ceases, the mixture is concentrated by distillation (1 atm) to a volume of ca. 50 mL. Methanol (200 mL) is added, and the distillation repeated (Note 41). The residue containing the product and diphenylprolinol is diluted with methanol (100 mL) and then loaded onto a 2.5 x 30-cm column packed with Amberlyst 15 (NH_4^+) (Note 42) at ca. 2.5 mL/min, collecting 40-mL fractions. The column is rinsed with methanol until the product is eluted (Note 43). The column is then rinsed with 1 M methanolic ammonia until the diphenylprolinol is eluted. The fractions containing the product (2-15) are combined, and then concentrated under reduced pressure (40°C, 100 mm) to give 26.2 g of crude product as a white crystalline solid (Note 44). The crude product is dissolved in hexane (260 mL) at 50-60°C in a 500-mL, round-bottomed flask fitted with a mechanical stirrer. The product is allowed to crystallize as the stirred mixture is slowly cooled to 20-25°C. The mixture is stirred for 4 hr at 20-22°C, filtered, and the cake washed with hexane (2 x 25 mL). The product is dried in a vacuum oven (30°C, 10 mm) to constant weight to yield 24.1-25.2 g (90-94%) of the title compound as a white crystalline solid (Notes 45 and 46).

2. Notes

1. The submitters used reagent grade tetrahydrofuran (EM Science) dried over 4 Å or 5 Å molecular sieves (Aldrich Chemical Company, Inc.), and determined the water content in the solvents by Karl Fisher titration (KF). The submitters report that the tetrahydrofuran should be dry (KF < 50 µg/mL) and *MUST NOT* contain any DMF (≤ 1 ppm). Also, the KF of the tetrahydrofuran/proline suspension should be < 50 µg/mL. The submitters used (S)-proline obtained from Ajinomoto while the checkers used (S)-proline from Aldrich Chemical Company, Inc. The proline should be milled and/or delumped prior to use, if necessary, to insure complete reaction. The enantiomeric purity of the (S)-proline should be > 99.5%, and can be assayed via the procedure of Marfey.[2] The checkers used HPLC grade THF (Fisher Scientific Company) that was dried over 4 Å molecular sieves.

2. Phosgene (1.93 M in toluene) was obtained from Fluka Chemical Company. Neat phosgene (Matheson Gas Products) and triphosgene (Aldrich Chemical Company, Inc.) (98%) can also be used. Phosgene is an insidious poison so all manipulations with it must be performed in a fume hood with good ventilation. Any excess phosgene should be *CAREFULLY* decomposed in cold aqueous sodium hydroxide or aqueous ammonia. Since the addition is slightly exothermic, the rate of addition and external cooling is adjusted to maintain the internal temperature at 15-20°C.

3. The vacuum is decreased gradually in order to prevent bumping as the hydrogen chloride, excess phosgene, and tetrahydrofuran are removed. The checkers concentrated the reaction mixture using a rotary evaporator (a drying tube was placed in-line between the water aspirator and the rotary evaporator) under aspirator pressure (bath temp < 30°C), in an efficient fume hood The temperature must be maintained < 20°C during the concentration. At a volume of 80 mL the mixture is viscous but can still be stirred. The reaction can be assayed at this point by ^1H NMR; an aliquot withdrawn (ca. 30 mg) and dissolved in $CDCl_3$ (0.6 mL) giving ^1H NMR (250 MHz, $CDCl_3$) δ: 2.35 (s, toluene), 1.83-2.44 (m, 4 H, C-3-H_2 and C-4-H_2), 3.54-3.86 (m, 2 H, C-5-H_2), 4.47-4.52 (dd, 0.6 H, C-2-H rotamer), 4.58-4.61 (dd, 0.4 H, C-2-H rotamer), 7.12-7.29 (m, toluene), 10.6-10.8 (br s, CO_2H). The ^1H NMR spectrum should not contain resonances at δ 4.9 (dd, 0.4 H, C-2-H) and 4.7 (dd, 0.6 H, C-2-H) (i.e., those corresponding to the N-carbamoyl chloride, acid chloride derivative of proline).

4. Triethylamine obtained from J. T. Baker Inc. was dried over 4 Å or 5 Å molecular sieves to a KF < 50 µg/mL. The checkers used triethylamine (Fisher Scientific Company) dried over 4 Å molecular sieves. The reaction is slightly exothermic so the rate of addition and external cooling is adjusted to maintain the

internal temperature at 0-5°C. During the addition, a white precipitate of triethylammonium hydrochloride is formed.

5. The tetrahydrofuran solution of the proline-N-carboxyanhydride must be used immediately since on standing the material can polymerize and release CO_2. The reaction can be assayed at this point by ^{1}H NMR: (1.0 mL of the THF solution is concentrated under reduced pressure, and the residue dissolved in 0.6 mL of $CDCl_3$) δ: 1.83-2.39 (m, 4 H), 3.29-3.39 (m, 1 H, C-5-H), 3.70-3.81 (m, 1 H, C-5-H), 4.35-4.39 (dd, 1 H, C-2-H); ^{13}C NMR (62.9 MHz, $CDCl_3$) δ: 26.7, 27.2, 46.2, 62.8, 154.6, 168.8. The amount of triethylammonium hydrochloride, δ 3.12 (q, 6 H) and 1.4 (t, 9 H), should be less than 5 mol%.

6. Alternate work-up: The filtered tetrahydrofuran solution is concentrated under reduced pressure (20°C, 50 mm) to a volume of 80 mL. Dry hexane (KF < 50 μg/mL, 600 mL) is slowly added with stirring. The product should begin to crystallize during the addition; seed if necessary. The mixture is stirred at 20-25°C for 1 hr, then filtered. The cake is washed with dry hexane (2 x 25 mL). Filtration is performed in an enclosed, medium-frit Schlenk funnel, under an atmosphere of nitrogen (with careful exclusion of moisture). The cake is dried under reduced pressure (30°C, < 100 mm) for 2 hr to yield 70.5 g (95%) as a free-flowing white crystalline solid, mp 51.5-52°C. The product should be stored cool, protected from moisture. By microanalysis, the product contains 1-4 wt% of triethylammonium hydrochloride. It is also possible to concentrate the THF to precipitate the product as a white solid.

7. Phenylmagnesium chloride (2.0 M in tetrahydrofuran) was obtained from Aldrich Chemical Company, Inc. The vessel must be dry and flushed with nitrogen prior to the addition of the phenylmagnesium chloride.

8. The reaction should not be cooled lower than -10°C because the Grignard reagent can crystallize. This can cause the stir shaft to break. It is best to add a small

portion of the proline-N-carboxyanhydride, slurry the solution, and then cool to -10°C. Crystallization of the Grignard reagent does not affect the reaction yield.

9. The reaction is exothermic; therefore the rate of addition and external cooling are adjusted to maintain the internal temperature at -15 to -10°C.

10. The checkers allowed the reaction mixture to warm to ambient temperature overnight before quenching the reaction. If desired, the progress of the reaction may be monitored by the following HPLC assay. An aliquot (0.5 mL, accurately measured) is quenched into water (5 mL) containing 6 M aqueous sulfuric acid (200 µL). The solution is diluted to 100 mL with 1:1 H_2O/MeCN and analyzed by HPLC. The checkers did not monitor the reaction by HPLC. Column: 4.6 x 250-mm ZORBAX RX ; Eluent A: H_2O (0.01 M KH_2PO_4); Eluent B: MeCN; Linear Gradient: 80:20 (A:B) to 20:80 over 12 min; Flow Rate: 1.0 mL/min; Injection: 10.0 µL; Detection: UV (210 nm); Retention Times: Proline, N-benzamide 4.82 min; Diphenylprolinol 6.68 min; Benzene 11.45 min; Triphenylmethanol 14.66 min.

11. Sulfuric acid (Mallinckrodt Inc.) is diluted to the desired concentration with deionized water. Since the quench is exothermic the rate of addition and external cooling are adjusted to maintain the internal temperature below 20°C. During the quench a white precipitate of magnesium sulfate is formed. The amount of water and sulfuric acid used is important. Insufficient water results in the product (sulfate salt) precipitating with the magnesium sulfate. Excess water reduces the amount of magnesium removed. Insufficient sulfuric acid results in non-filterable gels of magnesium hydroxide. Excess sulfuric acid increases the solubility of the diphenylprolinol sulfate, therefore reducing the yield of isolated product.

12. A large sintered glass funnel is required because of the amount of magnesium sulfate solution produced (~700 mL). Analysis of the magnesium sulfate cake by HPLC detected < 1 g of diphenylprolinol.

13. During concentration, a precipitate of diphenylprolinol sulfate and triphenylmethanol is formed. *Caution: Benzene (43 g), formed during the quench of the excess phenylmagnesium chloride, is removed during the concentration.*

14. Water is used to remove excess sulfuric acid, and ethyl acetate to remove triphenylmethanol, benzophenone and the N-benzamide derivative of proline.

15. Assay for enantiomeric purity: To a magnetically stirred suspension of diphenylprolinol sulfate (30 mg) in tetrahydrofuran (1 mL) is added 1 M aqueous sodium hydroxide (210 µL). The mixture is stirred for 15 min, then (R)-Mosher acid chloride (20 µL) is added. By TLC (EM Si-60, 8:2 hexane/EtOAc, R_f diphenylprolinol = 0.05, R_f Mosher amide = 0.4) the reaction is complete in 1 hr. The reaction mixture is diluted into hexane (9 mL), then eluted through a Baker Silica SPE (1 g) column (previously washed with hexane). The column was eluted with additional 9:1 hexane/tetrahydrofuran (5 mL). The combined eluates were than analyzed by capillary GC: Column: 0.33 mm x 30 m DB-23 (J&W Scientific); Oven Temperature: 250°C; Injector/Detector Temperature: 275°C; Carrier Gas: Helium (21 lbs/in^2), ca. 30:1 split; Injection: 1 µL; Detection: FID; Retention Times: (R,R)-Mosher amide 25.9 min; (R,S)-Mosher amide 29.6 min. The enantiomeric purity is 99:1.

16. There are occasions when a small amount of material does not dissolve. This does not affect the reaction.

17. Filtration is necessary to remove a small amount of polymeric solid that interferes with the phase separation.

18. Toluene is removed with a rotary evaporator (water aspirator, 40°C bath temperature). If desired, a toluene flush (300 mL) may be employed to remove residual traces of water. Concentration serves both to remove THF and azeotropically dry the solution. The last traces of solvent are removed by placing the oil under high vacuum (ca. 1 mBar) at room temperature overnight.

19. The physical properties are as follows: lit. mp: 76.5-77.5°C;[2e] ^{1}H NMR (500 MHz, $CDCl_3$) δ: 1.57-1.79 (m, 4 H, C-3-H_2, C-4-H_2), 2.93-2.98 (m, 1 H, C-5-H), 3.02-3.07 (m, 1 H, C-5-H), 4.27 (t, J = 7.6, C-2-H), 7.16-7.21 (m, 2 H, Ar-H), 7.28-7.34 (m, 4 H, Ar-H), 7.53-7.72 (m, 4 H, Ar-H); ^{13}C NMR (125 MHz, $CDCl_3$) δ: 25.4, 26.3 (C-3, C-4), 46.7 (C-5), 64.5 (C-2), 125.5, 125.8, 126.3, 126.4, 127.9, 128.2, (C-2'-C-6', C-2"-C-6"), 145.4 (C-1"), 148.1 (C-1').

20. If desired, the diphenylprolinol may be assayed for enantiomeric purity as described in Note 15.

21. The glassware was oven-dried (110°C) before use. The 500-mL round-bottomed flask was graduated in 100-mL increments. The checkers used argon in place of nitrogen.

22. The xylenes (EM Science) were dried over 4 Å molecular sieves. Residual water content < 20 µg/mL by Karl Fisher titration. The checkers dried hexanes (Fisher Scientific Company), tetrahydrofuran, and xylenes over 4 Å molecular sieves. The checkers did not measure the residual water content of the solvents.

23. Trimethylboroxine (as a 50 wt% solution in THF) and borane-methyl sulfide (10 M) were obtained from Callery Chemical Company. The checkers employed 50 wt% trimethylboroxine/THF solution (prepared from neat trimethylboroxine (Aldrich Chemical Company, Inc.) and dry THF). The checkers used borane-methyl sulfide obtained from Aldrich Chemical Company, Inc.

24. The submitters report that a two-stage exotherm occurs during the addition. The first exotherm occurs as the trimethylboroxine reacts with the diphenylprolinol, and the second occurs as the intermediate bismethylboronic acid adduct crystallizes. Typically, the intermediate begins to crystallize after ca. one third of the trimethylboroxine is added. The reaction should not be run more concentrated. The checkers observed an exothermic reaction (warming from 20 to 45°C) after approximately one half to two thirds of the trimethylboroxine/THF solution was added.

25. The vessel is swept with a gentle stream of nitrogen during the distillation and subsequent operation to ensure the complete removal of the reaction by-products. The condenser should be air-cooled during the initial stages of the distillation to prevent it from being plugged with methylboronic acid. The checkers heated the stirred mixture rapidly to the boiling point (over ca. 20-30 min period) and used an argon sweep in place of nitrogen during the distillation.

26. The essentially pure free oxazaborolidine will slowly solidify on standing (mp 79-81°C). Although this material can be stored and used "as is" if scrupulously protected from adventitious moisture, the submitters recommend that it be converted to the more stable borane complex.

27. The physical properties are as follows: mp 80-83°C (corr); ^{1}H NMR (500 MHz, 0.15 M in $CDCl_3$) δ: 0.46 (s, 3 H, B-CH_3), 0.85-0.93 (m, 1 H, C-4-H), 1.64-1.85 (m, 3 H, C-4-H, C-5-H_2), 3.08-3.13 (m, 1 H, C-6-H), 3.38-3.43 (m, 1 H, C-6-H), 4.40-4.43 (dd, 1 H, J = 5.5, 9.8, C-3a-H), 7.23-7.43 (m, 8 H, Ar-H), 7.59-7.61 (m, 2 H, Ar-H); ^{11}B NMR (64.2 MHz, 0.15 M in $CDCl_3$) δ: 34.6; ^{13}C NMR (125 MHz, $CDCl_3$) δ: -5.7 (br, B-CH_3), 26.3 (C-5), 30.1 (C-4), 42.8 (C-6), 72.6 (C-3a), 87.7 (C-3), 126.1, 126.2, 126.5, 127.0, 127.6, 128.0 (C-3', C-3'', C-4', C4'', C-5', C-5''), 143.9, 147.5 (C-1', C-1''); IR (CCl_4) cm^{-1}: 3060, 3025, 2980, 2875, 1445, 1335, 1315, 1235, 995, 695; low-resolution EIMS: m/z 277.2 $[M]^+$. The solvent used for the NMR spectra must be dry and free of HCl or DCl (molecular sieves and/or anhydrous potassium carbonate).

28. During the addition, the oxazaborolidine-borane complex crystallizes. A slow addition of dry hexane favors the formation of larger crystals.

29. During the filtration it is important to keep the operation under an atmosphere of dry nitrogen to prevent moisture from condensing on the product. The submitters have also isolated the product in a 600-mL, sintered-glass funnel contained in a large bag maintained under an atmosphere of dry nitrogen. The checkers

performed the filtration in a 600-mL, medium-frit sintered glass funnel contained in a large plastic glove bag maintained under an argon atmosphere.

30. The three hexane washes are first used to rinse product remaining in the reaction vessel.

31. The physical properties area as follows: mp 124-127°C (dec); ^{1}H NMR (500 MHz, $CDCl_3$) δ: 0.77 (s, 3 H, B-CH_3), 1.0-1.9 (very br s, 3 H), 1.28-1.37 (m, 1 H, C-4-H), 1.56-1.68 (m, 1 H, C-5-H_2), 1.89-2.0 (m, 2 H, C-5-H_2), 3.19-3.24 (m, 1 H, C-6-H), 3.37-3.43 (m, 1 H, C-6-H), 4.66 (t, 1 H, J = 7.9, C-3a-H), 7.2-7.4 (m, 8 H, Ar-H), 7.6 (m, 2 H, Ar-H); ^{11}B NMR ($CDCl_3$) δ: 34.5 (oxazaborolidine-B), -14.5 (complexed H_3); ^{13}C NMR (125 MHz, $CDCl_3$) δ: -3.9 (br, B-CH_3), 24.9 (C-5), 31.4 (C-4), 57.7 (C-6), 76.2 (C-3a), 90.6 (C-3), 125.0 (C-2", C-6"), 125.4 (C-2', C-6'), 127.1 (C-4"), 127.3 (C-4'), 128.1 (C-3", C-5"), 128.2 (C-3', C-5'), 143.5 (C-1"), 144.6 (C-1'). Anal. Calcd for $C_{18}H_{23}B_2NO$: C, 74.29, H, 7.97; N, 4.81. Found: C, 74.34; H, 8.00; N, 4.69.

32. The product is stored at room temperature protected from moisture. A nitrogen atmosphere is recommended for long term storage. Unlike the free oxazaborolidine that readily reacts with and is decomposed by atmospheric moisture, the oxazaborolidine-borane complex is significantly more stable, allowing it to be handled briefly in the open.

33. The glassware was oven-dried (110°C) before use. The checkers used argon in place of nitrogen.

34. The checkers observed gas evolution upon dissolution of the (S)-oxazaborolidine-borane complex in dichloromethane (CH_2Cl_2) at room temperature.

35. Dichloromethane (Fisher Scientific Company) was dried over 4 Å molecular sieves before use (water content < 20 µg/mL by Karl Fisher titration). The checkers did not measure the residual water content of the solvent.

36. Borane-methyl sulfide (10 M) was obtained from Callery Chemical Company. The checkers purchased borane-methyl sulfide (10 M) from Aldrich Chemical Company, Inc.

37. The reaction temperature was controlled using a cooling bath of methanol/ice and sufficient dry ice to maintain the indicated temperature.

38. 1-Indanone (Aldrich Chemical Company, Inc., 99+%) was used as received.

39. The progress of the reaction was followed by capillary GC. An aliquot (50 μL) is quenched into methanol (5 mL) and then analyzed using a 30-m x 0.32-mm DB-23 (J&W Associates) column (He carrier, 15 lbs/in^2; ca. 30:1 split; oven temp: 50 to 150°C at 10°C/min; injector/detector temp: 250°C). 1-Indanol (t_R 10.9 min); 1-indanone (t_R 11.1 min).

40. *Caution: Hydrogen is evolved during the quench. The flask should be flushed with nitrogen. The reaction mixture is added at a rate to control the foaming.*

41. Methanol distillation serves to remove the majority of the boron species, as $B(OMe)_3$ and $MeB(OMe)_2$

42. The Amberlyst (NH_4+) resin column is prepared as follows: Amberlyst 15 (H^+) (56 g, 100 mL dry, Rohm & Haas Co.) is suspended in an open beaker containing methanol (100 mL). [*Caution*: *the slurry exotherms to ca. 40°C without external cooling, and expands to ca. 1.5 times its initial volume.*] The slurry is poured into a 2.5 x 30-cm column and is eluted with 1 M methanolic ammonia (ca. 1 L) until a sample of the eluent diluted 1:1 with water is basic. The resin is then eluted with methanol (ca. 0.5 L) until a sample of the eluent diluted 1:1 with water is neutral. Once prepared, the column can be reused multiple times.

43. The progress of the column is monitored by UV (260 nm) or capillary GC (Note 39).

44. Chiral HPLC assay (4.6 x 250-mm Chiralcel-OB, 90:10 Hexane/isopropyl alcohol, 0.5 mL/min, UV 254 nm): 1-indanone (t_R 29.6 min, < 0.1%), (R)-carbinol (t_R

10.3 min, 98.9%), (S)-carbinol (t_R 15.3 min, 1.1%). The sample was taken as a homogenous solution in methanol (prior to crystallization) to avoid enantiomeric enrichment. The checker's weight of crude 1-(R)-indanol (prior to recrystallization); was 26 g, ca. 90% ee. The checkers also recovered crude (S)-diphenylprolinol (1.9 g) from the column.

45. The physical properties are as follows: Mp 73.0-73.5°C [lit.[3] 72.0°C]; HPLC (Chiralcel-OB) (R)-carbinol (t_R 10.3 min, > 99.9%), (S)-carbinol (t_R 15.3 min, < 0.1%); $[\alpha]_D^{21}$ -45.5° (MeOH, *c* 1.184) [lit.[4] (S)-carbinol $[\alpha]_D^{25}$ +22.6° ($CHCl_3$, *c* 4.2) reported for the enantiomer]; ^1H NMR (500 MHz, $CDCl_3$) δ: 2.2-2.25 (br s, 1 H, OH), 1.91-1.98 (m, 1 H, C-2-H), 2.46-2.53 (m, 1 H, C-3-H), 2.81-2.87 (m, 1 H, C-3-H), 3.04-3.10 (m, 1 H, C-3-H), 5.23 (t, 1 H, J = 6.0, C-1-H), 7.27-7.45 (m, 4 H, Ar-H); ^{13}C NMR (125 MHz, $CDCl_3$) δ: 29.7, 35.7, 76.2, 124.1, 124.7, 126.5, 128.1, 143.2, 144.9. Checkers yield of (R)-2,3-dihydro-1H-inden-1-ol after recrystallization was 19.6-20.3 g (73-76%); mp (corr): 73-74°C, $[\alpha]_D^{29}$ -18.1° (MeOH, *c* 1.3).

46. From the mother liquors was obtained an additional 1.75 g (6.5%) of 1-indanol of lower enantiomeric purity (HPLC: 85:15 R/S). The checkers concentrated the mother liquors to afford additional 1-(R)-indanol (6.5-6.9 g) of lower enantiomeric purity (60-70% ee).

Waste Disposal Information

All toxic materials were disposed of in accordance with "Prudent Practices in the Laboratory"; National Academic Press; Washington, DC, 1996.

3. Discussion

A-B. Enantiomerically pure (R)- and (S)-α,α-diaryl-2-pyrrolidinemethanols are precursors to several useful asymmetric catalysts,[5a-e] including the corresponding oxazaborolidines.[5f-l] Although several routes have been reported for making these compounds, none were suitable for our purposes (poor yields, multiple steps, difficult isolations).[6-10] We therefore developed a practical two-step synthesis of (R)-α,α-diaryl-2-pyrrolidinemethanols from (R)- or (S)-proline based on the addition of proline-N-carboxyanhydride to phenylmagnesium chloride.[11,12] This investigation led to the development of a reliable procedure for oxazaborolidine preparation based on the reaction of trimethylboroxine with diphenylprolinol.[11] Using this the submitters prepared a series (see Table) of enantiomerically pure α,α-diphenyl-2-pyrrolidinemethanols and their corresponding B-alkyl- and B-aryloxazaborolidines beginning with either (R)- or (S)-proline.

The yield obtained for the preparation of (S)-α,α-diaryl-2-pyrrolidinemethanol was 67% at -10°C. It was found that addition of the Grignard reagent to the proline-N-carboxyanhydride resulted in a lower yield; therefore the proline-N-carboxyanhydride is added to the Grignard. The reaction is successful with meta- and para- substituted aryl Grignards, but with ortho-substituted aryl Grignard reagents only one equivalent adds, affording an unstable ketone intermediate too hindered for further addition. Alkyl Grignards do not afford the desired product so the submitters recommend the procedure described by Enders.[10]

TABLE

α,α-DIARYL-2-PYRROLIDINEMETHANOLS

Entry	R	Yield[a]	e.e.[b]
1	Phenyl	73%	99.4%
2	4-Fluorophenyl	89%	99.4%
3	4-Chlorophenyl	59%	99.4%
4	4-Methylphenyl	57%	99.4%
5	4-(CF_3)-Phenyl	46%	99.4%
6	4-tert-Butylphenyl	50%	99.4%
7	4-Methoxyphenyl	53%	99.4%
8	3-Chlorophenyl	62%	99.4%
9	3,5-Dichlorophenyl	68%	99.4%
10	3,5-Dimethylphenyl	60%	99.4%
11	2-Naphthyl	64%	99.4%

[a]Isolated yield. [b]Enantiomeric excess (e.e.) determined by capillary GC analysis of the (R)-α-methoxy-α-(trifluoromethyl)phenylacetic acid amide derivative.

The methods reported for the preparation of the oxazaborolidine include the reaction of diphenylprolinol with methylboronic acid 1) in toluene at 23°C for 1.5 hr with 4 Å molecular sieves present and 2) in toluene at reflux for 3 hr using a Dean-Stark trap, both followed by evaporation of solvent and molecular distillation (0.1 mm, 170°C).[5f] An alternate method involved heating a toluene solution of 2-naphthylprolinol and methylboronic acid at reflux for 10 hr using a Soxhlet extractor containing 4 Å molecular sieves.[5i] These methods afforded erratic results. The submitters therefore developed an alternate synthesis.

During their investigations they isolated the methylboronic acid adduct and the water adduct of the catalyst. The submitters found, by NMR experiments, that these were the compounds giving rise to spurious signals that were attributed to a "dimer". It was shown that the methylboronic acid adduct could be synthesized directly by treating 0.667 equiv of methylboronic acid with (S)-α,α-diphenyl-2-pyrrolidine-methanol. Heating at reflux in toluene yielded the catalyst free of any spurious signals. The catalyst prepared this way reproducibly afforded high levels of enantioselection. The oxazaborolidine must, however, be protected from moisture in order to retain high levels of enantioselection.

C. The reported procedure provides a practical preparation of (S)-tetrahydro-1-methyl-3,3-diphenyl-1H,3H-pyrrolo[1,2-*c*][1,3,2]oxazaborole and conversion to its more stable borane complex.[13] The oxazaborolidine-borane complex has also been prepared by treatment of a toluene solution of the free oxazaborolidine with gaseous diborane followed by recrystallization from a dichloromethane-hexane bilayer.[14] This and other chiral oxazaborolidines have been used to catalyze the enantioselective reduction of prochiral ketones.[15] The yield and enantioselectivity of reductions using catalytic amounts of the oxazaborolidine-borane complex are equal to or greater than those obtained using the free oxazaborolidine.[13]

D. The use of chiral oxazaborolidines as enantioselective catalysts for the reduction of prochiral ketones, imines, and oximes, the reduction of 2-pyranones to afford chiral biaryls, the addition of diethylzinc to aldehydes, the asymmetric hydroboration, the Diels-Alder reaction, and the aldol reaction has recently been reviewed.[15b,d] The yield and enantioselectivity of reductions using stoichiometric or catalytic amounts of the oxazaborolidine-borane complex are equal to or greater than those obtained using the free oxazaborolidine.[13] The above procedure demonstrates the catalytic use of the oxazaborolidine-borane complex for the enantioselective reduction of 1-indanone. The enantiomeric purity of the crude product is 97.8%. A single recrystallization from hexane then affords enantiomerically pure (> 99.8% ee) 1-indanol in 90-94% overall yield.

1. Merck Research Laboratories, Department of Process Research, Rahway, NJ 07065.

2. Marfey, P. *Carlsberg Res. Commun.* **1984**, *49,* 591-596.

3. Top, S.; Meyer, A.; Jaouen, G. *Tetrahedron Lett.* **1979**, 3537-3540.

4. Kabuto, K.; Imuta, M.; Kempner, E. S.; Ziffer, H. *J. Org Chem.* **1978**, *43*, 2357-2361.

5. Enantioselective addition of diethylzinc to aldehydes: (a) Soai, K.; Ookawa, A.; Ogawa, K.; Kaba, T. *J. Chem. Soc., Chem. Commun.* **1987**, 467-468; (b) Soai, K.; Ookawa, A.; Kaba, T.; Ogawa, K. *J. Am. Chem. Soc.* **1987**, *109*, 7111-7115. Enantioselective addition of methyltitanium diisopropoxides to aldehydes: (c) Takahashi, H.; Kawabata, A.; Higashiyama, K. *Chem. Pharm. Bull.* **1987**, *35*, 1604-1607. Enantioselective reduction of ketones: (d) Kraatz, U. German Patent DE 3 609 152, 1987; *Chem. Abstr.* **1988**, *108*, 56111C; (e) Corey, E. J.; Bakshi, R. K.; Shibata, S. *J. Am. Chem. Soc.* **1987**, *109*, 5551-5553; (f) Corey, E. J.; Bakshi, R. K.; Shibata, S.; Chen, C.-P.; Singh, V. K. *J. Am. Chem. Soc.* **1987**, *109*, 7925-7926; (g) Corey, E. J.; Shibata, S.; Bakshi, R. K. *J. Org. Chem.* **1988**, *53*, 2861-2863; (h) Corey, E. J.; Reichard, G. A. *Tetrahedron Lett.* **1989**, *30*, 5207-5210; (i) Corey, E. J.; Link, J. O. *Tetrahedron Lett.* **1989**, *30*, 6275-6278; (j) Corey, E. J.; Bakshi, R. K. *Tetrahedron Lett.* **1990**, *31*, 611-614; (k) Rama Rao, A. V.; Gurjar, M. K.; Sharma, P. A.; Kaiwar, V. *Tetrahedron Lett.* **1990**, *31*, 2341-2344; (l) Jones, T. K.; Mohan, J. J.; Xavier, L. C.; Blacklock, T. J.; Mathre, D. J.; Sohar, P.; Jones, E. T. T.; Reamer, R. A.; Roberts, F. E.; Grabowski, E. J. J. *J. Org. Chem.* **1991**, *56*, 763-769.

6. Addition of lithiated N-nitrosopyrrolidine to benzophenone to give racemic α,α-diphenyl-2-pyrrolidinemethanol (58-60% yield based on benzophenone): (a) Seebach, D.; Enders, D.; Renger, B. *Chem. Ber.* **1977**, *110*, 1852-1865; (b) Enders, D.; Pieter, R.; Renger, B.; Seebach, D. *Org. Synth., Coll. Vol. VI* **1988,** 542-549.

7. Addition of phenylmagnesium chloride to methyl pyroglutamate followed by reduction with borane to give racemic α,α-diphenyl-2-pyrrolidinemethanol (3 steps, 51% yield from pyrroglutamic acid) that was then resolved as its O-acetylmandelate salt to give (R)- or (S)-α,α-diphenyl-2-pyrrolidinemethanol (2 steps, 3 recrystallizations, 30% yield from racemic α,α-diphenyl-2-pyrrolidinemethanol ref. 5f).

8. Addition of (S)-proline ethyl ester to phenylmagnesium chloride to give (S)-α,α-diaryl-2-pyrrolidinemethanol (2 steps, 21% yield from (S)-proline): (a) Roussel-UCLAF French Patent FR M3638, 1965; *Chem. Abstr.* **1969**, *70*, 106375m An earlier report (26% yield) did not indicate the enantiomeric purity of the product; (b) Kapfhammer, J.; Matthes, A. *Z. physiol. Chem.* **1933**, *223*, 43-52. In the submitters hands, addition of (S)-proline methyl ester hydrochloride to phenylmagnesium chloride, afforded (S)-α,α-diphenyl-2-pyrrolidinemethanol in 20% yield and 80% e.e. See also ref. 5d.

9. Addition of N-(benzyloxycarbonyl)-(S)-proline methyl ester to phenyl-magnesium chloride to give (S)-α,α-diphenyl-2-pyrrolidinemethanol (3 steps, 5 isolations, 0-50% yield from (S)-proline): ref. 5e and 5g. It should be noted that 6-10 equiv of the Grignard reagent are required to drive this reaction to completion. Quenching the excess phenylmagnesium bromide affords benzene--an environmental and health hazard.

10. Addition of N-benzyl-(S)-proline ethyl ester to phenylmagnesium chloride followed by catalytic hydrogenolysis affords (S)-α,α-diphenyl-2-

pyrrolidinemethanol (4 steps, 3 isolations, 51% yield from (S)-proline: Enders, D.; Kipphardt, H.; Gerdes, P.; Brena-Valle, L. J.; Bhushan, V. *Bull. Soc. Chim. Belg.* **1988**, *97* , 691-704. This method has also been used to prepare other (S)-α,α-diaryl-2-pyrrolidinemethanol: ref. 5i.

11. Mathre, D. J.; Jones, T. K.; Xavier, L. C.; Blacklock, T. J.; Reamer, R. A.; Mohan, J. J.; Jones, E. T. T.; Hoogsteen, K.; Baum, M. W.; Grabowski, E. J. J. *J. Org. Chem.* **1991**, *56*, 751-762.
12. (a) See ref 8; (b) Blacklock, T. J.; Jones, T. K.; Mathre, D. J.; Xavier, L. C., U.S. Patent 5 039 802, 1991; *Chem. Abstr.* **1991**, *115*, 232503n; (c) For an alternate source of (R)-diphenylprolinol see: Kerrick, S. T.; Beak, P. *J. Am. Chem. Soc.* **1991**, *113*, 9708-9710; Nikolic, N. A.; Beak, P. *Org. Synth.* **1996**, *74*, 23.
13. (a) Mathre, D. J.; Thompson, A. S.; Douglas, A. W.; Hoogsteen, K.; Carroll, J. D.; Corley, E. G.; Grabowski, E. J. J. *J. Org. Chem.* **1993**, *58*, 2880-2888; (b) Blacklock, T. J.; Jones, T. K.; Mathre, D. J.; Xavier, L. C., U.S. Patent 5 189 177, 1993: See U.S. Patent 5 039 802; *Chem. Abstr.* **1991**, *115*, 232503n.
14. Corey, E. J.; Azimioara, M.; Sarsha, S. *Tetrahedron Lett.* **1992**, *33*, 3429-3430.
15. (a) Itsuno, S.; Ito, K.; Hirao, A.; Nakahama, S. *J. Chem. Soc., Chem. Commun.* **1983**, 469-470; (b) see ref 5f. For recent reviews see: (c) Singh, V. K. *Synthesis* **1992**, 605-617; (d) Wallbaum, S.; Martens, J. *Tetrahedron: Asymmetry* **1992**, *3*, 1475-1504.

Appendix
Chemical Abstracts Nomenclature (Collective Index Number); (Registry Number)

(S)-Tetrahydro-1-methyl-3,3-diphenyl-1H,3H-pyrrolo[1,2-*c*][1,3,2]oxazaborole-borane complex: Boron, trihydro(tetrahydro-1-methyl-3,3-diphenyl-1H,3H-pyrrolo[1,2-*c*][1,3,2]oxazaborole-N^7)-, [l-4-(3aS-cis)]- (12); (112022-90-9)

(S)-Proline-N-carboxyanhydride: 1H,3H-Pyrrolo[1,2-*c*]oxazole-1,3-dione, tetrahydro-, (S)- (9); (45736-33-2)

(S)-(-)-Proline: L-Proline (8,9); (147-85-3)

Phosgene: HIGHLY TOXIC. Carbonic dichloride (9); (75-44-5)

Triethylamine (8); Ethanamine, N, N-diethyl- (9); (121-44-8)

(S)-1,1-Diphenylprolinol: 2-Pyrrolidinemethanol, α,α-diphenyl-, (S)- (12); (112068-01-6)

Trimethylboroxine: Boroxin, trimethyl- (8,9); (823-96-1)

(S)-Diphenylprolinol sulfate: 2-Pyrrolidinemethanol, α,α-diphenyl-, (S)-, sulfate (2:1) (salt) (12); (131180-44-4)

Borane-methyl sulfide complex: Methyl sulfide, compd, with borane (1:1) (8); Borane, compd. with thiobis[methane] (1:1) (9); (13292-87-0)

(R)-2,3-Dihydro-1H-inden-1-ol: 1-Indanol, (R)- (8); 1H-Inden-1-ol, 2,3-dihydro-, (R)- (9); (697-64-3)

1-Indanone (8); 1H-Inden-1-one, 2,3-dihydro- (9); (83-33-0)

1,2,3-TRIPHENYLCYCLOPROPENIUM BROMIDE

(Cyclopropenylium, triphenyl-, bromide)

Ph—≡—Ph + $PhCHCl_2$ + 2 t-BuOK ⟶

[Ph, Ot-Bu, Ph, Ph cyclopropene] $\xrightarrow{HBr}$ triphenylcyclopropenium Br^-

Submitted by Ruo Xu and Ronald Breslow.[1]

Checked by Rebecca Calvo and Robert K. Boeckman, Jr.

1. Procedure

1,2,3-Triphenylcyclopropenium bromide. A flame-dried, 500-mL, three-necked, round-bottomed flask, equipped with a magnetic stirring bar and a reflux condenser fitted with an argon inlet vented through a mineral oil bubbler, glass stopper, and rubber septum, is flushed with argon and charged with a solution of diphenylacetylene (3.92 g, 0.022 mol) (Note 1) and potassium tert-butoxide (9.97 g, 0.089 mol) in dry benzene (100 mL) (Note 2). Efficient magnetic stirring is initiated, and 5.8 mL of α,α-dichlorotoluene (7.2 g, 0.045 mol) (Note 3) is added dropwise over ~5 min via a syringe under argon (Note 4). The reaction mixture is then heated under reflux for 3 hr during which time the precipitate dissolves (Note 5). After the reaction mixture is cooled to room temperature, 100 mL of water is added to remove inorganic salts. The organic layer is separated and the aqueous layer is extracted with portions of ether (2 x 50-mL). The organic layers are combined, dried over magnesium sulfate, and filtered. The filtrate is saturated with anhydrous hydrogen bromide (Note 6), whereupon a light yellow precipitate forms that is collected to afford 6.8-7.2 g (89-93%)

of essentially pure triphenylcyclopropenium bromide (Note 7). The material prepared in this manner provides a satisfactory elemental analysis,[2] but it can be further purified by recrystallization from acetonitrile, if desired.

2. Notes

1. Diphenylacetylene, potassium tert-butoxide, and α,α-dichlorotoluene were purchased from Aldrich Chemical Company, Inc. and used without further purification.

2. Benzene is obtained from the Fisher Scientific Company and purified by distillation from CaH_2.

3. An excess of α,α-dichlorotoluene and potassium tert-butoxide is necessary to ensure high conversion of diphenylacetylene. When 1 equiv of α,α-dichlorotoluene and 2 equiv of potassium tert-butoxide are used, only 60% conversion of diphenylacetylene is achieved and the remainder is recovered during the workup. Nonetheless, the yield corrected for recovered diphenylacetylene is still more than 90%.

4. The reaction mixture is observed to bubble, fume, become brown in color, and deposit a precipitate. When the reaction is carried out in the air rather than under argon, the conversion is only 50%.

5. Slightly lower conversion is consistently realized when the reflux period is shortened to 1 hr.

6. When the usual preparative procedure was interrupted before the addition of water, and the inorganic material was removed by filtration, concentration of the solution and addition of dry hexane caused crystallization of 1,2,3-triphenylcyclopropenyl tert-butyl ether as white prisms.[2] The checkers did not observe reprecipitation of the salts upon cooling the reaction mixture.

7. The crude product is quite pure and has the following spectral characteristics: ^1H NMR (300 MHz, CD_3NO_2) δ: 7.99 (t, 6 H, J = 8), 8.12 (t, 3 H, J = 8), 8.71 (d, 6 H, J = 8); ^{13}C NMR (75 MHz, CD_3NO_2) δ: 119.65, 130.18, 135.40, 138.18, 155.26; IR ($CHCl_3$) cm^{-1}: 3382, 1712, 1594, 1505, 1410. On heating it decomposes without a defined melting point.

Waste Disposal Information

All toxic materials were disposed of in accordance with "Prudent Practices in the Laboratory"; National Academic Press; Washington, DC, 1996.

3. Discussion

This is the procedure first reported in 1961.[2] It involves generation of phenylchlorocarbene, or more likely a related carbenoid, which adds to the acetylene to form triphenylcyclopropenium chloride. This chloride reacts with tert-butoxide ion to form the covalent tert-butyl ether, which can be isolated. With water, this ether hydrolyzes to bis(triphenylcyclopropenyl) ether, but either of these compounds is converted to the ionic bromide salt with HBr.

According to this procedure, with proper choice of reaction conditions, a sizeable amount of diphenylacetylene can be converted quantitatively to triphenylcyclopropenium bromide in a few hours. The reaction is of wide generality and can be applied to p-anisylphenylacetylene and to di-p-anisylacetylene, or with p-anisal chloride instead of α,α-dichlorotoluene, to prepare p-methoxy derivatives of the title compound.[2] Since the initial preparation of this derivative of the cyclopropenyl cation by a less efficient procedure,[3] many aryl-, alkyl- and heteroatom-substituted derivatives of this simplest cyclic aromatic system have been synthesized,[4] including the parent cyclopropenyl cation itself.[5] They have been used for various physical studies,[6] and for the preparation of derivatives such as covalent cyclopropenes and metal coordination complexes.[7,8]

The cyclopropenyl cation is the simplest aromatic system, and thus of some theoretical interest. In addition, the chemistry of cyclopropene derivatives is full of interesting rearrangements to other novel structures,[9] reflecting the great strain energy of the cyclopropene ring.

1. Department of Chemistry, Columbia University, New York, New York 10027.
2. Breslow, R.; Chang, H. W. *J. Am. Chem. Soc.* **1961**, *83*, 2367.
3. Breslow, R. *J. Am. Chem. Soc.* **1957**, *79*, 5318.
4. (a) Breslow, R.; Yuan, C. *J. Am. Chem. Soc.* **1958**, *80*, 5991; (b) Breslow, R.; Höver, H.; Chang, H. W. *J. Am. Chem. Soc.* **1962**, *84*, 3168; (c) Yoshida, Z. I. *Top. Curr. Chem.* **1973**, *40*, 47.
5. Breslow, R.; Groves, J. T. *J. Am. Chem. Soc.* **1970**, *92*, 984.
6. Wasielewski, M. R.; Breslow, R. *J. Am. Chem. Soc.* **1976**, *98*, 4222.
7. Pienta, N. J.; Kessler, R. J.; Peters, K. S.; O'Driscoll, E. D.; Arnett, E. M.; Molter, K. E. *J. Am. Chem. Soc.* **1991**, *113*, 3773.
8. Hughes, R. P.; Klaeui, W.; Reisch, J. W.; Mueller, A. *Organometallics* **1985**, *4*, 1761.
9. Breslow, R.; Boikess, R.; Battiste, M. *Tetrahedron Lett.* **1960**, *26*, 42.

Appendix

Chemical Abstracts Nomenclature (Collective Index Number); (Registry Number)

1,2,3-Triphenylcyclopropenium bromide: Cyclopropenylium, triphenyl-, bromide (8,9); (4919-51-5)

Diphenylacetylene: Acetylene, diphenyl- (8); Benzene, 1,1'-(1,2-ethynediyl)bis- (9); (501-65-5)

Potassium tert-butoxide: tert-Butyl alcohol, potassium salt (8); 2-Propanol, 2-methyl-, potassium salt (9); (865-47-4)

α,α-Dichlorotoluene: Toluene, α,α-dichloro- (8); Benzene, (dichloromethyl)- (9); (98-87-3)

Hydrogen bromide: Hydrobromic acid (8,9); (10035-10-6)

PYRIDINE-DERIVED TRIFLATING REAGENTS: N-(2-PYRIDYL)–TRIFLIMIDE AND N-(5-CHLORO-2-PYRIDYL)TRIFLIMIDE

(Methanesulfonamide, 1,1,1-trifluoro-N-2-pyridinyl-N-[(trifluoromethyl)sulfonyl]-] and Methanesulfonamide, N-(5-chloro-2-pyridinyl)-1,1,1-trifluoro-N-[(trifluoromethyl)sulfonyl]-)

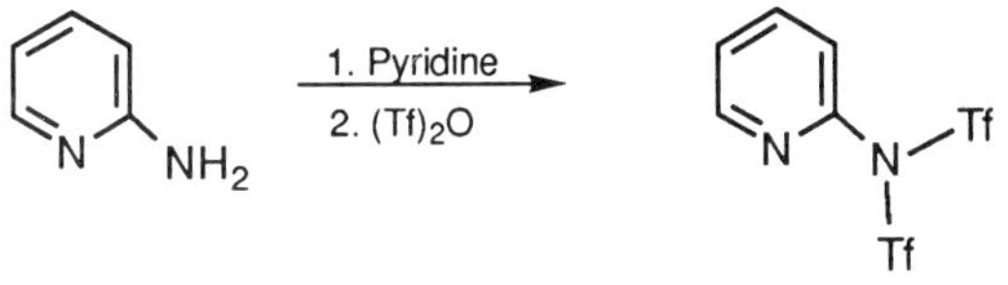

1. Pyridine
2. $(Tf)_2O$

NH_2 N N Tf Tf

1

Cl

1. Pyridine
2. $(Tf)_2O$

NH_2 N N Tf Tf

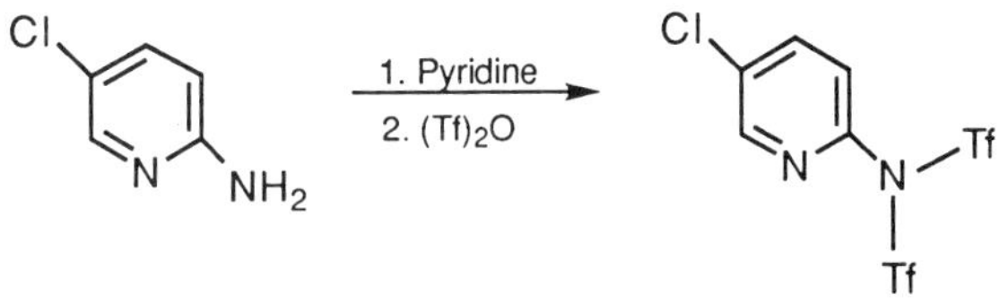

2

Submitted by Daniel L. Comins, Ali Dehghani, Christopher J. Foti, and Sajan P. Joseph.[1]

Checked by Maria A. Cichy and Amos B. Smith, III.

1. Procedure

A. N-(2-Pyridyl)triflimide (**1**). A 2-L, two-necked, round-bottomed flask equipped with a mechanical stirrer (Note 1) and a rubber septum is charged with 2-aminopyridine (19.859 g, 0.211 mol) (Note 2) and pyridine (35.04 g, 35.88 mL, 0.443 mol) (Note 3) in 800 mL of dichloromethane (CH_2Cl_2) (Note 4) under an argon atmosphere. The reaction mixture is cooled to -78°C and a solution of triflic anhydride (125 g, 74.54 mL, 0.443 mol) (Note 5) in 150 mL of CH_2Cl_2 is added dropwise via a cannula over 3.5 hr with vigorous stirring. After the solution is stirred for 2 hr at -78°C, the cooling bath is removed and stirring is continued at room temperature for 19 hr. The reaction mixture is quenched with 50 mL of cold water and the layers are separated. The aqueous layer is extracted with CH_2Cl_2 (4 x 50 mL). The combined organic extracts are washed with cold aqueous 10% sodium hydroxide (1 x 150 mL), cold water (1 x 100 mL), brine (1 x 100 mL) and dried over magnesium sulfate. After filtration, the solvent is removed under vacuum to give 69 g of the crude product. After Kugelrohr distillation (Note 6), 61 g (81%) of pure N-(2-pyridyl)triflimide (bp 85-100°C/0.25 mm, mp 41-42°C) (Note 7) is obtained as a white solid.

B. N-(5-Chloro-2-pyridyl)triflimide (**2**). Using the same procedure as described above, 2-amino-5-chloropyridine (27.13 g, 0.211 mol) (Note 8) is converted to N-(5-chloro-2-pyridyl)triflimide (mp 47-48°C) (61.84 g, 75%, bp 88-100°C/0.15 mm) (Note 9).

2. Notes

1. A magnetic stirrer can be used with a large stirrer bar and 1.5 L of dichloromethane.

2. 2-Aminopyridine was purchased from Aldrich Chemical Company, Inc., and used without further purification.

3. Anhydrous pyridine was purchased from Aldrich Chemical Company, Inc., and kept over 3 Å molecular sieves for two days prior to use.

4. Anhydrous dichloromethane was purchased from Aldrich Chemical Company, Inc., and used without further purification.

5. Triflic anhydride was purchased from Aldrich Chemical Company, Inc., and used as such.

6. Sometimes a second distillation is needed to obtain pure compound.

7. The spectral properties of N-(2-pyridyl)triflimide are as follows: IR (nujol) cm^{-1}: 1590, 1570, 1460, 1220, 1215, 1120, 1040, 990, 940, 910, 880, 735, 710; ^{1}H NMR (300 MHz, $CDCl_3$) δ: 7.46-7.55 (m, 2 H), 7.91-7.97 (dt, 1 H, J = 8.06, 2.2), 8.63 (dd, 1 H, J = 4.4, 1.46); ^{13}C NMR (75 MHz, $CDCl_3$) δ: 112.80, 117.10, 121.39, 125.50, 125.69, 126.79, 139.71, 145.98, 150.31.

8. 2-Amino-5-chloropyridine was purchased from Aldrich Chemical Company, Inc., and used without further purification.

9. The spectral properties of N-(5-chloro-2-pyridyl)triflimide are as follows: IR (nujol) cm^{-1}: 1570, 1460, 1230, 1215, 1125, 1010, 925, 905, 745, 730; ^{1}H NMR (300 MHz, $CDCl_3$) δ: 7.42 (d, 1 H, J = 8.8), 7.90 (dd, 1 H, J = 8.8, 2.2), 8.58 (d, 1 H, J = 2.93); ^{13}C NMR (75 MHz, $CDCl_3$) δ: 112.77, 117.06, 121.36, 125.66, 126.18, 135.84, 139.33, 143.82, 149.31.

Waste Disposal Information

All toxic materials were disposed of in accordance with "Prudent Practices in the Laboratory"; National Academic Press; Washington, DC, 1996.

3. Discussion

Vinyl triflates are important intermediates, since they can be used as synthetic precursors for vinyl cations and alkylidene carbenes, and as substrates for regiospecific coupling reactions.[2,3] Vinyl triflates are also valuable intermediates in a mild, two-step procedure for the deoxygenation of ketones.[4,5] These new triflating reagents are highly reactive and easy to prepare and handle.[6] When compared with other triflating reagents, the vinyl triflate can in most cases be made at lower temperatures, and any excess reagent and by-products can be removed by washing with cold aqueous 5% sodium hydroxide solution. The utility of the pyridine-derived triflating reagents is illustrated by the examples in the Table.[6] Recently reagent **2** has been used in the total syntheses of (-)-porantheridine[7] and trans-decahydroquinoline alkaloid (+)-219 A.[8]

1. Department of Chemistry, North Carolina State University, Raleigh, NC 27695-8204.
2. For reviews see: (a) Stang, P. J. *Acc. Chem. Res.* **1978**, *11*, 107; (b) Stang, P. J.; Hanak, M.; Subramanian, L. R. *Synthesis* **1982**, 85.
3. (a) Scott, W. J.; Peña, M. R.; Swärd, K.; Stoessel, S. J.; Stille, J. K. *J. Org. Chem.* **1985**, *50*, 2302; (b) Tamaru, Y.; Ochiai, H.; Nakamura, T.; Yoshida, Z. *Tetrahedron Lett.* **1986**, *27*, 955; (c) Wulff, W. D.; Peterson, G. A.; Bauta, W. E.; Chan, K-S.; Faron, K. L.; Gilbertson, S. R.; Kaesler, R. W.; Yang, D. C.; Murray, C. K. *J. Org. Chem.* **1986**, *51*, 277; (d) Cacchi, S.; Morera, E.; Ortar, G. *Synthesis* **1986**, 320; (e) McCague, R. *Tetrahedron Lett.* **1987**, *28,* 701; (f) Arcadi, A.; Burini, A.; Cacchi, S.; Delmastro, M.; Marinelli, F.; Pietroni, B. *Synlett* **1990**, 47, and references therein; (g) for recent applications in total syntheses of natural products, see: Corey, E. J.; Kigoshi, H. *Tetrahedron Lett.* **1991**, *32,* 5025; Smith, A. B., III; Sulikowski, G. A.; Sulikowski, M. M.; Fujimoto, K. *J. Am. Chem. Soc.* **1992**, *114*, 2567.
4. (a) Jigajinni, V. B.; Wrightman, R. H. *Tetrahedron Lett.* **1982**, *23*, 117; (b) Subramanian, L. R.; Martinez, A. G.; Fernandes, A. H.; Alvarez, R. M. *Synthesis* **1984**, 481.

5. The submitters successfully used this reduction procedure in their recent synthesis of (±)-pumiliotoxin C, see: Comins, D. L.; Dehghani, A. *Tetrahedron Lett.* **1991**, *32*, 5697.
6. Comins, D. L.; Dehghani, A. *Tetrahedron Lett.* **1992**, *33*, 6299.
7. Comins, D. L.; Hong, H. *J. Am. Chem. Soc.* **1993**, *115*, 8851.
8. Comins, D. L.; Dehghani, A. *J. Org. Chem.* **1995**, *60*, 794.

TABLE[6]

PREPARATION OF VINYL TRIFLATES FROM KETONE ENOLATES

Ketone	Base; Reagent; Conditions	Product	Yield, %
(4-tert-butylcyclohexanone)	NaHMDS; **2** -78°C, 2 hr	OTf	92
Ph	LDA; **1** -78°C, 3 hr	OTf Ph	76
O	LDA; **2** -78°C, 2 hr	OTf	77
Ph Ph	LDA; **2** -78°C, 3 hr	OTf Ph Ph	79
O	LDA; **1** -78°C, 2 hr; -23°C, 4 hr; -16°C, 12 hr	OTf	76
O	LDA; **2** -78°C, 2 hr	OTf	86
O Ph N Ph CO_2Ph	LDA; **2** -78°C, 2 hr	OTf Ph N Ph CO_2Ph	88
O N CO_2Bn	L-Selectride; **2** -23°C, 2 hr	OTf N CO_2Bn	80

Appendix

Chemical Abstracts Nomenclature (Collective Index Number); (Registry Number)

N-(2-Pyridyl)triflimide: Methanesulfonamide, 1,1,1-trifluoro-N-2-pyridinyl-N-[(trifluoromethyl)sulfonyl]- (13); (145100-50-1)

N-(5-Chloro-2-pyridyl)triflimide: Methanesulfonamide, N-(5-chloro-2-pyridinyl)-1,1,1-trifluoro-N-[(trifluoromethyl)sulfonyl]- (13); (145100-51-2)

2-Aminopyridine: Pyridine, 2-amino- (8); 2-Pyridinamine (9); (504-29-0)

Pyridine (8,9); (110-86-1)

Trifluoromethanesulfonic anhydride: Methanesulfonic acid, trifluoro-, anhydride (8,9); (358-23-6)

2-Amino-5-chloropyridine: Pyridine, 2-amino-5-chloro- (8); 2-Pyridinamine, 5-chloro- (9); (1072-98-6)

BIS(TRIMETHYLSILYL) PEROXIDE (BTMSPO)

(Silane, dioxybis[trimethyl-)

A. $N(CH_2CH_2)_3N + 2H_2O_2 \xrightarrow[0°C]{THF} N(CH_2CH_2)_3N \cdot 2H_2O_2$

B. $N(CH_2CH_2)_3N \cdot 2H_2O_2 + N(CH_2CH_2)_3N + 4Me_3SiCl \xrightarrow[0°C]{CH_2Cl_2} 2Me_3SiOOSiMe_3$
$2N(CH_2CH_2)_3N$
$2HCl$

Submitted by P. Dembech,[1] A. Ricci, G. Seconi, and M. Taddei.

Checked by Christopher S. Brook, Mark Guzman, and Amos B. Smith, III.

1. Procedure

Caution! The possibility of explosions when handling bis(trimethylsilyl) peroxide (BTMSPO) has been reported,[2] especially in the presence of metal needles, cannulas, etc., although the product was prepared using a different procedure that included the use of 85% H_2O_2 and pyridine. Our method never gave rise to explosions in the general practice of an organic chemistry laboratory. Nevertheless all the reactions were carried out in safety cupboards and behind blast shields (Note 1).

A. Diazabicyclooctane hydrogen peroxide (DABCO·$2H_2O_2$) complex. A 1-L, two-necked flask, fitted with a dropping funnel, 3-cm stirring bar, and a thermometer, is charged with diazabicyclooctane (DABCO) (28.05 g, 0.25 mol) (Note 2) dissolved in commercial grade tetrahydrofuran (THF) (375 mL) (Note 3). The flask is cooled to 0°C (internal temperature, ice-salt bath), and hydrogen peroxide (49 mL of a 35% solution, 0.5 mol) (Note 4) is slowly added at such a rate that the temperature remains constant, under vigorous stirring. A precipitate appears immediately and, after completing the addition of hydrogen peroxide, stirring is continued at 0°C for 1 hr. The mixture is then filtered through a Büchner funnel and the collected precipitate washed with cold THF (3 x 50 mL) and dried under reduced pressure (2 mm, 35-40°C) in a flask for 2 hr. The yield of DABCO·$2H_2O_2$ obtained is 28.3 g (63%), m.p. 102-105°C (sealed tube) with decomposition, softness at 80°C (lit.[3] m.p. 112°C) (Note 5).

B. Bis(trimethylsilyl) peroxide (BTMSPO). A 2-L, three-necked flask, containing a 3-cm stirring bar, and a thermometer, dropping funnel, and an argon inlet adapter, is maintained under a slight pressure of argon. The flask is charged with the DABCO·$2H_2O_2$ complex (28.3 g, 0.157 mol) and DABCO (28 g, 0.25 mol). Dry dichloromethane (700 mL) (Note 6) is added and the mixture is cooled to 0°C. Chlorotrimethylsilane (80 mL, 0.628 mol) (Note 7) is added dropwise, maintaining the temperature at 0°C, and the resulting mixture is stirred for 5 hr at room temperature. The mixture is filtered through a sintered Büchner funnel and the precipitate is washed with pentane (2 x 25 mL). The solution, transferred to a 2-L flask fitted with a vacuum-jacketed, packed column (70 cm long, 4-cm i.d.) and a distillation head equipped with a dry ice/acetone cold finger (Note 8), is concentrated to ca. 80 mL under reduced pressure (Note 9). Pentane (25 mL) is added to the mixture, the additional precipitate is filtered off, and the solution is transferred to a smaller flask, (500 mL). It is concentrated under reduced pressure with the distillation apparatus previously employed to give a residue of 40.1 g (71% yield) of BTMSPO as a clear colorless liquid. It is shown (Note 10) from multinuclear NMR to be contaminated by ca. 8% of hexamethylsiloxane (Note 11).

For synthetic purposes the BTMSPO obtained can be used without further purification; on the other hand, distillation (40°C/30 mm) or column chromatography (Florisil - eluent pentane) does not seem to improve the grade of purity (Note 12). The compound can be stored under nitrogen in a refrigerator for months without any appreciable decomposition.

2. Notes

1. Because of the reported[4] instability of the DABCO·$2H_2O_2$ complex, the checkers found it prudent to use new glassware for each run.

2. Diazabicyclooctane (DABCO) was used as purchased from Aldrich Chemical Company, Inc.

3. In the preparation of the complex the use of THF, miscible with water, produced a fine precipitate.

4. Hydrogen peroxide, 35% m/v, 120 volumes, from Aldrich Chemical Company, Inc., was used without titration.

5. The submitters ran this procedure on twice the scale (i.e., 0.5 mol of DABCO) in 59% yield. The complex is hygroscopic and can be stored in a desiccator overnight, but decomposes over long storage.

6. DABCO is previously dried at 20-27°C (room temperature) under reduced pressure (2 mm) Dichloromethane was used as purchased from Aldrich Chemical Company, Inc., (99.9%).

7. Trimethylchlorosilane was used as purchased from Aldrich Chemical Company, Inc.

8. If the solvent is removed with a rotary evaporator, considerable loss of product results from codistillation with dichloromethane. The checkers found that if this exact distillation set-up was not used the yield of BTMSPO decreased substantially (30-40%) because of the codistillation of CH_2Cl_2 and BTMSPO (observed in distillate by ^1H NMR). The distillation set-up requires use of a "walk-in" hood (see diagram).

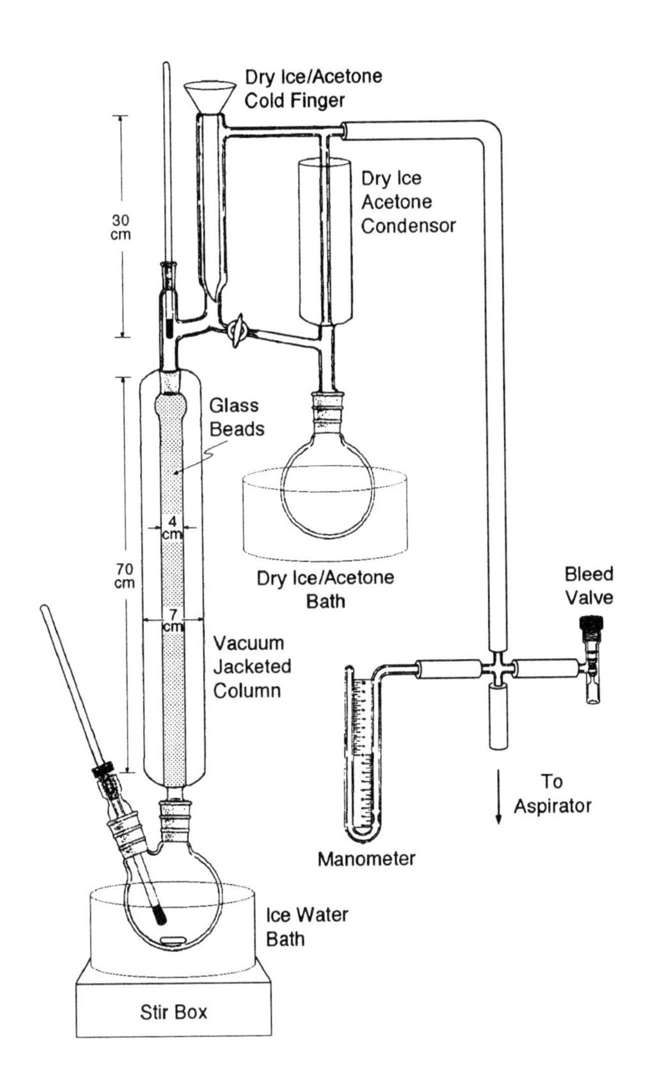

9. Important points of this procedure are the temperature of the pot, which never must exceed 0°C, the temperature of the cooled head of the column, which

should vary in the range of -5 - 0°C, and the vacuum, which throughout the distillation has been kept in the range of 90-100 mm (90-100 drops/min).

These conditions apply to CH_2Cl_2 removal, whereas for pentane distillation a pressure of about 200 mm is recommended. To remove the last traces of solvent at the very end of the fractionation, the pot temperature was raised to 15-20°C and the pressure decreased to 40 mm.

10. The submitters ran this procedure on twice the scale (i.e., 0.28 mol of DABCO·$2H_2O_2$) and report a yield of 94%. The checkers found that the final stillpot residue still contained some DABCO·HCl salt that had to be filtered to obtain pure BTMSPO. Filtration was done via a fluted filter paper.

11. A typical NMR spectroscopic analysis was as follows: ^{1}H NMR (500 MHz, $CDCl_3$) δ: 0.187 ppm; ^{13}C NMR (125 MHz, $CDCl_3$) δ: -1.40

12. Recently another procedure has appeared,[4] dealing with the large scale preparation of BTMSPO. The costs and the yield of the two procedures are comparable, calculated on the amount of H_2O_2-base complex used, but the present procedure avoids the *potential* danger of violent decomposition[2] of the silylated peroxide during its distillation.

Waste Disposal Information

All toxic materials were disposed of in accordance with "Prudent Practices in the Laboratory"; National Academic Press; Washington, DC, 1996.

3. Discussion

The procedure described here, based on a modification of that of A. G. Davies,[5] provides the first large scale preparation of BTMSPO, a reagent widely used for synthetic purposes over the last few years. The present method is characterized by high reliability; the low concentration of H_2O_2, high yields, and the safe procedure are further major advantages.

BTMSPO has been mainly employed as a reagent for the transformation of carbanions into the corresponding trimethylsiloxy or hydroxy derivatives,[6] and in the hydroxylation of neutral aromatic systems[7] (in the presence of TfOH).

It has, for example, been used in the preparation of 3-trimethylsiloxyfuran,[8] and to prepare stereochemically pure E- or Z-silyl enol ethers starting from the corresponding E- or Z-vinyl bromides[9] or vinylstannanes.[6]

It can also be used as a mild aprotic oxidizing agent (soluble in all common organic solvents), for the transformation of sulfides into the corresponding sulfoxides,[10] phosphines into phosphine oxides,[11] in the oxidation of alcohols to aldehydes[12] (as co-oxidant of CrO_3·Py), in the Baeyer-Villiger reaction[13] (in the presence of $TfOSiMe_3$). It has also been employed in the transformation of acetylenes to the corresponding iodoacetylenes.[14]

1. C.N.R.-Istituto dei Composti del Carbonio contenenti Eteroatomi e loro Applicazioni, Via P. Gobetti 101, 40129 Bologna, Italy.
2. Neumann, H.; Seebach, D. *Chem. Ber.* **1978**, *111*, 2785.
3. Oswald, A. A.; Guertin, D. L. *J. Org. Chem.* **1963**, *28*, 651.
4. Jackson, W. P. *Synlett*, **1990**, 536.
5. Cookson, P. G.; Davies, A. G.; Fazal, N. *J. Organomet. Chem.* **1975**, *99*, C31.
6. (a) Camici, L.; Dembech, P.; Ricci, A.; Seconi, G.; Taddei, M. *Tetrahedron* **1988**, *44*, 4197; (b) Dembech, P.; Guerrini, A.; Ricci, A.; Seconi, G.; Taddei, M. *Tetrahedron* **1990**, *46*, 2999.
7. Olah, G. A.; Ernst, T. D. *J. Org. Chem.* **1989**, *54*, 1204.
8. Camici, L.; Ricci, A.; Taddei, M. *Tetrahedron Lett.* **1986**, *27*, 5155.

9. Davis, F. A.; Sankar Lai, G.; Wei, J. *Tetrahedron Lett.* **1988**, *29*, 4269.

10. (a) Brandes, D.; Blaschette, A. *J. Organometal. Chem.* **1974**, *73*, 217; (b) Curci, R.; Mello, R.; Troisi, L. *Tetrahedron* **1986**, *42*, 877.

11. (a) Wozniak, L.; Kowalski, J.; Chojnowski, J. *Tetrahedron Lett.* **1985**, *26*, 4965;. (b) Hayakawa, Y.; Uchiyama, M.; Noyori, R. *Tetrahedron Lett.* **1986**, *27*, 4195.

12. Kanemoto, S.; Matsubara, S.; Takai, K.; Oshima, K.; Utimoto, K.; Nozaki, H. *Bull. Chem. Soc. Jpn.* **1988**, *61*, 3607.

13. Suzuki, M.; Takada, H.; Noyori, R. *J. Org. Chem.* **1982**, *47*, 902.

14. Ricci, A.; Taddei, M.; Dembech, P.; Guerrini, A.; Seconi, G. *Synthesis* **1989**, 461.

Appendix

Chemical Abstracts Nomenclature (Collective Index Number); (Registry Number)

Bis(trimethylsilyl) peroxide: 3,4-Dioxa-2,5-disilahexane, 2,2,5,5-tetramethyl- (8); Silane, dioxybis[trimethyl- (9); (5796-98-5)

1,4-Diazabicyclo[2.2.2]octane (8,9); (280-57-9)

Hydrogen peroxide (8,9); (7722-84-1)

Chlorotrimethylsilane: Silane, chlorotrimethyl- (8,9); (75-77-4)

SYNTHESIS OF EPOXIDES USING DIMETHYLDIOXIRANE: trans-STILBENE OXIDE

(Dioxirane, dimethyl- and Oxirane, 2,3-diphenyl-, trans)

A. Me₂C=O + Oxone $\xrightarrow[H_2O]{NaHCO_3}$ dimethyldioxirane

B. trans-PhCH=CHPh + dimethyldioxirane $\xrightarrow[rt]{acetone}$ trans-stilbene oxide

Submitted by Robert W. Murray and Megh Singh.[1]

Checked by Thomas G. Marron, Lance A. Pfeifer and William R. Roush.

1. Procedure

Caution! Dimethyldioxirane is a volatile peroxide and should be treated as such. The preparation and all reactions of the dioxirane should be carried out in a hood.

A. Dimethyldioxirane (See Figure 1). A 2-L, three-necked, round-bottomed flask containing a mixture of water (80 mL), acetone (50 mL, 0.68 mol), and sodium bicarbonate (96 g), is equipped with a magnetic stirring bar and a pressure equalizing addition funnel containing water (60 mL) and acetone (60 mL, 0.82 mol) (Note 1). A solid addition flask containing Oxone (180 g, 0.29 mol) is attached to the reaction vessel via a rubber tube (Note 2). An air condenser (20 cm length) loosely packed with glass wool is attached to the reaction vessel. The outlet of the air condenser is connected to a 75 x 350-mm Dewar condenser filled with dry ice-acetone that is connected to a receiving flask (100 mL) cooled in a dry ice-acetone bath. The receiving flask is also connected in series to a second dry ice-acetone cold trap, a trap containing a potassium iodide solution, and a drying tube. A gas inlet tube is connected to the reaction flask and a stream of nitrogen gas is bubbled through the reaction mixture (Note 3). The Oxone is added in portions (10-15 g) while the acetone-water mixture is simultaneously added dropwise (Note 4). The reaction mixture is stirred vigorously throughout the addition of reagents (ca. 30 min). A yellow solution of dimethyldioxirane in acetone collects in the receiving flask. Vigorous stirring is continued for an additional 15 min while a slight vacuum (ca. 30 mm, water aspirator) is applied to the cold trap (Note 5). The yellow dioxirane solution (62-76 mL, Note 6) is dried over sodium sulfate (Na_2SO_4), filtered and stored in the freezer (-25°C) over Na_2SO_4. The dioxirane content of the solution is assayed using phenyl methyl sulfide and the GLC method (Notes 7 and 8). Generally concentrations in the range of 0.07-0.09 M are obtained.

B. trans-Stilbene oxide. To a magnetically stirred solution of trans-stilbene (0.724 g, 4.02 mmol) (Note 9) in 5 mL of acetone is added a solution of 0.062 M dimethyldioxirane in acetone (66 mL, 4.09 mmol) at room temperature. The progress of the reaction is followed by GLC (Note 10), which analysis indicates that trans-stilbene is converted to the oxide in 6 hr (Note 11). Removal of solvent on a rotary evaporator gives a white crystalline solid. The solid is dissolved in dichloromethane (CH_2Cl_2) (30 mL) and dried with anhydrous Na_2SO_4. The drying agent is filtered off and washed with CH_2Cl_2. The solvent is removed on a rotary evaporator. Remaining solvent is removed under reduced pressure to give an analytically pure sample of the oxide (0.788 g, 100% yield). Recrystallization from aqueous ethanol gives white plates/prisms, mp 69-70°C (Notes 12, 13).

2. Notes

1. A mechanical stirrer may also be used.

2. The submitters used Oxone supplied by DuPont, whereas the checkers used Oxone purchased from Aldrich Chemical Company, Inc.

3. The submitters recommended that a stream of helium be passed through the reaction system during the course of the experiment. The checkers substituted nitrogen for helium with no decrease in yield . In addition, on several occasions the checkers did not use a gas purge and there was no decrease in yield. Therefore, use of a gas purge is viewed as optional, not mandatory. It is noted that other investigators have reported dimethyldioxirane preparations that do not require use of a gas purge.[2]

4. The procedure followed was generally that contained in the original publication.[3] See an alternative procedure, by Adam.[2]

5. The submitters recommended that the distillation be performed at ca. 30 mm. By using this procedure the checkers obtained an average of 67 mL (range: 62-76 mL) of dioxirane solution with an average concentration of 0.077 M (range: 0.068-0.087 M) over 10 repetitions of the procedure. When the distillation was performed at 80 mm for up to 90 min, greater volumes of dimethyldioxirane solution were obtained (84-89 mL), but with a corresponding decrease in the reagent concentration (0.053-0.066 M).

6. The submitters reported that 80-90 mL of dimethyldioxirane solution is obtained.

7. Determination of dimethyldioxirane concentration by the GLC method is as follows: A standard solution of thioanisole (phenyl methyl sulfide) is prepared. The solution is usually 0.2 M in acetone, but other concentrations may be used. It is important to keep the sulfide in excess so that oxidation by the dioxirane will produce largely or exclusively the sulfoxide and not the sulfone.

A standard solution of an internal standard (dodecane or hexadecane) is also prepared in acetone. This solution should be at the same concentration as that of the sulfide.

To determine the dioxirane concentration 1 mL each of the dioxirane, sulfide, and internal standard solutions are combined in a vial. The GLC analysis is then carried out using the following: Column: DB 210; temp 1 = 60°C, time 1 = 5 min, rate 1 = 20°/min; temp 2 = 200°C, time 2 = 5 min. The analysis is conducted on 1 µL of solution. The analysis is made quantitative by determining the response factors of the sulfide and internal standard in the usual manner. The dimethyldioxirane concentration is determined by measuring the sulfide concentration before and after adding the dioxirane. Under these conditions the following retention times are observed: dodecane, 7.15 min; sulfide, 8.2 min; sulfoxide, 12.9 min.

8. Generally concentrations in the range of 0.07-0.09 M are obtained. The submitters reported a concentration range of 0.05-0.10 M.

9. trans-Stilbene was purchased from Eastman Organic Chemical Co. The purity of the sample was 99%. Sample purity was also checked by GLC and GC-MS. The GC-MS analysis suggests that the sample contains bibenzyl (<1%) as an impurity.

10. Gas chromatographic conditions are as follows: Column DB-210 (30 m x 0.318 mm x 0.5 µm, fused silica capillary column), column temp 1 = 100°C, time 1 = 5 min, rate = 20°/min; temp 2 = 200°C, time 2 = 7 min, injector temp 250°C, detector temp 250°C, inlet P, 24 psi, retention times: trans-stilbene 11.4 min, trans-stilbene oxide 11.9 min.

11. In 2 hr, 96% conversion had occurred.

12. The checkers obtained 790 mg (100%) of crude trans-stilbene oxide that was recrystallized from aqueous ethanol to give 744 mg (95%; two crops) of analytically pure trans-stilbene oxide.

13. trans-Stilbene oxide has the following properties: ^{1}H NMR (300 MHz, $CDCl_3$) δ: 3.86 (s, 2 H, 2 X -CH-), 7.26-7.45 (m, 10 H, 2 X C_6H_5-); ^{13}C NMR (75 MHz, $CDCl_3$) δ: 62.81 (-CH-); 125.4 (C-4, Ar), 128.19 (C-3,5, Ar), 128.44 (C-2,6, Ar), 136.99 (C-1, ipso C of phenyl); IR ($CHCl_3$) cm^{-1}: 3076, 3036, 2989, 1603, 1497, 1457, 870, 698; high resolution mass spectrum, Calcd. for $C_{14}H_{12}O$, 196.0888. Found 196.0896. Anal. Calcd. for $C_{14}H_{12}O$: C, 85.68; H, 6.16. Found: C, 85.76; H, 6.05.

Waste Disposal Information

All toxic materials were disposed of in accordance with "Prudent Practices in the Laboratory"; National Academic Press; Washington, DC, 1996.

3. Discussion

For most epoxidations dimethyldioxirane (DMD) is the reagent of choice. The reaction is usually carried out at room temperature or below and in neutral solution. The reaction is stereospecific, proceeds rapidly, and generally in essentially quantitative yield. The procedure is remarkably convenient. In many cases removal of the solvent gives the pure product. The reaction is applicable to a variety of unsaturated systems (Table). The data given in the Table also compare yields with the commonly used epoxidation reagent m-chloroperbenzoic acid (MCPBA). In almost every case use of DMD gave a higher yield than did MCPBA. In many cases the difference is dramatic (see entries 1 and 2, for example). The use of MCPBA frequently leads to opening of sensitive epoxides under the acidic conditions of the reaction. This problem is conveniently avoided when using DMD. In a growing number of cases DMD has successfully given an epoxide when other methods fail. A particularly pleasing example[4] of this is the preparation of the 8,9-epoxide of the mycotoxin, aflatoxin B_1.

Linear free energy relationship studies[5,6] have demonstrated that DMD is an electrophilic reagent. This property is demonstrated in the Table where substrates with electron-withdrawing substituents require longer reaction times (compare entries 1 and 2 with 4, for example). This is particularly noticeable in entry 8 where the substrate contains a strong electron-withdrawing substituent. The previously reported[5] faster rates for DMD epoxidation of cis-alkenes compared to their trans stereoisomers is seen in the longer reaction times required for the trans isomers (compare entries 1 and 2, and 3 and 4). This difference in rates is taken as support for a spiro transition state[5] for epoxidation. A number of reviews[7-11] of the chemistry of DMD, including its use as an epoxidizing reagent, have been published.

Acknowledgment We gratefully acknowledge support of this work by the National Institute of Environmental Health Sciences through Grant No. ES01984.

1. Department of Chemistry, University of Missouri-St. Louis, St. Louis, MO 63121.
2. See, for example, Adam, W.; Bialas, J.; Hadjiarapoglou, L. *Chem. Ber.* **1991**, *124*, 2377.
3. Murray, R. W.; Jeyaraman, R. J. *J. Org. Chem.* **1985**, *50*, 2847.
4. Baertschi, S. W.; Raney, K. D.; Stone, M. P.; Harris, T. M. *J. Am. Chem. Soc.* **1988**, *110*, 7929.
5. Baumstark, A. L.; Vasquez, P. C. *J. Org. Chem.* **1988**, *53*, 3437.
6. Murray, R. W.; Shiang, D. L. *J. Chem. Soc., Perkin Trans 2* **1990**, 349.
7. Murray, R. W. *Chem. Rev.* **1989**, *89*, 1187.
8. Murray, R. W. In "Molecular Structure and Energetics, Vol. 6: Modern Models of Bonding and Delocalization"; Liebman, J. F.; Greenberg, A., Eds.; VCH: New York, 1988; pp. 311-351.
9. Curci, R. *Adv. Oxygenated Processes* **1990**, *2*, 1-59.

10. Adam, W.; Curci, R.; Edwards, J. O. *Acc. Chem. Res.* 1989, *22*, 205.

11. Adam, W.; Hadjiarapoglou, L. P.; Curci, R.; Mello, R. In "Organic Peroxides"; Ando, W., Ed.; J. Wiley and Sons: West Sussex, England, 1992; p. 195.

12. Curtin, D. Y.; Kellom, D. B. *J. Am. Chem. Soc.* **1953**, *75*, 6011.

13. Bissing, D. E.; Speziale, A. J. *J. Am. Chem. Soc.* **1965**, *87*, 2683.

14. Imuta, M.; Ziffer, H. *J. Org. Chem.* **1979**, *44*, 1351.

15. Marshall, P. A.; Prager, R. H. *Aust. J. Chem.* **1977**, *30*, 151.

16. Valente, V. R.; Wolfhagen, J. L. *J. Org. Chem.* **1966**, *31*, 2509.

Appendix

Chemical Abstracts Nomenclature (Collective Index Number); (Registry Number)

Dimethyldioxirane: Dioxirane, dimethyl- (10); (74087-85-7)

Oxone: Peroxymonosulfuric acid, monopotassium salt, mixt. with dipotassium sulfate and potassium hydrogen sulfate (9); (37222-66-5)

trans-Stilbene: Stilbene, (E)- (8); Benzene, 1,1'-(1,2-ethendiyl)bis-, (E)- (9); (103-30-0)

trans-Stilbene oxide: Oxirane, 2,3-diphenyl-, trans, (9); (1439-07-2)

TABLE

EPOXIDATION OF OLEFINS WITH DIMETHYLDIOXIRANE

Entry	Olefin	Equiv.	Time hr	Product	Yield(a,ref)
1	$C_6H_5(H)C=C(H)C_6H_5$	1	6	$C_6H_5(H)C(O)C(H)C_6H_5$	100 (55,[12] 90[13])
2	$C_6H_5(H)C=C(H)C_6H_5$	1	8	$C_6H_5(H)C(O)C(H)C_6H_5$	99 (52,[12] 55[13])
3	$CH_3CH_2CH_2(H)C=C(H)CH_2CH_2CH_3$	1	3	$CH_3CH_2CH_2(H)C(O)C(H)CH_2CH_2CH_3$	74 (70[13])
4	$CH_3CH_2CH_2(H)C=C(H)CH_2CH_2CH_3$	1	1	$CH_3CH_2CH_2(H)C(O)C(H)CH_2CH_2CH_3$	81 (60[13])
5		1	20 min	O	97
6	$C_6H_5(H)C=C(H)CH_3$	1	1	$C_6H_5(H)C(O)C(H)CH_3$	98 (90,[14]b)
7	$C_6H_5(H)C=C(CH_3)CH_3$	1	1	$C_6H_5(H)C(O)C(CH_3)CH_3$	99 (95[15])

TABLE (contd.)

EPOXIDATION OF OLEFINS WITH DIMETHYLDIOXIRANE

Entry	Olefin	Equiv.	Time hr	Product	Yield(a,ref)
8	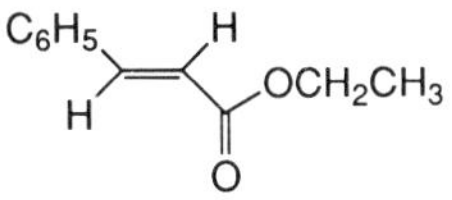 C_6H_5, H / H, $C(=O)OCH_2CH_3$	2	24	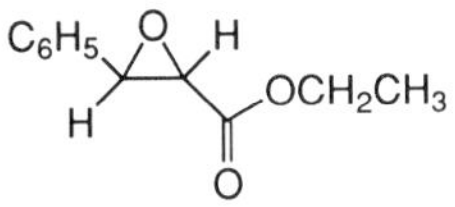 C_6H_5, H (epoxide, O) / H, $C(=O)OCH_2CH_3$	100 (47[16])
9	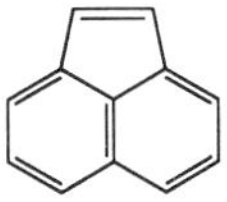 [acenaphthylene]	1	1	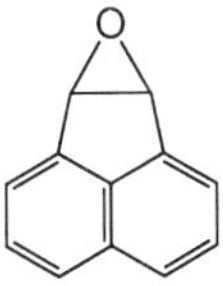 [acenaphthylene oxide]	97 (100,[14]b)
10	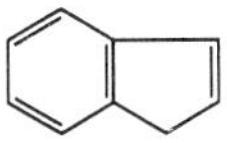 [indene]	1	1	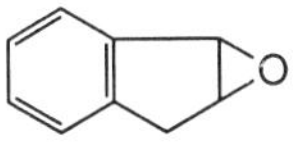 [indene oxide]	96 (91[14])

(a) % Yield by MCPBA oxidation; (b) NMR yield.

Figure 1

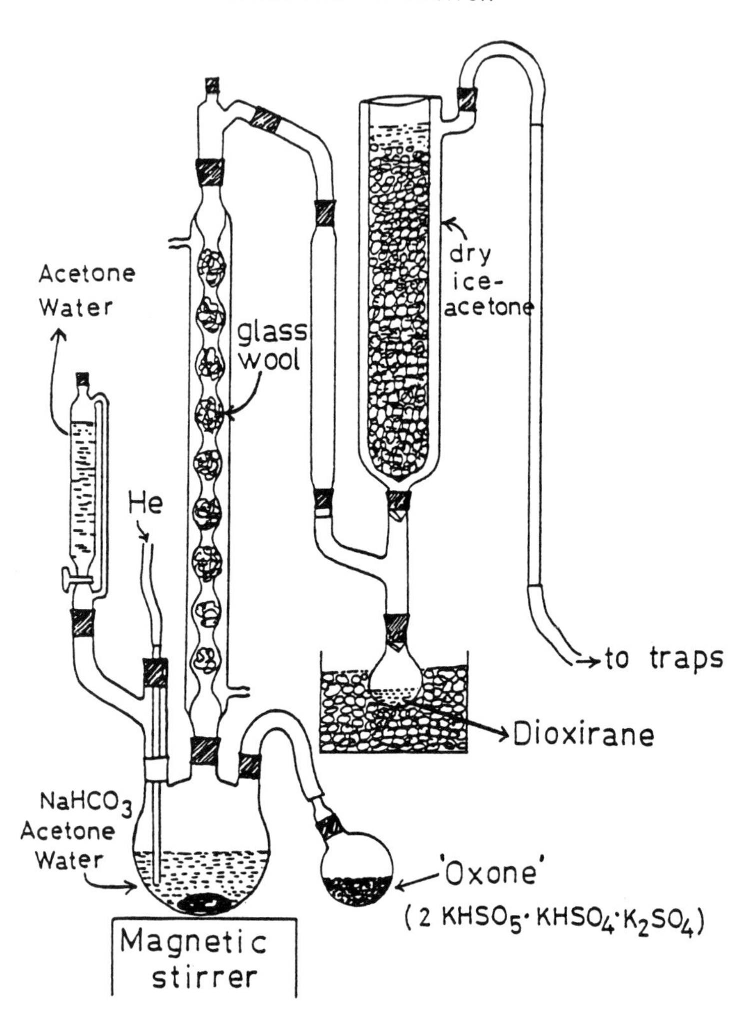

DIOXIRANE GENERATOR
Acetone
Water
glass
wool
dry
ice-
acetone
He
to traps
Dioxirane
NaHCO3
Acetone
Water
'Oxone'
(2 KHSO5·KHSO4·K2SO4)
Magnetic
stirrer

PREPARATION AND REACTIONS OF 2-tert-BUTYL-1,1,3,3-TETRAMETHYLGUANIDINE: 2,2,6-TRIMETHYLCYCLOHEXEN-1-YL IODIDE

A. $(CH_3)_2N-C(=O)-N(CH_3)_2$ → 1. $(Cl_3CO)_2CO$, toluene, 0°C; 2. t-$BuNH_2$ → $(CH_3)_2N-C(=NBu^t)-N(CH_3)_2$ **1**

B. 2,2,6-trimethylcyclohexanone hydrazone + I_2 → [1], Ether → 2,2,6-trimethylcyclohexen-1-yl iodide

Submitted by Derek H. R. Barton, Mi Chen, Joseph Cs. Jászberényi, and Dennis K. Taylor.[1]

Checked by Richard A. Hartz and Amos B. Smith, III.

1. Procedure

Caution! These reactions, which involve toxic reagents, must be carried out (including workup) in an efficient fume hood.

A. 2-tert-Butyl-1,1,3,3-tetramethylguanidine (**1**). To an oven-dried, 500-mL, three-necked, round-bottomed flask, equipped with a nitrogen inlet with gas bubbler, magnetic stirring bar, thermometer, condenser, and a 250-mL dropping funnel, are added triphosgene (14.8 g, 0.05 mol) (Note 1), and anhydrous toluene (120 mL) (Note 2). The mixture is kept under argon and cooled to ≈10°C with the aid of an external ice bath. A solution of N,N,N',N'-tetramethylurea (18.0 mL, 0.15 mol) (Note 3) in dry toluene (50 mL) is slowly added to the mixture over 30 min (Note 4). After the addition is complete, the mixture is allowed to warm to ambient temperature, and stirring of the mixture is continued for an additional hour. During this time a white precipitate forms (Note 5). tert-Butylamine (47.3 mL, 0.45 mol) (Note 6) is slowly added to the mixture over 30 min (Note 7). After the addition is complete, the mixture is heated under reflux for 5 hr and then cooled to room temperature. Anhydrous ether (200 mL) (Note 8) is added and the white precipitate is quickly removed by filtration (Note 9). The precipitate is washed with a further quantity of anhydrous ether (300 mL) (Note 10) and immediately dissolved in aqueous 25% sodium hydroxide solution (100 mL). The mixture is then extracted with three portions of ether (300 mL). The combined organic layers are dried (potassium carbonate), filtered, and the solvent is removed under reduced pressure. The resulting colorless liquid is purified by distillation (bp 88-89°C/36 mm) to afford 18.7 g (73%) of 2-tert-butyl-1,1,3,3-tetramethylguanidine **1** (Note 11).

B. 2,2,6-Trimethylcyclohexen-1-yl iodide (**2**). To an oven-dried, 500-mL, three-necked, round-bottomed flask, equipped with a nitrogen inlet with gas bubbler, magnetic stirring bar, and a 250-mL dropping funnel, are added 2,2,6-trimethylcyclohexanone hydrazone (4.6 g, 0.03 mol) (Note 12), anhydrous ether (100 mL) (Note 8), and 2-tert-butyl-1,1,3,3-tetramethylguanidine (**1**) (46.25 g, 0.27 mol). The mixture is kept under argon at ambient temperature and an ethereal solution (100 mL) of iodine (15.25 g, 0.06 mol) is added to the mixture over 40 min with vigorous stirring. (Note 13). After the addition is complete, stirring is continued for an additional 30 min. The ether is removed under reduced pressure (Note 14) and the residue is heated at 90°C for 30 min (Note 15) under an inert atmosphere. The reaction mixture is allowed to attain ambient temperature. Ether (100 mL) is added and the organic phase is washed twice with 2 N hydrochloric acid (30 mL), aqueous sodium thiosulfate solution (30 mL), aqueous sodium bicarbonate solution (30 mL) and saturated sodium chloride solution (30 mL). The organic phase is dried (sodium sulfate) and the solvent is removed under reduced pressure to afford crude iodide (**2**). Purification of **2** can be achieved by flash chromatography (Note 16) affording pure iodide (**2**) (6.34 g, 85%) as a colorless oil. (Note 17).

2. Notes

1. Triphosgene was purchased from the Aldrich Chemical Company, Inc., and used as received.

2. Toluene is distilled from calcium hydride (CaH_2) under argon just prior to use.

3. N,N,N',N'-Tetramethylurea was purchased from the Aldrich Chemical Company, Inc., and purified by distillation prior to use.

4. Although no major temperature increase is observed, the reaction proceeds best with slow addition.

5. This salt is the corresponding Vilsmeier salt. See reference 2.

6. tert-Butylamine was purchased from the Aldrich Chemical Company, Inc., and dried prior to use by distillation from CaH_2 under argon.

7. No major temperature increase is observed.

8. Diethyl ether is distilled from sodium under argon just prior to use.

9. The white precipitate should be collected as quickly as possible to avoid hydrolysis to the starting urea. The precipitate turns pale yellow if hydrolysis is occurring. Additional ether (300 mL) may be needed to ensure complete transfer of the solids to the filtration apparatus.

10. The filtrate must be colorless, indicating that all impurities have been removed.

11. Distillation is not neccessary if the solids are washed correctly. Spectral data for **1** are as follows: IR (neat) cm^{-1}: 1620; ^1H NMR ($CDCl_3$) δ: 1.22 (s, 9 H), 2.67 (s, 12 H). **1** should be stored under argon in the refrigerator to prevent hydrolysis. The purity is estimated to be ≈95% by NMR and TLC analysis. The impurity is the starting urea and could not be avoided.

12. The hydrazone of 2,2,6-trimethylcyclohexanone is prepared according to a procedure outlined in reference 3. To a solution of absolute ethanol (37 mL), hydrazine (26.0 g, 25.40 mL), and triethylamine (6.8 g, 9.43 mL) is added 2,2,6-trimethylcyclohexanone (6.3 g, 7.0 mL). The mixture is heated to 100°C for 2-3 days, cooled to ambient temperature, and the solvent removed under reduced pressure. Recrystallization of the residue from hexanes affords the hydrazone as white needles (4.85 g, 70%, mp 48-49°C). The spectra are as follows: ^1H NMR ($CDCl_3$) δ: 1.0-1.2 (m, 9 H), 1.40-1.92 (m, 6 H), 2.95 (m, 1 H), 4.51 (s, 2 H); ^{13}C NMR ($CDCl_3$) δ: 17.18, 17.39, 26.51, 28.92, 29.48, 31.68, 37.61, 40.43, 162.47.

13. Vigorous stirring is required as large quantities of precipitate form during the addition.

14. The nitrogen inlet is removed and replaced with a line to a water pump.

15. Heating causes most of the solids to liquify. The temperature refers to the outside oil bath temperature.

16. Flash chromatography was performed on standard Silica Gel 60Å, (230-400 mesh) with hexane, R_f = 0.75.

17. Spectral data for **2** are as follows: ^1H NMR ($CDCl_3$) δ: 1.09 (s, 6 H), 1.53-1.72 (m, 4 H), 1.87 (s, 3 H), 2.12 (t, 2 H); ^{13}C NMR ($CDCl_3$) δ: 19.40, 31.08, 31.56, 33.69, 37.86, 39.51, 117.36, 137.70. Compound **2** deteriorates rapidly at ambient temperature, but is stable for several weeks if stored under argon in the refrigerator. The purity is estimated to be >98% by NMR and TLC analysis. No microanalytical data was obtained because of the instability of **2**.

Waste Disposal Information

All toxic materials were disposed of in accordance with "Prudent Practices in the Laboratory"; National Academic Press; Washington, DC, 1996.

3. Discussion

The procedure described here allows the convenient preparation of large quantities of the strong, non-nucleophilic base 2-tert-butyl-1,1,3,3-tetramethylguanidine (**1**). This reagent provides an inexpensive alternative to the amidine bases, 1,5-diazabicyclo[4.3.0]non-5-ene (DBN) and 1,8-diazabicyclo[5.4.0]undec-7-ene (DBU), which suffer from being easily alkylated.[4] Additionally, the hazards of using phosgene in the previous preparations of **1**[2,4,5] have been greatly reduced by employing triphosgene as a phosgene equivalent.[6]

The synthetic utility of this base (**1**) was demonstrated in the preparation of vinyl iodides in high yields from simple ketohydrazones and iodine (Table), a process that normally gives mixtures of vinyl iodides and geminal diiodides if less hindered bases are employed.[5] This base has also been used in the elimination of sulfonic acids from the corresponding sulfonates, the alkylation of compounds containing active methylene groups, the conversion of hydrazones to vinyl selenides, and the preparation of esters from sterically hindered acids.[4,5]

Other inexpensive, sterically hindered guanidine bases have also been synthesized and their reactivity is comparable to that described here.[2,4]

TABLE
PREPARATION OF VINYL IODIDES FROM HYDRAZONES

Entry	Hydrazone	Vinyl Iodide	%Yield
1	N-NH2	I	85
2	N-NH2, MeO	I, MeO	70
3	N-NH2	I	91
4	N-NH2, HO	I, HO	95

1. Department of Chemistry, Texas A&M University, College Station, TX 77843-3255, USA.

2. Barton, D. H. R.; Elliott, J. D.; Géro, S. D. *J. Chem. Soc., Chem. Comm.* **1981**, 1136.

3. Di Grandi, M. J.; Jung, D. K.; Krol, W. J.; Danishefsky, S. J. *J. Org. Chem.* **1993**, *58*, 4989.

4. Barton, D. H. R.; Elliott, J. D.; Géro, S. D. *J. Chem. Soc., Perkin Trans. 1* **1982**, 2085.

5. Barton, D. H. R.; Bashiardes, G.; Fourrey, J.-L. *Tetrahedron* **1988**, *44*, 147; Barton, D. H. R.; O'Brien, R. E.; Sternhell, S. *J. Chem. Soc.* **1962**, 470; Pross, A.; Sternhell, S. *Aust. J. Chem.* **1970**, *23*, 989.

6. Eckert, H.; Forster, *Angew. Chem., Int. Ed. Engl.* **1987**, *26*, 894.

Appendix
Chemical Abstracts Nomenclature (Collective Index Number); (Registry Number)

2-tert-Butyl-1,1,3,3-tetramethylguanidine: Guanidine, 2-tert-butyl-1,1,3,3-tetramethyl- (8); Guanidine, N''-(1,1-dimethylethyl)-N,N,N',N'-tetramethyl- (9); (29166-72-1)

Triphosgene: Carbonic acid, bis(trichloromethyl) ester (8,9); (32315-10-9)

N,N,N',N'-Tetramethylurea: Urea, tetramethyl- (8,9); (632-22-4)

tert-Butylamine: HIGHLY TOXIC (8); 2-Propanamine, 2-methyl- (9); (75-64-9)

Iodine (8,9); (7553-56-2)

Hydrazine: HIGHLY TOXIC. CANCER SUSPECT AGENT (8,9); (302-01-2)

Triethylamine (8); Ethanamine, N,N-diethyl- (9); (121-44-8)

2,2,6-Trimethylcyclohexanone: Cyclohexanone, 2,2,6-trimethyl- (8,9); (2408-37-9)

DIETHYL (DICHLOROMETHYL)PHOSPHONATE. PREPARATION AND USE IN THE SYNTHESIS OF ALKYNES: (4-METHOXYPHENYL)ETHYNE

(Phosphonic acid, (dichloromethyl)-, diethyl ester to prepare Benzene, 1-ethylnyl-4-methoxy-)

A. $(C_2H_5O)_2P(O)-CCl_3$ → (1. iPrMgCl, THF/ether, -78°C; 2. EtOH, H_2O, HCl) → $(C_2H_5O)_2P(O)-CHCl_2$ **1**

B. $(C_2H_5O)_2P(O)-CHCl_2$ **1** + $CH_3O-C_6H_4-CHO$ → (1. LDA/THF, -78°C → 0°C; 2. BuLi, -78°C → 0°C; 3. H_2O, HCl) → $CH_3O-C_6H_4-C{\equiv}CH$ **2**

Submitted by Angela Marinetti and Philippe Savignac.[1]
Checked by Catherine Dubuisson and Louis S. Hegedus.

1. Procedure

A. *Diethyl (dichloromethyl)phosphonate,* **1**. An oven-dried, 1-L, four-necked, round-bottomed flask (or a 1-L, three-necked flask with a Claisen head fitted in a side neck) is fitted with an efficient mechanical stirrer, thermometer, reflux condenser with a bubbler, and a 200-mL pressure-equalizing addition funnel with a nitrogen inlet.

Under a gentle flow of nitrogen the flask is charged with 400 mL of tetrahydrofuran (THF) (Note 1). The addition funnel is charged with a 1.9 M solution of isopropylmagnesium chloride in diethyl ether (83 mL, 0.158 mol) (Note 2). The mixture is cooled to -78°C with a dry ice/acetone bath (Note 3), and isopropylmagnesium chloride is added over a few minutes with stirring. At this temperature, a solution of diethyl (trichloromethyl)phosphonate (38.3 g, 0.150 mol) (Note 4) in tetrahydrofuran (50 mL) is added dropwise over 15 min. The resulting solution is stirred for an additional 15 min at -78°C producing a clear orange solution. Anhydrous ethanol (12 g, 0.260 mol) in tetrahydrofuran (15 mL) is added dropwise at -78°C, producing a clear yellow solution. The resulting mixture is stirred for a few minutes, then allowed to warm slowly to -40°C. The reaction mixture is poured into a beaker containing a stirred mixture of 3 M hydrochloric acid (70 mL), and an equal volume of crushed ice and dichloromethane (70 mL). The yellow color initially dissipates, but the solution becomes yellow orange upon warming to room temperature. The organic layer is separated, and the aqueous layer is extracted with dichloromethane (2 x 60 mL). The extracts are combined and dried over anhydrous magnesium sulfate. After the filtration, the solvents are evaporated on a rotary evaporator. The bright yellow, crude liquid (36.3 g) is transferred to a pear-shaped flask fitted with a 10-cm Vigreux column and distilled under reduced pressure to give 26.6 g (80%) of diethyl (dichloromethyl)phosphonate, **1,** as a pale yellow liquid, bp 115-119°C/9 mm, >90% pure by ^{1}H, ^{13}C and ^{31}P NMR spectroscopy (Notes 5 and 6).

B. *(4-Methoxyphenyl)ethyne,* **2**. An oven-dried, 1-L, four-necked, round-bottomed flask is fitted as above, flushed with nitrogen, and charged with a solution of butyllithium in hexane (1.56 M, 92 mL, 0.143 mol) (Note 7). The solution is cooled with stirring to -20°C with a dry ice/acetone bath and a solution of diisopropylamine (15.1 g, 0.149 mol) (Note 8) in tetrahydrofuran (220 mL) (Note 1), is added dropwise over 15 min. The resulting clear solution is cooled to -78°C and treated by the dropwise addition of a solution of **1** (30 g, 0.136 mol) and 4-methoxybenzaldehyde (18.1 g, 0.133 mol) (Note 8) in tetrahydrofuran (60 mL) over 30 min. The resulting brown solution is stirred at -78°C for an additional 30 min, then allowed to warm slowly to 0°C over 60 min. The resulting mixture is then cooled to -78°C and a solution of butyllithium (1.56 M in hexane, 183 mL, 0.285 mol) (Note 7) is added dropwise over 20 min. The resulting brown solution is stirred at -78°C for an additional 30 min, then allowed to warm slowly to 0°C over 60 min. At this temperature, the reaction mixture is quenched by the dropwise addition of 3 M hydrochloric acid to pH 5-6 (125-130 mL). The brown color has practically disappeared and the solution is yellow orange. The organic layer is separated and the aqueous layer is extracted with diethyl ether (3 x 50-mL). The extracts are combined, washed with water (3 x 10 mL), and dried over anhydrous magnesium sulfate. Magnesium sulfate is removed by filtration and the solvents are evaporated on a rotary evaporator. The residue is dissolved in hexane (200 mL) and filtered after 30 min. Solvent is again completely evaporated under reduced pressure. Crude product **2** thus obtained is purified by column chromatography on silica gel (Note 9) to afford 11.0 g (63%) of pure (4-methoxyphenyl)ethyne as a colorless liquid, bp 70 - 72°C (3 mm) that gives white crystals on standing in the freezer (Note 10).

2. Notes

1. Tetrahydrofuran is obtained from S.D.S. Company and is purified by distillation from sodium and benzophenone.

2. Isopropylmagnesium chloride (2 M in diethyl ether) is available from Aldrich Chemical Company, Inc., and is standardized before use by titration against a solution of benzyl alcohol in toluene with cuproine (2,2'-biquinoline).

3. An alternative cooler was also used by the submitters consisting of a Dewar partially filled with liquid nitrogen.

4. Diethyl (trichloromethyl)phosphonate (97%) either purchased from Aldrich Chemical Company, Inc., or prepared according to a described procedure[2] is used.

5. The crude liquid could be distilled using a Kugelrohr apparatus. The main fraction (27.5 g, 83% yield) consists of a clear orange liquid that was collected at 145-155°C (oven temperature)/9-10 mm. ^{1}H NMR (200 MHz, $CDCl_3$) indicates that this material is of the same purity as that from conventional distillation.

6. The product displays the following spectroscopic data: ^{31}P NMR (81 MHz, $CDCl_3$) δ: +10.9; ^{1}H NMR (200 MHz, $CDCl_3$) δ: 1.29 (t, 6 H, CH_3), 4.22 (dq, 4 H, CH_2), 5.6 (d, 1 H, $^{2}J_{H\text{-}P}$ = 2, CH); ^{13}C NMR (50 MHz, $CDCl_3$) δ: 16.2 (d, $J_{C\text{-}P}$ = 5.9, CH_3), 60.7 (d, $J_{C\text{-}P}$ = 178, CH), 65.0 (d, $J_{C\text{-}P}$ = 7.4, CH_2) ppm.

7. Butyllithium (1.6 M solution in hexane) is available from Janssen Chimica or Aldrich Chemical Company, Inc., and is standardized before use by titration against a solution of benzyl alcohol in toluene with cuproine (2,2'-biquinoline) or with (±)-2-butanol and 1,10-phenanthroline in diethyl ether.

8. Diisopropylamine (99%) and 4-methoxybenzaldehyde (98%) were purchased from Aldrich Chemical Company, Inc., and used without purification.

9. The crude product is purified by chromatography (200 g of silica gel, Silitech 32-63 μm, purchased from ICN Biomedicals; column diameter, 5.5 cm). Elution is performed first with hexanes (200 mL) and then with hexanes-diethyl ether (20:1). Alternatively, the submitters purified the product by distillation from a pear-shaped flask using an 8-cm fractionating column.

10. The product displays the following spectroscopic data: ^{1}H NMR (200 MHz, $CDCl_3$) δ: 3.02 (s, 1 H), 3.80 (s, 3 H), 6.8 ($\underline{AA}$'BB', 2 H), 7.4 (AA'$\underline{BB'}$, 2 H); ^{13}C NMR (50, MHz, $CDCl_3$) δ : 55.2 (OCH_3), 75.8 (C$\underline{C}$H), 83.6 ($\underline{C}$CH), 113.9 [CH (Ar)], 114.0 ($\underline{C}$-C≡CH), 133 [CH (Ar)], 159.9 ($\underline{C}OCH_3$) ppm.

Waste Disposal Information

All toxic materials were disposed of in accordance with "Prudent Practices in the Laboratory"; National Academic Press; Washington, DC, 1996.

3. Discussion

The procedure described in Part A represents a convenient synthesis of diethyl (dichloromethyl)phosphonate,**1**, for large-scale preparations (up to 1 mol). The use of isopropylmagnesium chloride instead of butyllithium[3,4] reduces the amount of by-products,[5] simplifies the purification step, and improves the yield. An alternative synthetic method for **1** is chlorination of diethyl (chloromethyl)phosphonate;[6,7] however, yields and selectivities are lower than in the procedure described here. Diethyl (dichloromethyl)phosphonate, **1**, or the corresponding lithiated derivative, are useful intermediates in organic synthesis: 1,1-dichloroalkenes[3,4,6] as well as terminal alkynes[4] have been prepared.

Part B shows the use of **1** as starting material for the synthesis of (4-methoxyphenyl)ethyne. The generation of the phosphorylated carbanion is performed through metalation with lithium diisopropylamide (LDA). A mixture containing the phosphorus reagent **1** and the aldehyde is added directly to LDA, in order to trap the unstable phosphorylated, lithiated carbanion, thus preventing decomposition and side reactions. The formation of (4-methoxyphenyl)ethyne described here is an example of a general synthetic method for the conversion of aldehydes into acetylenes, on a large scale, by means of diethyl (dichloromethyl)phosphonate, **1**. The methodology is well suited for the synthesis of a wide variety of acetylenic compounds, such as $C_9H_{19}C{\equiv}CH$, $(C_2H_5)_2CHC{\equiv}CH$, $C_6H_5CH{=}CHC{\equiv}CH$,[4] and analogous terminal alkynes. Several methods for the preparation of alkynes based on phosphorus reagents have been reported previously. Each of these procedures can be regarded as an extension of the Wittig-Horner olefin synthesis. In the final step the olefins are converted into the corresponding alkynes. The first method that employed the PPh_3 - CBr_4 couple (or PPh_3 - CBr_4 - zinc dust) was introduced by Corey and Fuchs[8] (1972). The amounts of PPh_3 involved (2 or 4 equiv and of PPh_3O formed during the reaction, are obstacles for large scale synthesis. Dimethyl diazomethylphosphonate first prepared by Seyferth[9] in 1971 was shown, by Colvin and Hamill[10] in 1977, then by Gilbert and Weerasooriya[11] in 1982, to be an effective reagent for the transformation of aldehydes, alkyl aryl ketones or diaryl ketones into alkynes. However, this reagent is rather difficult to prepare (four steps, 46% yield), and very hazardous to handle (explosive decomposition on distillation), and therefore is unsuitable for preparative scale synthesis.

1. Hétéroéléments et Coordination, URA CNRS 1499, DCPH, Ecole Polytechnique, 91128 Palaiseau Cedex, France.
2. Kosolapoff, G. M. *J. Am. Chem. Soc.* **1947**, *69*, 1002
3. Seyferth, D.; Marmor R. S. *J. Organometal. Chem.* **1973**, *59*, 237
4. Villieras, J.; Perriot, P.; Normant J. F. *Synthesis* **1975**, 458
5. Lowen, G. T.; Almond, M. R. *J. Org. Chem.* **1994**, *59*, 4548.
6. Savignac, Ph.; Dreux, M.; Coutrot, Ph. *Tetrahedron Lett.* **1975**, 609
7. Lee, K.; Shin, W.S.; Oh, D. Y. *Synth. Comm.* **1991**, *21*, 1657
8. Corey, E. J.; Fuchs, P. L. *Tetrahedron Lett.* **1972**, 3769
9. Seyferth, D.; Marmor, R. S.; Hilbert, P. *J. Org. Chem.* **1971**, *36*, 1379.
10. Colvin, E. W.; Hamill, B. J. *J. Chem. Soc., Perkin Trans*. I **1977**, 869.
11. Gilbert, J. C.; Weerasooriya, U. *J. Org. Chem.* **1982**, *47*, 1837.

Appendix

Chemical Abstracts Nomenclature (Collective Index Number); (Registry Number)

Diethyl (dichloromethyl)phosphonate: Phosphonic acid, (dichloromethyl)-, diethyl ester (8,9); (3167-62-2)

(4-Methoxyphenyl)ethyne: Anisole, p-ethynyl- (8); Benzene, 1-ethynyl-4-methoxy- (9); (768-60-5)

Isopropylmagnesium chloride: Magnesium, chloroisopropyl- (8); Magnesium, chloro (1-methylethyl)- (9); (1068-55-9)

Diethyl (trichloromethyl)phosphonate: Phosphonic acid, (trichloromethyl)-, diethyl ester (8,9); (866-23-9)

Butyllithium: Lithium, butyl- (8,9); (109-72-8)

Diisopropylamine (8); 2-Propanamine, N-(1-methylethyl)- (9); (108-18-9)

4-Methoxybenzaldehyde: p-Anisaldehyde (8); Benzaldehyde, 4-methoxy- (9); (123-11-5)

2,3-DIBROMO-1-(PHENYLSULFONYL)-1-PROPENE AS A VERSATILE REAGENT FOR THE SYNTHESIS OF FURANS AND CYCLOPENTENONES: 2-METHYL-4-[(PHENYLSULFONYL)METHYL]FURAN AND 2-METHYL-3-[(PHENYLSULFONYL)METHYL]-2-CYCLOPENTEN-1-ONE

(Benzene, [(2,3-dibromo-1-propenyl)sulfonyl]-, Furan, 2-methyl-4-[(phenylsulfonyl)methyl]-, and 2-Cyclopenten-1-one, 2-methyl-3-[(phenylsulfonyl)methyl]-)

A. H—≡—CH_2OH —[1) PhSCl, Et_3N; 2) H_2O_2, HOAc]→ $CH_2{=}C{=}C(SO_2Ph)H$

B. $CH_2{=}C{=}C(SO_2Ph)H$ —[Br_2, HOAc]→ $BrCH_2C(Br){=}CH(SO_2Ph)$

C. $BrCH_2C(Br){=}CH(SO_2Ph)$ —[$CH_3COCH_2COCH_3$; CH_3ONa, CH_3OH]→ 2-methyl-4-(CH_2SO_2Ph)furan

D. $BrCH_2C(Br){=}CH(SO_2Ph)$ —[$CH_3COCH(CH_3)COCH_3$; CH_3ONa, CH_3OH]→ 2-methyl-3-(CH_2SO_2Ph)-2-cyclopenten-1-one

Submitted by Scott H. Watterson, Zhijie Ni, Shaun S. Murphree, and Albert Padwa.[1]

Checked by Huguette Vanlierde, Amy Dekoker, and Leon Ghosez.

1. Procedure

A. 1-(Phenylsulfonyl)-1,2-propadiene. In a 1-L, three-necked, round-bottomed flask equipped with a magnetic stirring bar, 100-mL addition funnel, and a septum fitted with a nitrogen inlet, are placed 10.5 mL (0.18 mol) of propargyl alcohol (Note 1), 30.1 mL (0.22 mol) of triethylamine, and 700 mL of dichloromethane. The flask is cooled with an ice-water bath and 26.0 g (0.18 mol) of phenylsulfenyl chloride (Note 2) in 50 mL of dichloromethane is added under a nitrogen atmosphere from the addition funnel. After the solution is warmed to room temperature and stirred for 3 hr, the mixture is washed with water (2 x 200-mL) and then concentrated under reduced pressure. The resulting residue (ca. 37 g) is dissolved in 100 mL of acetic acid in a 500-mL, round-bottomed flask fitted with a magnetic stirring bar, an addition funnel, and a thermometer. The solution is heated to 95°C and then 40 mL of hydrogen peroxide (30-35%) (Note 3) is slowly added over a 20-min period maintaining the internal temperature in the flask below 95°C. The mixture is heated for an additional 10 hr, maintaining the external temperature at 95°C (Note 4). After being cooled to 25°C, the mixture is poured into 200 mL of water, and extracted with dichloromethane (3 x 100 mL). The organic layer is washed with water (2 x 100 mL) followed by saturated sodium bicarbonate solution (100 mL), dried over magnesium sulfate, and filtered. The filtrate is concentrated to give 28.1 g (68%) of 1-(phenylsulfonyl)-1,2-propadiene as an oil, which is used in the next step (Notes 5, 6).

B. 2,3-Dibromo-1-(phenylsulfonyl)-1-propene. To a solution containing 25.4 g (0.14 mol) of 1-(phenylsulfonyl)-1,2-propadiene in 100 mL of acetic acid in a 250-mL, round-bottomed flask fitted with a dropping funnel is added 8.0 mL (0.14 mol) of bromine over a period of 30 min. The solution is stirred at room temperature for 8 hr, poured into 200 mL of water, and extracted with dichloromethane (3 x 100 mL). The dichloromethane layer is washed with 50 mL of a 1.0 M aqueous sodium thiosulfate solution, water (2 x 100 mL) and saturated sodium bicarbonate solution (100 mL). The organic phase is dried over anhydrous magnesium sulfate, filtered under reduced pressure through a 60-mL fritted funnel containing 4.0 g of Celite topped with 6.0 g of silica gel. The mixture is concentrated under reduced pressure to give 41.5 g (87%) of 2,3-dibromo-1-(phenylsulfonyl)-1-propene (Note 7), which is used in the next step without further purification.

C. 2-Methyl-4-[(phenylsulfonyl)methyl]furan. Into a flame-dried, 1-L, round-bottomed flask equipped with an addition funnel fitted with a nitrogen inlet are placed 20.0 g (58.8 mmol) of 2,3-dibromo-1-(phenylsulfonyl)-1-propene and 6.1 mL (59.4 mmol) of 2,4-pentanedione in 300 mL of methanol (Note 8). The flask is blanketed with nitrogen and cooled using an ice-water bath. To this mixture is added 140 mL (70.0 mmol) of a 0.5 N methanolic solution of sodium methoxide (Note 9) dropwise over 30 min. The solution is stirred at 25°C for 12 hr and cooled to 0°C. Another 140 mL (70.0 mmol) of the 0.5 N methanolic solution of sodium methoxide is added over 20 min. After being stirred at 25°C for 10 hr, the reaction is quenched by the addition of 50 mL of saturated ammonium chloride solution. The solvent is removed with a rotary evaporator at aspirator vacuum and the resulting residue is taken up in 100 mL of water and 200 mL of dichloromethane. The aqueous layer is extracted with dichloromethane (2 x 200 mL). The organic layer is washed with water (100 mL) and brine (100 mL) and then dried over anhydrous magnesium sulfate. After filtration and concentration, the residue is recrystallized from ethyl acetate to give 5.6 g (59%) of 2-methyl-4-[(phenylsulfonyl)methyl]furan (Notes 6, 10).

D. 2-Methyl-3-[(phenylsulfonyl)methyl]-2-cyclopenten-1-one. In a flame-dried, 1-L, round-bottomed flask equipped with an addition funnel under a nitrogen atmosphere are placed 20.0 g (58.8 mmol) of 2,3-dibromo-1-(phenylsulfonyl)-1-propene and 7.0 mL (60.0 mmol) of 3-methyl-2,4-pentanedione in 300 mL of methanol (Note 8). The flask is cooled using an ice-water bath. To this mixture is added 140 mL

(70.0 mmol) of a 0.5 N methanolic solution of sodium methoxide (Note 9) dropwise over 30 min. The solution is stirred at 25°C for 5 hr and cooled to 0°C. Another 140-mL (70.0 mmol) portion of the 0.5 N methanolic solution of sodium methoxide is added over 20 min. After being stirred at 25°C for 10 hr, the reaction is quenched by the addition of 50 mL of saturated ammonium chloride solution. The solvent is removed with a rotary evaporator at aspirator vacuum and the resulting residue is taken up in 100 mL of water and 200 mL of dichloromethane. The aqueous layer is extracted with dichloromethane (2 x 200 mL). The organic layer is washed with water (100 mL) and brine (100 mL) and then dried over anhydrous magnesium sulfate. After filtration and concentration, the residue is recrystallized from ethyl acetate to give 6.5 g (44%) of 2-methyl-3-[(phenylsulfonyl)methyl]-2-cyclopenten-1-one (Notes 6, 11).

2. Notes

1. Propargyl alcohol, 2,4-pentanedione and 3-methyl-2,4-pentanedione were purchased from Aldrich Chemical Company, Inc., and were used without further purification. Dichloromethane and triethylamine were distilled from calcium hydride prior to use. Methanol was dried and distilled from magnesium-iodine.

2. Phenylsulfenyl chloride was prepared according to the procedure of Barrett, A. G. M.; Dhanak, D.; Graboski, G. G.; Taylor, S. J. *Org. Synth., Coll. Vol. VIII* **1993**, 550.

3. Hydrogen peroxide (30-35%), glacial acetic acid, and bromine were purchased from Fisher Scientific Company and were used without further purification.

4. For the next 30-60 min the internal temperature should not be allowed to exceed 95°C.

5. The submitters were able to obtain most of this product as a crystalline solid that could be recrystallized from ether at -20°C, mp 44-45°C [lit.[2] mp 44-45°C]. It has

the following spectral properties: ^{1}H NMR (300 MHz, $CDCl_3$) δ: 5.40 (d, 2 H, J = 6.3), 6.21 (t, 1 H, J = 6.3), 7.47-7.87 (m, 5 H); ^{13}C NMR (75 MHz, $CDCl_3$) δ: 84.0, 100.7, 127.3, 129.0, 133.4, 140.9, 209.2.

6. The checkers were not able to purify the crude oil. As a result their overall yield of final product was 40%. The submitters obtained a 71% overall yield using the crystalline material.

7. This product contains a mixture of E- and Z-isomers in a 7:3 ratio, which could be separated by flash chromatography on silica gel eluting with hexane:ethyl acetate (4:1). The spectral properties of both isomers are as follows: E-isomer: mp 62-63°C (recrystallized from ether); ^{1}H NMR (300 MHz, $CDCl_3$) δ: 4.96 (s, 2 H), 6.78 (s, 1 H), 7.50-8.15 (m, 5 H); ^{13}C NMR (75 MHz, $CDCl_3$) δ: 29.8, 127.8, 129.6, 133.7, 134.4, 137.6, 139.6; Anal. Calcd for $C_9H_8Br_2O_2S$: C, 31.79; H, 2.37. Found: C, 31.86; H, 2.36. Z-Isomer (oil); ^{1}H NMR (300 MHz, $CDCl_3$) δ: 4.20 (s, 2 H), 7.31 (s, 1 H), 7.50-7.95 (m, 5 H); ^{13}C NMR (75 MHz, $CDCl_3$) δ: 35.7, 127.9, 128.5, 129.1, 133.8, 134.0, 139.4.

8. Both isomers gave similar yields. If the crystalline E-isomer was used, it took about 30 min for the solid to dissolve.

9. This solution was prepared from sodium and methanol and could be stored in a plastic bottle over a period up to several months without any effect on the reaction.

10. The product has the following spectral properties; mp 91-92°C; ^{1}H NMR (300 MHz, $CDCl_3$) δ: 2.22 (s, 3 H), 4.10 (s, 2 H), 5.91 (s, 1 H), 7.01 (s, 1 H), 7.50-7.95 (m, 5 H); ^{13}C NMR (75 MHz, $CDCl_3$) δ: 12.9, 53.0, 106.9, 112.5, 127.9, 128.3, 133.1, 137.3, 140.3, 152.6. Anal. Calcd for $C_{12}H_{12}O_3S$: C, 61.00; H, 5.12. Found: C, 60.91; H, 5.12.

11. The product has the following spectral properties; mp 166-167°C; ^{1}H NMR (300 MHz, $CDCl_3$) δ: 1.26 (s, 3 H), 2.45 (m, 2 H), 2.73 (m, 2 H), 4.15 (s, 2 H), 7.40-7.95 (m, 5 H); ^{13}C NMR (75 MHz, $CDCl_3$) δ: 7.2, 29.4, 33.7, 57.9, 127.5, 128.9, 133.7,

137.7, 142.4, 155.8, 207.8. Anal. Calcd. for $C_{13}H_{14}O_3S$: C, 62.38, H, 5.64. Found: C, 62.14, H, 5.47.

Waste Disposal Information

All toxic materials were disposed of in accordance with "Prudent Practices in the Laboratory"; National Academic Press; Washington, DC, 1996.

3. Discussion

Many methods have been devised for the formation of multicyclic furans[3] and cyclopentenones[4] because of their importance in organic synthesis. The procedure described here provides a simple and general approach for the construction of 2-methyl-4-[(phenylsulfonyl)methyl]furan and 2-methyl-3-[(phenylsulfonyl)methyl]-2-cyclopenten-1-one using 2,3-dibromo-1-(phenylsulfonyl)-1,2-propene (DBP) as the key reagent.[5] Addition of bromine to 1-(phenylsulfonyl)-1,2-propadiene proceeds smoothly at 25°C and can be controlled so that the reaction may be terminated after 1 equiv of bromine is consumed. The resulting dibromide is a stable, crystalline solid, requiring no special precautions to prevent decomposition.[6] This dibromosulfone can be viewed as a multielectrophilic reagent with great potential as a nucleophilic acceptor for sequential addition. Functionalized allylic reagents that contain both a leaving group and a π-activating substituent have been used extensively in organic synthesis.[7-10] These substituted 1-propenes have been referred to as multicoupling reagents.[8] Because of the molecular weight and stability of the phenylsulfonyl group, the carbon backbone of DBP is very small, without the drawback of volatility or thermal lability seen in other synthetic intermediates with the same carbon skeleton. The synthetic potential of DBP is demonstrated here by taking advantage of two properties of the molecule: 1) the ability of the phenylsulfonyl group to activate the double bond toward Michael addition with soft dicarbonyl anions and 2) the facility with which both bromides can be displaced.

Treatment of 1,3-dicarbonyl compounds with DBP in a methoxide/methanol system affords 2-alkyl-4-[(phenylsulfonyl)methyl]furans, where reaction proceeds by initial addition-elimination on the vinyl sulfone moiety. In contrast, silyl enol ethers in the presence of silver tetrafluoroborate resulted in products derived from S_N2 displacement at the allylic site.[11] Anions derived from 1,3-dicarbonyls substituted at the C-2 position are found to induce a complete reversal in the mode of ring closure.[12] The major products obtained are 3-[(phenylsulfonyl)methyl]-substituted cyclopentenones. The internal displacement reaction leading to the furan ring apparently encounters an unfavorable $A_{1,3}$-interaction in the transition state when a substituent group is present at the 2-position of the dicarbonyl compound. This steric interaction is not present in the transition state leading to the cyclopentenone ring.

Since DBP can react with a variety of β-dicarbonyl anions, a wide assortment of furans and cyclopentenones is available. In addition to its ease of removal,[13] the pendant sulfone also offers a convenient and versatile site for further elaboration (via alkylation[14] or Julia coupling[15]). This strategy toward furans and cyclopentenones can clearly be applied to more complex targets.

1. Department of Chemistry, Emory University, Atlanta, GA 30322.
2. Stirling, C. J. M. *J. Chem. Soc.* **1964**, 5863; Cinquini, M.; Colonna, S.; Cozzi, F.; Stirling, C. J. M. *J. Chem. Soc., Perkin Trans. I* **1976**, 2061.
3. Dean, F. M. in "Advances in Heterocyclic Chemistry"; Katritzky, A. R., Ed.; Academic Press: New York, 1982; Vol. 30, pp. 167-238; Lipshutz, B. H. *Chem. Rev.* **1986**, *86*, 795.
4. Pauson, P. L. *Tetrahedron* **1985**, *41*, 5855; Ramaiah, M. *Synthesis* **1984**, 529.

5. Padwa, A.; Austin, D. J.; Ishida, M.; Muller, C. L.; Murphree, S. S.; Yeske, P. E. *J. Org. Chem.* **1992**, *57*, 1161; Padwa, A.; Ishida, M.; Muller, C. L.; Murphree, S. S.; *J. Org. Chem.* **1992**, *57*, 1170.
6. The stereochemical assignment rests on an X-ray single-crystal structure analysis of the E-isomer.[5]
7. Nelson, R. P.; Lawton, R. G. *J. Am. Chem. Soc.* **1966**, *88*, 3884.
8. Knochel, P.; Seebach, D. *Nouv. J. Chim.* **1981**, *5*, 75; Seebach, D.; Knochel, P. *Helv. Chim. Acta* **1984**, *67*, 261; Knochel, P.; Normant, J. F. *Tetrahedron Lett.* **1984**, *25*, 1475; **1984**, *25*, 4383; **1985**, *26,* 425.
9. Trost, B. M.; Lavoie, A. C. *J. Am. Chem. Soc.* **1983**, *105*, 5075.
10. Speckamp, W. N.; Dijkink, J.; Huisman, H. O. *J. Chem. Soc. C* **1970**, 196.
11. Padwa, A.; Ishida, M. *Tetrahedron Lett.* **1991**, *32*, 5673.
12. Murphree, S. S.; Muller, C. L.; Padwa, A. *Tetrahedron Lett.* **1990**, *31*, 6145.
13. Smith, A. B., III; Hale, K. J. *Tetrahedron Lett.* **1989**, *30*, 1037.
14. Magnus, P. D. *Tetrahedron* **1977**, *33*, 2019.
15. Julia, M.; Stacino, J.-P. *Tetrahedron* **1986**, *42*, 2469.

Appendix

Chemical Abstracts Nomenclature (Collective Index Number); (Registry Number)

2,3-Dibromo-1-(phenylsulfonyl)-1-propene: Benzene, [(2,3-dibromo-1-propenyl)sulfonyl]- (12); (132604-65-0)

1-(Phenylsulfonyl)-1,2-propadiene: Benzene, (1,3-propanedienylsulfonyl)- (9); (2525-42-0)

Propargyl alcohol: 2-Propyn-1-ol (8,9); (107-19-7)

Triethylamine (8); Ethanamine, N,N-diethyl- (9); (121-44-8)

Phenylsulfenyl chloride (8,9); (931-59-9)

Hydrogen peroxide (8,9); (7722-84-1)

Bromine (8,9); (7726-95-6)

2-Methyl-4-[(phenylsulfonyl)methyl]furan: Furan, 2-methyl-4-[(phenylsulfonyl)methyl]- (12); (128496-98-0)

2,4-Pentanedione (8,9); (123-54-6)

Sodium methoxide (8,9); (124-41-4)

2-Methyl-3-[(phenylsulfonyl)methyl]-2-cyclopenten-1-one (12); 2-Cyclopenten-1-one, 2-methyl-3-[(phenylsulfonyl)methyl]- (12); (132604-66-1)

PHENYL VINYL SULFIDE

(Benzene, (ethenylthio)-)

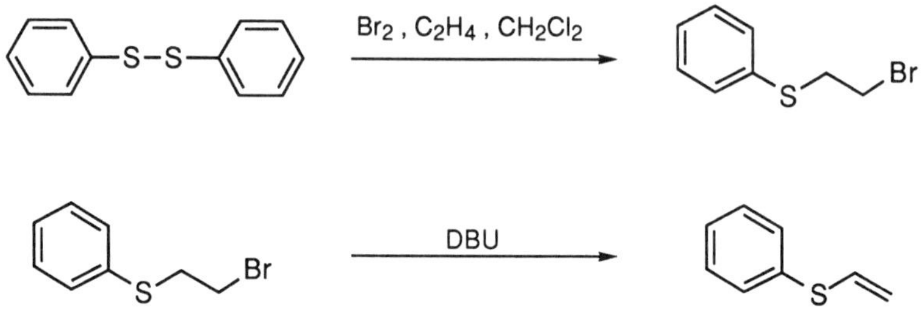

Br_2, C_2H_4, CH_2Cl_2

DBU

Submitted by Daniel S. Reno[1a] and Richard J. Pariza.[1b]

Checked by David A. Barda and William R. Roush.

1. Procedure

Caution! The intermediate, 1-phenylthio-2-bromoethane, produced in the first step of this one pot reaction sequence is a strong alkylating agent, and some bromine escapes from the reaction vessel. Therefore the reaction should be run in a properly operating ventilation hood, and care must be exercised to avoid exposure to these substances.

A 2-L, three-necked, round-bottomed flask fitted with a reflux condenser, an addition funnel, a magnetic stirring bar, thermometer, and nitrogen inlet is charged with diphenyl disulfide (200 g, 917 mmol) and dichloromethane (320 mL) (Note 1). The addition funnel is charged with bromine (161 g, 52 mL, 1.01 mol). After the diphenyl disulfide dissolves, the nitrogen inlet is replaced with a calcium sulfate-packed drying tube, and the flask is fitted with a gas-dispersion tube. Ethylene (73.1 g, 2.61 mol) (Notes 2, 3) is slowly bubbled into the solution through a gas dispersion tube, and the bromine is added in 2-3-mL portions over 5 hr (Notes 4, 5). After addition of the bromine and ethylene is complete, the drying tube is replaced with the nitrogen inlet, and the flask is fitted with a clean addition funnel. The addition funnel is charged with 1,8-diazabicyclo[5.4.0]undec-7-ene, DBU, (306 g, 300 mL, 2.01 mol) (Note 6). DBU is added at a rate such that the temperature of the reaction mixture does not exceed 55°C. After the DBU is added the reaction mixture is maintained at about 50°C for 15-18 hr. A 1.0 M ammonium hydroxide solution (600 mL) is added to the reaction mixture, and the mixture is transferred to a separatory funnel. The layers are separated, and the aqueous layer is extracted with 300 mL of dichloromethane. The organic fractions are combined, washed with water (600 mL), and dried with 10 g of magnesium sulfate. The mixture is filtered, and the solvent is evaporated under reduced pressure. Distillation of the residue affords 162-184 g (65-74%) of phenyl vinyl sulfide, bp 80-84°C/11-12 mm (Notes 7, 8, 9 and 10). The purity is greater than 98% by GC analysis (Note 11, 12).

2. Notes

1. All the materials used in this process were obtained from the Aldrich Chemical Company, Inc., and were used as received.

2. Initiation of the ethylene addition should precede the introduction of the first portion of bromine by 2-4 min. Ethylene is then added continuously at a slow rate until all the bromine is consumed (see Note 4).

3. The submitters reported that use of less than 1.41 equiv (2.62 mol) of ethylene reduces the yield of product. The checkers used a 145-174-µm fritted gas dispersion tube, and up to 1.84 equiv of ethylene (3.38 mol) was required to consume all the bromine (see Note 4). However, on one occasion the checkers obtained excellent results (83% yield) using only 1.26 equiv of ethylene (2.30 mol). In this

instance, a 25-50-µm fritted gas dispersion tube was used, which permitted a slower and more efficient rate of ethylene introduction.

4. The first 2-3-mL portion of bromine is added until the reaction mixture is intensely violet. Subsequent portions of bromine are added when the reaction mixture fades to an amber color. After all the bromine has been added, ethylene addition is continued until the color fades to amber.

5. A competing process involving halogenation of the phenyl substituent occurs when the bromine concentration becomes too high. The slow addition of bromine specified here minimizes the competing aromatic bromination.

6. The submitters report that use of less than this amount of DBU reduces the yield of phenyl vinyl sulfide.

7. The checkers obtained 204-207 g (82-84%) of phenyl vinyl sulfide, bp 83-84°C/11-12 mm.

8. Phenyl vinyl sulfide has the following spectral properties: EI MS 136; ^{1}H NMR (400 MHz, $CDCl_3$) δ: 5.35 (d, 1 H, J = 16.9), 5.36 (d, 1 H, J = 9.8), 6.55 (dd, 1 H, J = 9.8, 16.9), 7.22-7.40 (m, 5H); ^{13}C NMR (100 MHz, $CDCl_3$) δ: 115.4, 127.1, 129.1, 130.4, 131.8, 134.2.

9. Phenyl vinyl sulfide prepared using this procedure is stable at room temperature under a nitrogen atmosphere for months. The submitters have kept samples for over nine months at ambient temperature without any visible degradation.

10. This procedure was used by the submitters to prepare 97 kg of phenyl vinyl sulfide from 100 kg of diphenyl disulfide (78% yield).

11. The GC analysis was performed with a Hewlett-Packard, HP-1 column (10 m x 0.53 mm x 2.65 µm). The temperature program was as follows: initial temperature, 50°C; initial time, 2.0 min; rate, 20°C/min; final temperature, 250°C; final time, 8 min.

12. The checkers identified the two major impurities as p-bromophenyl vinyl sulfide (see Note 5) and 2-chloroethyl phenyl sulfide by GCMS analysis.

Waste Disposal Information

All toxic materials were disposed of in accordance with "Prudent Practices in the Laboratory"; National Academic Press; Washington, DC, 1996.

3. Discussion

Phenyl vinyl sulfide possesses a number of synthetically useful attributes. It participates as an electron-rich alkene in 1+2,[2] 2+2,[3] 3+2,[4] and 4+2[5] cycloaddition reactions. Deprotonation of phenyl vinyl sulfide with strong base affords an α-metallated sulfide that reacts with electrophiles.[6] The metallation-electrophile sequence and the cycloaddition reactions afford products amenable to further synthetic manipulation via the sulfide functionality. Furthermore, phenyl vinyl sulfide is a convenient precursor to the synthetically useful phenyl vinyl sulfoxide and phenyl vinyl sulfone.[7]

The procedure described here affords phenyl vinyl sulfide in a high yield using common reagents and mild conditions. The material obtained via this procedure is stable at room temperature under a nitrogen atmosphere for months. As indicated in Note 10, this process is readily scaled up. Other methods either afford lower yields,[7,8,9] less stable product,[7] or require more extreme conditions.[10,11]

1. (a) Process Development, Chemical and Agricultural Products Division, Abbott Laboratories, North Chicago, IL 60064; (b) C&P Associates, 43323 North Oak Crest Lane, Zion, IL 60099.
2. Weber, A.; Sabbioni, G.; Galli, R.; Stampfli, U.; Neuenshwander, M. *Helv. Chim. Acta* **1988**, *71*, 2026.

3. Brückner, R.; Huisgen, R. *Tetrahedron Lett.* **1990**, *31*, 2561; Sugimura, H.; Osumi, K. *Tetrahedron Lett.* **1989**, *30*, 1571; Pabon, R. A.; Bellville, D. J.; Bauld, N. L. *J. Am. Chem. Soc.* **1984**,*106*, 2730 and references cited therein.
4. Singleton, D. A.; Church, K. M. *J. Org. Chem.* **1990**, *55*, 4780; Tsuge, O.; Kanemasa, S.; Sakamoto, K.; Takenaka, S. *Bull. Chem. Soc. Jpn.* **1988**, *61*, 2513.
5. Gupta, R. B.; Franck, R. W.; Onan, K. D.; Soll, C. E. *J. Org. Chem.* **1989**, *54*, 1097 and references cited therein.
6. Kanemasa, S.; Kobayashi, H.; Tanaka, J.; Tsuge, O. *Bull. Chem. Soc. Jpn.* **1988**, *61*, 3957; Jacobsen, E. J.; Levin, J.; Overman, L. E.; *J. Am. Chem. Soc.* **1988**, *110*, 4329; Reich, H. J.; Willis, W. W. Jr.; Clark, P. D.; *J. Org. Chem.* **1981**, *46*, 2775 and references cited therein.
7. Paquette, L. A.; Carr, R. V. C.; *Org. Synth., Coll. Vol. VII* **1990**, 453.
8. Smith, M. B. *Synth. Commun.* **1986**, *16*, 85.
9. Bartels, B.; Hunter, R.; Simon, C. D.; Sny, G. D. *Tetrahedron Lett.* **1987**, *28*, 2985.
10. Brandsma, L.; Verkruijsse, H. D.; Schade, C.; von Ragué Schleyer, P. *J. Chem. Soc., Chem. Commun.* **1986**, 260.
11. Prior to submission of this manuscript a new, efficient synthesis of phenyl vinyl sulfide was published: Brace, N. O. *J. Org. Chem.* **1993**, *58*, 4506.

Appendix

Chemical Abstracts Nomenclature (Collective Index Number); (Registry Number)

Phenyl vinyl sulfide: Sulfide, phenyl vinyl (8); Benzene, (ethenylthio)- (9); (1822-73-7)

1-Phenylthio-2-bromoethane: Sulfide, 2-bromoethyl phenyl (8); Benzene, [(2-bromoethyl)thio]- (9); (4837-01-8)

Bromine (8,9); (7726-95-6)

Diphenyl disulfide: Phenyl disulfide (8); Disulfide, diphenyl (9); (882-33-7)

Ethylene (8); Ethene (9); (74-85-1)

1,8-Diazabicyclo[5.4.0]undec-7-ene: Pyrimido[1,2-a]azepine, 2,3,4,6,7,8,9,10-octahydro- (8,9); (6674-22-2)

NITROACETALDEHYDE DIETHYL ACETAL

(Ethane, 1,1-diethoxy-2-nitro-)

$$O_2N\text{-}CH_3 \quad + \quad HC(OC_2H_5)_3 \xrightarrow[\sim 90°C]{ZnCl_2} O_2NCH_2CH(OC_2H_5)_2$$

1

Submitted by V. Jäger and P. Poggendorf.[1]

Checked by E. Jnoff and Leon Ghosez.

1. Procedure

CAUTION! Distillation of nitromethane and reactions using it as a solvent or a reactant at an elevated temperature, as well as reactions of nitroalkanes in general, should be conducted behind a safety shield. In one instance a minor deflagration was observed upon erroneously aerating the distillation residue while it was still hot. The apparatus, therefore, should only be ventilated after cooling to ambient temperature, and nitrogen, not air, is recommended for this purpose.

A 500-mL, round-bottomed flask, equipped with a magnetic stirring bar, 20-cm Vigreux column, column head, Claisen distilling head, and thermometer, is charged with 89.3 g (602 mmol) of triethyl orthoformate (Note 1), 180 g (2.95 mol) of nitromethane (Note 2), and 5.00 g (36.6 mmol) of anhydrous zinc chloride. The solution is heated to 90°C (oil bath temperature, Note 3). After 16 hr (overnight) ca. 30 mL of ethanol is collected (Note 4). The remaining mixture, a brown suspension, is cooled to room temperature, and filtered by suction through a sintered glass funnel. The brown liquid obtained is distilled from a 100-mL, round-bottomed flask through a 20-cm Vigreux column at reduced pressure. First, excess nitromethane is removed (bp ca. 30°C/35 mm), then the fraction boiling at 58-60°C/1 mm (Note 3) is collected to afford 39-41 g (40-42% yield) of nitroacetaldehyde diethyl acetal **1** as a colorless liquid (Notes 4, 5).

2. Notes

1. All reagents were purchased from Fluka Feinchemikalien GmbH, Neu-Ulm, Germany and used without further purification.

2. Nitromethane, 98% purity, was used.

3. The rate of heating depends on the type of Vigreux column used, in this case a 20-cm, silver-plated Vigreux column with 4-cm outer and 1 1/2-cm inner diameter. The bath temperature should not be raised above 110°C when using this type of column, to avoid co-distillation of nitromethane (bp at 760 mm, 101°C). Ethanol should be distilled off at a rate of about 10 drops/min.

4. The spectroscopic properties of nitroacetaldehyde diethyl acetal are as follows: IR (film) ν_{max} cm^{-1}: 2970 (CH), 2920 (CH), 2880 (CH), 1550 (N=O), 1365 (N=O), 1340, 1120 (C-O), 1060; ^{1}H NMR (250 MHz, $CDCl_3$) δ: 1.15 (t, 6 H, $CH_2C\underline{H}_3$), 3.56 and 3.67 (2 q, 4 H, J = 7.1, 2 $C\underline{H}_2CH_3$), 4.44 ("d", 2 H, J = 5.8, CH_2NO_2), 5.09 ("t", 1 H, J = 5.8, CH); ^{13}C NMR (63 MHz, $CDCl_3$) δ: 15.0 (q, $CH_2\underline{C}H_3$), 63.3 (t, $\underline{C}H_2CH_3$), 76.8 (t, $\underline{C}H_2NO_2$), 98.7 [d, $\underline{C}H(OEt)_2$].

5. This reaction was carried out in the submitter's group more than 30 times, with yields ranging from 32-41% (lit.[2]: 49%). The purity of this material was repeatedly determined by gas chromatographic analysis to be >98%. GLC analysis: Column PS 086/.30 mm x 20 m glass capillary, 95:5 methyl/phenylsilicone. Program T_1, 40°C (1 min), rate 10°C/min; T_2, 300°C, 0.4 mm hydrogen pressure; R_t = 9.37 min.

Waste Disposal Information

All toxic materials were disposed of in accordance with "Prudent Practices in the Laboratory"; National Academic Press; Washington, DC, 1996.

3. Discussion

This procedure describes the preparation of nitroacetaldehyde diethylacetal **1** according to the method of René and Royer.[2]

Two other procedures for the preparation of **1** are known, i.e., by treatment of chloroacetaldehyde diethylacetal with silver nitrite,[3] and by the reaction of α,α-diethoxymethyltriethylammonium tetrafluoroborate and nitromethane.[4] These alternate methods are less suited and less economic for preparation of **1** on a large scale. The dimethyl acetal has been obtained by treatment of 1-chloro-2-nitroethene with sodium methoxide.[5]

Diethylacetal **1** has been used to obtain various other acetals by transacetalization,[6,14c] such as the dimethyl, the ethyleneglycol, and the neopentylglycol acetal (2,2-dimethyl-1,3-propylidene acetal).

Aliphatic nitro compounds are highly versatile building blocks in organic synthesis[7,8] (see Scheme 1). For example, the nitroaldol addition (Henry reaction)[9] leads to the formation of 1,2-nitro alcohols, **2**, which are easily transformed into 1,2-amino alcohols, **3**, by reduction, and into α-hydroxycarbonyl compounds, **4**, by hydrolysis[10] (Nef reaction). The former process, mostly using nitromethane, has been widely employed in carbohydrate chemistry.[11]

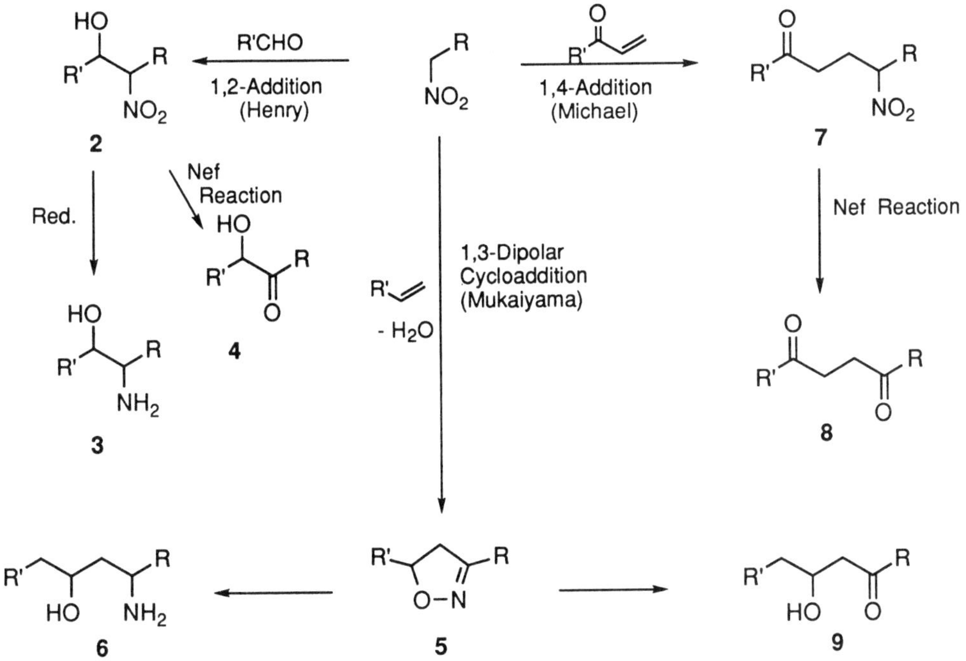

Scheme 1

Dehydration of primary nitro compounds (Mukaiyama reaction)[12] affords nitrile oxides, which may dimerize to yield furoxans, or otherwise be trapped by suitable dipolarophiles such as double or triple bond systems, leading to the formation of various heterocyclic systems, **5**.[13] The latter have been used for further derivatization in the heterocyclic series, or in "return" as precursors of acyclic products after ring cleavage,[7d,14] for example, 1,3-amino alcohols **6** or β-hydroxycarbonyl compounds, **9**.

Both nitroaldol[15] and 1,3-dipolar cycloaddition products (e.g., isoxazolines, from alkenes)[13,14] have shown that nitroacetaldehyde diethyl acetal **1** constitutes a versatile C_2 building block in organic synthesis, notably what concerns amino sugar target structures. Recent work both on nitroaldol and nitroalkane derived dipolar additions has concentrated on the study and elaboration of stereoselective C-C forming steps with nitroalkanes.

Further uses of nitroalkanes are in 1,4-additions (Michael reaction) to α,β-unsaturated carbonyl compounds and the like. Recent reports deal with transformations of 1,4-nitro ketones, **7**, into 1,4-keto aldehydes, **8**, and cyclization to cyclopentenones.[16]

1. Institut für Organische Chemie und Isotopenforschung der Universität Stuttgart, Pfaffenwaldring 55, D-70569 Stuttgart, Germany.
2. René, L.; Royer, R. *Synthesis* **1981**, 878.
3. Losanitsch, M. S. *Ber. Dtsch. Chem. Ges.* **1909**, *42*, 4044.
4. Kabusz, S.; Tritschler, W. *Synthesis* **1971**, 312.
5. Francotte, E.; Verbruggen, R.; Viehe, H. G.; van Meerssche, M.; Germain, G.; Declercq, J.-P. *Bull. Soc. Chim. Belg.* **1978**, *87*, 693.
6. Müller, R. Dissertation, Universität Würzburg, 1992.
7. (a) Schickh, O. von.; Apel, G.; Padeken, H. G.; Schwarz, H. H.; Segnitz, A. In "Methoden der Organischen Chemie (Houben-Weyl)", 4th ed.; Müller, E., Ed.; Georg Thieme Verlag: Stuttgart, 1971; Vol. X. Part 1, pp. 1-462; (b) Behnisch, R.; Behnisch, P.; Mattmer, R. In "Methoden der Organischen Chemie (Houben-Weyl)", 4th ed.; Klamann, D., Ed.; Georg Thieme: Stuttgart, 1992; Vol. E 16d, Part 1, pp. 142-254; (c) Nielsen, A. T. In "The Chemistry of the Nitro and Nitroso Groups"; Feuer, H., Ed.; Wiley: New York, 1969, pp. 349-486; (d) Torssell, K. B. G. "Nitrile Oxides, Nitrones, and Nitronates in Organic Synthesis"; VCH: New York, 1987

8. (a) Baer, H. H. *Adv. Carbohydr. Chem. Biochem.* **1969**, *24*, 67; (b) Wade, P. A.; Giuliano, R. M. In "Nitro Compounds; Recent Advances in Synthesis and Chemistry"; Feuer, H.; Nielsen, A. T., Eds.; VCH: Weinheim, 1990, pp. 137-265.

9. Henry, L. *C. R. Hebd. Acad. Sci.* **1895**, *120*, 1265.

10. Noland, W. E. *Chem. Rev.* **1955**, *55*, 137.

11. (a) Sowden, J. C. *Adv. Carbohydr. Chem.* **1951**, *6*, 291; (b) Sowden, J. C. *Methods Carbohydr. Chem.* **1962**, *1*, 132; (c) Whistler, R. L.; BeMiller, J. N. *Methods Carbohydr. Chem.* **1962**, *1*, 137; (d) Sowden, J. C.; Oftedahl, M. L. *Methods Carbohydr. Chem.* **1962**, *1*, 235; (e) Paulsen, H.; Brieden, M.; Sinnwell, V. *Liebigs Ann. Chem.* **1985**, 113; (f) Baer, H. H. *J. Am. Chem. Soc.* **1962**, *84*, 83; (g) Lichtenthaler, F. W.; Nakagawa, T. *Chem. Ber.* **1968**, *101*, 1846.

12. Mukaiyama, T.; Hoshino, T. *J. Am. Chem. Soc.* **1960**, *82*, 5339.

13. (a) Jäger, V.; Grund, H.; Buβ, V.; Schwab, W.; Müller, I.; Schohe, R.; Franz, R.; Ehrler, R. *Bull. Soc. Chim. Belg.* **1983**, *92*, 1039; (b) Jäger, V.; Müller, I.; Schohe, R.; Frey, M.; Ehrler, R.; Häfele, B.; Schröter, R. *Lect. Heterocycl. Chem.* **1985**, *8*, 79.

14. (a) Jäger, V.; Grund, H. *Angew. Chem.* **1976**, *88*, 27; *Angew. Chem. Int. Ed. Engl.* **1976**, *15*, 50; (b) Jäger, V.; Müller, I. *Tetrahedron* **1985**, *41*, 3519 and references therein; (c) Jäger, V.; Schohe, R. *Tetrahedron* **1984**, *40*, 2199.

15. (a) Wehner, V.; Jäger, V. *Angew. Chem.* **1990**, *102*, 1180; *Angew. Chem. Int. Ed. Engl.* **1990**, *29*, 1169; (b) Jäger, V.; Raczko, J.; Steuer, B.; Peters, K.; Wehner, V.; Öhrlein, R.; Poggendorf, P. *XVIth International Carbohydrate Symposium*, Paris, **1992**, Abstr. A227.

16. (a) Rosini, G.; Ballini, R.; Sorrenti, P. *Tetrahedron* **1983**, *39*, 4127; (b) Rosini, G.; Ballini, R.; Petrini, M.; Sorrenti, P. *Tetrahedron* **1984**, *40*, 3809.

Appendix

Chemical Abstracts Nomenclature (Collective Index Number); (Registry Number)

Nitroacetaldehyde diethyl acetal: Ethane, 1,1-diethoxy-2-nitro- (11); (34560-16-2)

Triethyl orthoformate: Orthoformic acid, triethyl ester (8); Ethane, 1,1',1"-[methylidynetris(oxy)]tris- (9); (122-51-0)

Nitromethane: Methane, nitro- (8,9); (75-52-5)

Zinc chloride (8,9); (7646-85-7)

CYCLOPENTANONE ANNULATION VIA CYCLOPROPANONE DERIVATIVES: (3aβ,9bβ)-1,2,3a,4,5,9b-HEXAHYDRO-9b-HYDROXY-3a-METHYL-3H-BENZ[e]INDEN-3-ONE

A. CH_3, O; $BrMgN(i\text{-}Pr)_2$, Et_2O; CH_3, OMgBr (in solution)

B. CH_3, OMgBr + EtO OMgBr → H_3C, OH, O

C. H_3C, OH, O; NaH, Et_2O → H_3C, O, OH

Submitted by Michael J. Bradlee and Paul Helquist.[1]

Checked by Steve Berberich and Stephen F. Martin.

1. Procedure

A. Magnesium enolate of 2-methyl-1-tetralone. Into a 500-mL, round-bottomed flask equipped with a magnetic stirring bar is placed 6.72 g of magnesium bromide diethyl etherate ($MgBr_2 \cdot Et_2O$) (26.0 mmol) under nitrogen (Note 1). Anhydrous ether, 350 mL, (Note 2) is added with a syringe, and the mixture is stirred at 25°C for a few minutes to dissolve the solid. Using an air-tight syringe, 11.2 mL of a 3.0 M solution of methylmagnesium bromide in ether (34 mmol, Note 3) is added to the slightly turbid solution at 25°C over a period of a few minutes. N,N-Diisopropylamine, 4.46 mL (34 mmol, Note 4) is added over 5 min, and the mixture is stirred for 16-18 hr, during which time a fine white precipitate forms (Note 5). The mixture is cooled to 0°C, and a solution of 5.28 g of 2-methyl-1-tetralone (33 mmol, Note 6) in 20 mL of anhydrous ether is added with a cannula over a 10-min period. The mixture becomes a bright yellow solution during the addition, and after 45 min of additional stirring, it becomes pale yellow.

B. 2-(1-Hydroxycyclopropyl)-2-methyl-1-tetralone. Simultaneously with the above operations, a separate 2-L, three-necked, round-bottomed flask is equipped with a mechanical stirrer, reflux condenser, nitrogen inlet, and a septum, and the flask is placed under a nitrogen atmosphere. Anhydrous ether, 600 mL and 20 mL of a 3.0 M solution of methylmagnesium bromide in ether (60 mmol, Note 3) are added to the flask with a cannula and a syringe, respectively, and the solution is cooled to 0°C. A solution of 6.12 g of cyclopropanone ethyl hemiketal[2] (60.0 mmol, Note 7) in 20 mL of anhydrous ether is added with vigorous stirring over several minutes with a cannula. During this addition, gas evolution is observed, and a fine white precipitate forms. After the mixture is stirred for 15 min, the solution of the enolate of 2-methyl-1-tetralone is added using a wide gauge cannula (ca. 1 mm id) over several minutes. The resulting turbid, light yellow solution is heated at reflux for 4 hr. The resulting clear, yellow solution is cooled to 25°C and diluted with 200 mL of ether. The organic solution is washed with saturated aqueous ammonium chloride (3 x 200 mL) and saturated aqueous sodium chloride (200 mL) and dried over anhydrous magnesium sulfate. The drying agent is removed by filtration and the filtrate is concentrated by rotary evaporation under reduced pressure to give 7.92 g of crude 2-(1-hydroxycyclopropyl)-2-methyl-1-tetralone (Note 8) as a clear yellow oil that is used in the next step without further purification.

C. (3aβ,9bβ)-1,2,3a,4,5,9b-Hexahydro-9b-hydroxy-3a-methyl-3H-benz[e]inden-3-one.[3] The yellow oil from Part B is dissolved in 100 mL of anhydrous ether in a 100-mL round-bottomed flask at 25°C under nitrogen. Into a separate 250-mL flask containing a magnetic stirring bar are placed 4.0 g of powdered sodium hydride (166 mmol, Note 9), and 100 mL of anhydrous ether at 25°C under nitrogen. The solution of the yellow oil is added with stirring over a period of a few minutes using a cannula. The mixture is stirred at 25°C for 3 hr, during which time it becomes yellow-orange. The reaction is quenched by adding 100 mL of ice-cold saturated aqueous ammonium chloride and 100 mL of ether. The organic layer is washed with saturated aqueous ammonium chloride (3 x 100 mL) and saturated aqueous sodium chloride (100 mL), dried over anhydrous magnesium sulfate, filtered, and concentrated by rotary evaporation under reduced pressure. The crude product is obtained as 5.8 g of yellow-orange crystals. The solid is dissolved in about 150 mL of warm ether, and 100 mL of hexanes is added. The mixture is concentrated to ca. 100 mL in a heating bath (or on a steam bath), and the resulting cloudy solution is cooled slowly to 0°C. An initial crop of 4.73 g (67%) of pale yellow crystals is obtained and the mother liquor is decanted from the solid and concentrated by rotary evaporation under reduced pressure to ca. 5 mL to give 0.39 g (5%) of pale yellow crystals as a second crop. The mother liquor may then be concentrated and the residue purified by flash chromatography[4] on silica gel using 1:1 ether/hexane to give 0.29 g (4%) of additional product (R_f = 0.33) as yellow crystals after washing with warm 1:1 ether/hexane. The total combined yield of (3aβ,9bβ)-1,2,3a,4,5,9b-hexahydro-9b-hydroxy-3a-methyl-3H-benz[e]inden-3-one is 5.41 g (76%) (Note 10).

2. Notes

1. The $MgBr_2 \cdot Et_2O$ was obtained from Aldrich Chemical Company, Inc., as a white solid in the form of a powder mixed with lumps. The solid was transferred to the 250-mL, round-bottomed flask in a nitrogen-filled glove box and carefully crushed with a glass stirring rod until the solid was a uniform powder. The flask was then equipped with a septum, taken out of the glove box, and attached to a standard nitrogen/vacuum double manifold system to maintain an inert, dry atmosphere of prepurified nitrogen in the flask throughout the remainder of the reaction sequence.

2. Anhydrous diethyl ether and anhydrous tetrahydrofuran (THF) were obtained from Fisher Scientific Company. They were redistilled under nitrogen from dark blue or purple solutions of sodium benzophenone ketyl immediately prior to use. All transfers of these and other anhydrous materials were performed with syringes or stainless steel cannulas while carefully maintaining a nitrogen atmosphere.

3. Methylmagnesium bromide was purchased from Aldrich Chemical Company, Inc., as a solution in ether, and the solution was used as obtained without titration.

4. N,N-Diisopropylamine was purchased from Aldrich Chemical Company, Inc., and distilled under nitrogen from calcium hydride prior to use.

5. The submitters observed the fine white precipitate after 12 hr of stirring.

6. 2-Methyl-1-tetralone was purchased from Aldrich Chemical Company, Inc., and used without further purification.

7. Cyclopropanone ethyl hemiketal is contaminated with 8-9% of the cyclopropanone methyl hemiketal as a result of exchange with the solvent.[2] The calculated number of mmoles of total cyclopropanone hemiketal is based upon this composition.

8. ^{1}H NMR analysis of the crude product indicated greater than 90% conversion of the starting 2-methyl-1-tetralone to 2-(1-hydroxycyclopropyl)-2-methyl-1-tetralone: IR ($CHCl_3$) cm^{-1}: 3485 (O-H), 3067 (aromatic C-H), 2972 (aliphatic C-H), 1670 (C=O); ^{1}H NMR (300 MHz, $CDCl_3$) δ: 0.49-0.72 (m, 4 H, cyclopropyl $C\underline{H}_2C\underline{H}_2$), 1.20 (s, 3 H, $C\underline{H}_3$), 1.61-1.71 (dt, 1 H, J = 13.4, 4.5, $CH\underline{H}$), 2.03-2.15 (m, 1 H, $C\underline{H}H$), 2.88-3.04 (m, 2 H, benzylic $C\underline{H}_2$), 3.68 (s, 1 H, $O\underline{H}$), 7.15-7.96 (m, 4 H, aromatic $C\underline{H}$); ^{13}C NMR (75 MHz, $CDCl_3$) δ: 9.0, 10.4 (cyclopropyl $\underline{C}H_2\underline{C}H_2$), 18.2 ($\underline{C}H_3$), 25.5 ($\underline{C}H_2$), 31.0 (benzylic $\underline{C}H_2$), 48.0 ($\underline{C}C=O$), 59.3 (cyclopropyl $\underline{C}OH$), 126.7, 127.8, 128.7, 131.7 (aromatic $\underline{C}H$), 133.6, 143.4 (ipso aromatic $\underline{C}$), 204.0 ($\underline{C}=O$).

9. Sodium hydride was obtained as a powder of 95% purity grade from Aldrich Chemical Company, Inc. A nitrogen-filled glove box was used to weigh the reagent and transfer it into the reaction flask.

10. Yields of the product ranged from 67 to 81% over a series of several runs. For (3aβ,9bβ)-1,2,3a,4,5,9b-hexahydro-9b-hydroxy-3a-methyl-3H-benz[e]inden-3-one, mp 117-118°C; IR ($CHCl_3$) cm^{-1}: 3466 (O-H), 3066, 3070 (aromatic C-H), 2976 (aliphatic C-H), 1736 (C=O); ^{1}H NMR (300 MHz, $CDCl_3$) δ: 1.14 (s, 3 H, $C\underline{H}_3$), 1.58-1.67 (m, 1 H), 1.80 (s, 1 H, $O\underline{H}$), 1.87-1.99 (m, 1 H), 2.20-2.41 (m, 3 H), 2.48-2.61 (m, 1 H), 2.66-2.76 (m, 1 H), 2.84-2.97 (m, 1 H), 7.08-7.68 (m, 4 H, aromatic $C\underline{H}$); ^{13}C NMR (75 MHz, $CDCl_3$) δ: 14.5 ($\underline{C}H_3$), 25.0 ($\underline{C}H_2$), 29.1 ($\underline{C}H_2COH$), 34.9 ($\underline{C}H_2C=O$), 36.1 (benzylic $\underline{C}H_2$), 52.5 ($\underline{C}CH_3$), 79.8 (benzylic $\underline{C}OH$),126.6, 126.9, 127.5, 128.5 (aromatic $\underline{C}H$), 134.6, 140.5 (ipso aromatic $\underline{C}$), 204.5 ($\underline{C}=O$); CIMS, m/e (rel intensity) 217 (10%, MH^+), 199 (100%, MH^+ - 18); Anal. Calcd for $C_{14}H_{16}O_2$: C, 77.75; H, 7.46. Found: C, 78.00; H, 7.60.

Waste Disposal Information

All toxic materials were disposed of in accordance with "Prudent Practices in the Laboratory"; National Academic Press; Washington, DC, 1996.

3. Discussion

A large number of methods have been developed for the construction of five-membered carbocyclic systems. These methods are too numerous to review here, but a number of key references are cited.[5]

The method that is used in the present preparation is based upon the *formal concept* of employing cyclopropanone in an aldol condensation with an enolate of another ketone. The resulting aldolate exhibits characteristics of a "homoenolate" and may undergo a second carbonyl addition reaction to form a five-membered ring as suggested in the following generalized pathway:

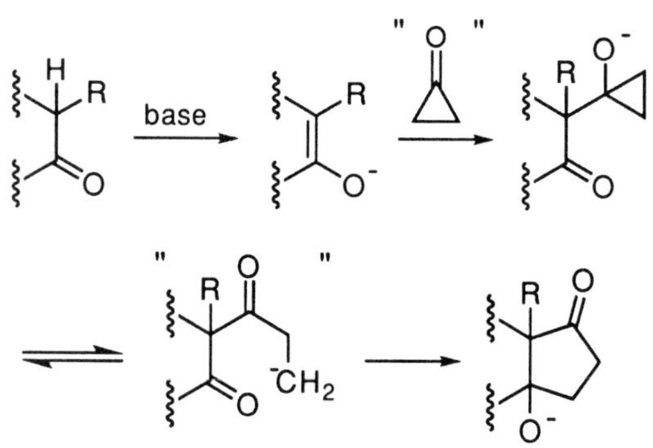

The homoenolate is not necessarily an intermediate in this pathway, but rather, it is shown only to illustrate the basic concept.

Cyclopropanone itself is a very unstable compound that has been isolated only at low temperature. Its chemistry has been studied thoroughly during the past several years.[6] Various forms of homoenolates have also been investigated, and their applications in synthesis are being actively developed.[7] The combination of cyclopropanone and homoenolate chemistry employed here is most closely related to studies of Narasimhan[8] and follows from a series of studies in the laboratory of the submitters.[3,9] Whether free cyclopropanone or the homoenolate form of the aldol adduct are produced as actual species in the above pathway is uncertain, but at least the formal concept of their hypothetical intermediacy and expected reactivity patterns have proven useful in the design of the annulation method employed in the present preparation. The cyclopropanone hemiketal used in this work could conceivably undergo ethoxide elimination to produce cyclopropanone in situ which would then serve as a very reactive acceptor for nucleophilic addition,[2,6,10] but other mechanistic pathways may also be consistent with the observed reaction sequence.

An attractive feature of this procedure is the directness with which the annulation of 3-hydroxycyclopentanone systems onto preexisting ketone skeletons can be accomplished to give usefully functionalized products. However, this method does have important limitations in its scope. The most common difficulty is that the "cyclopropanone" condensation with a wide range of ketone enolates often occurs in only low to modest yields. A two-fold excess of the cyclopropanone hemiketal is therefore used to provide at least some improvement in the yields of this step. In contrast, the subsequent "homoenolate" cyclization generally occurs in quite acceptable yields. A simpler representative example that illustrates the contrasting yields in the two steps is the following:[9b]

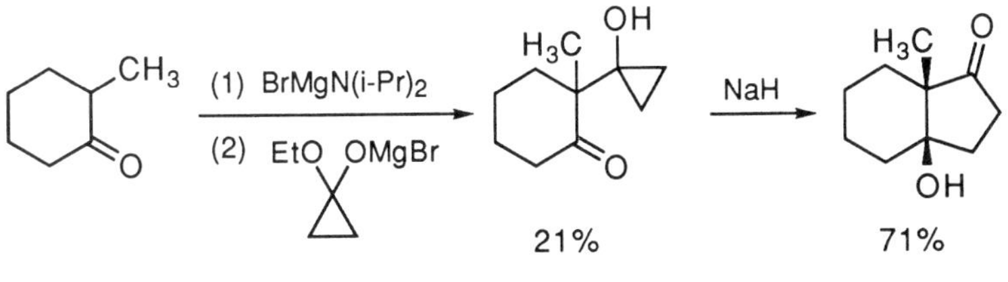

CH_3 (1) $BrMgN(i\text{-}Pr)_2$ (2) EtO OMgBr → H_3C OH 21% → NaH → H_3C O OH 71%

On the other hand, the parent, unsubstituted cyclohexanone enters into an apparently much more complex reaction pathway leading to the formation of a tricyclic, cycloheptanone-containing product.[9a] Also, cycloheptanones as starting materials give annulation products that undergo a subsequent retro-aldol/re-aldol sequence to give rearranged hydrazulenones as the final isolated products. In fact, the best cases of the present cyclopentanone annulation sequence are limited to benzo-fused cyclohexanones bearing an additional α-alkyl substituent. In addition to the preparation described here, another important example is the following (yields are not optimized as in the present preparation):[9b]

CH_3 O (1) $BrMgN(i\text{-}Pr)_2$ (2) EtO OMgBr (74%) (3) NaH H_3C O OH 78%

Although the limitations in the cases that provide good yields may appear to be very restrictive, this last example is suggestive of the potential applications of this procedure. Rather obvious possibilities that follow from this case include derivatives of the equilinin steroidal hormone system,[11] and of the steroidal cardiotonic agents, e.g., the cardenolides, digitoxigenin, bufadienolides, etc.[12]

1. Department of Chemistry and Biochemistry, University of Notre Dame, Notre Dame, Indiana 46556.
2. Salaün, J.; Marguerite, J. *Org. Synth.* **1985**, *63*, 147; *Org. Synth. Coll. Vol. VII* **1990**, 131.
3. Bradlee, M. J., M.Sc. Thesis, University of Notre Dame, 1991.
4. Still, W. C.; Kahn, M.; Mitra, A. *J. Org. Chem.* **1978**, *43*, 2923.
5. (a) Paquette, L. A. *Topics Curr. Chem.* **1979**, *79*, 41; (b) Trost, B. M. *Chem. Soc. Rev.* **1982**, *11*, 141; (c) Ramaiah, M. *Synthesis* **1984**, 529; (d) Paquette, L. A.

Topics Curr. Chem. **1984**, *119*, 1; (e) Hudlicky, T.; Kutchan, T. M.; Naqvi, S. M. *Org. React.* **1985**, *33*, 247; (f) Posner, G. H. *Chem. Rev.* **1986**, *86*, 831; (g) Ho, T.-L. "Carbocycle Construction in Terpene Synthesis"; VCH: Weinheim, 1988; (h) Hudlicky, T.; Price, J. D. *Chem. Rev.* **1989**, *89*, 1467; (i) Schick, H.; Eichhorn, I. *Synthesis*, **1989**, 477; (j) Haufe, G.; Mann, G. "Chemistry of Alicyclic Compounds: Structure and Chemical Transformations", Studies in Organic Chemistry 38; Elsevier: Amsterdam, 1989; (k) Schore, N. E. *Org. React.* **1991**, *40*, 1; (l) Welker, M. E. *Chem. Rev.* **1992**, *92*, 97.

6. (a) Salaün, J. *Chem. Rev.* **1983**, *83* , 619; (b) Wasserman, H. H.; Berdahl, D. R.; Lu, T.-J. In "The Chemistry of the Cyclopropyl Group", Rappoport, Z., Ed.; Wiley: Chichester, 1987; Part 2, Chapter 23; (c) Salaün, J. In "The Chemistry of the Cyclopropyl Group", Rappoport, Z., Ed.; Wiley: Chichester, 1987; Part 2, Chapter 13.
7. Leading references to "homoenolate" chemistry: (a) Werstiuk, N. H. *Tetrahedron* **1983**, *39*, 205-268; (b) Hoppe, D. *Angew. Chem., Int. Ed. Engl.* **1984**, *23*, 932-948; (c) Nakamura, E.; Oshino, H.; Kuwajima, I. *J. Am. Chem. Soc.* **1986**, *108*, 3745; (d) Fukuzawa, S.-i.; Sumimoto, N.; Fujinami, T.; Sakai, S. *J. Org. Chem.* **1990**, *55*, 1628; (e) Kuwajima, I.; Nakamura, E. *Topics Curr. Chem.* **1990**, *155*, 1-39.
8. (a) Narasimhan, N. S.; Patil, P. A. *Tetrahedron Lett.* **1986**, *27* , 5133; (b) Narasimhan, N. S.; Sunder, N. M. *Tetrahedron Lett.* **1988**, *29* , 2985.
9. (a) Carey, J. T.; Knors, C.; Helquist, P. *J. Am. Chem. Soc.* **1986**, *108*, 8313; (b) Carey, J. T.; Helquist, P. *Tetrahedron Lett.* **1988**, *29*, 1243; (c) Reydellet, V.; Helquist, P. *Tetrahedron Lett.* **1989**, *30*, 6837; (d) Carey, J. T. Ph.D. Dissertation, University of Notre Dame, 1988; (e) Reydellet, V. M.Sc. Thesis, University of Notre Dame, 1990.
10. Brown, H. C.; Rao, C. G. *J. Org. Chem.* **1978**, *43* , 3602.

11. See "Natural Products Chemistry", Nakanishi, K.; Goto, T.; Ito, S.; Natori, S.; Nozoe, S., Eds.; Academic Press: New York, 1974; Vol. 1, pp 442-445 and 485-487.

12. See "Natural Products Chemistry", Nakanishi, K.; Goto, T.; Ito, S.; Natori, S.; Nozoe, S., Eds.; Academic Press: New York, 1974; Vol. 1, pp 462-476.

Appendix

Chemical Abstracts Nomenclature (Collective Index Number); (Registry Number)

Magnesium bromide diethyl etherate: Magnesium, dibromo(ethyl ether)- (8); Magnesium, dibromo[1,1'-oxybis[ethane]]- (9); (29858-07-9)

Methylmagnesium bromide: Magnesium, bromomethyl- (8,9); (75-16-1)

Diisopropylamine (8); 2-Propanamine, N-(1-methylethyl)- (9); (108-18-9)

2-Methyl-1-tetralone: 1(2H)-Naphthalenone, 3,4-dihydro-2-methyl- (8,9); (1590-08-5)

Cyclopropanone ethyl hemiketal: Cyclopropanol, 1-ethoxy- (8,9); (13837-45-1)

Sodium hydride (8,9); (7646-69-7)

[3+2]-ANIONIC ELECTROCYCLIZATION USING 2,3-BIS(PHENYLSULFONYL)-1,3-BUTADIENE: trans-4,7,7-TRICARBOMETHOXY-2-PHENYLSULFONYLBICYCLO[3.3.0]OCT-1-ENE

(Benzene, 1,1'-[[1,2-bis(methylene)-1,2-ethanediyl]bis(sulfonyl)]bis-)

A. $HOCH_2-C\equiv C-CH_2OH \xrightarrow[NEt_3]{PhSCl}$ $CH_2=C(SOPh)-C(SOPh)=CH_2$

B. $CH_2=C(SOPh)-C(SOPh)=CH_2 \xrightarrow[AcOH]{H_2O_2}$ $CH_2=C(SO_2Ph)-C(SO_2Ph)=CH_2$ **1**

C. $CH_2(CO_2CH_3)_2$ → 1) NaH; 2) $CH_3OOC-CH=CH-CH_2Br$ → $CH_3OOC-CH=CH-CH_2-CH(CO_2CH_3)_2$

D. $CH_3OOC-CH=CH-CH_2-CH(CO_2CH_3)_2$ → 1) NaH; 2) $CH_2=C(SO_2Ph)-C(SO_2Ph)=CH_2$ → bicyclo[3.3.0]octene bearing SO_2Ph, CO_2CH_3, CO_2CH_3, CO_2CH_3

Submitted by Albert Padwa,[1] Scott H. Watterson, and Zhijie Ni.

Checked by David Young, Michael Rupp, Dan Kuzmich, and David J. Hart.

1. Procedure

A. 2,3-Bis(phenylsulfinyl)-1,3-butadiene. In a 2-L, three-necked, round-bottomed flask equipped with a mechanical stirrer, 25-mL dropping funnel, and a nitrogen inlet, are placed 7.75 g (0.09 mol) of 2-butyne-1,4-diol (Note 1), 37.6 mL (0.27 mol) of triethylamine, and 700 mL of dichloromethane. After the diol is completely dissolved, the mixture is cooled to -78°C. To this solution is added 26.0 g (0.18 mol) of phenylsulfenyl chloride (Note 2) over a 30-min interval. The mixture is gradually warmed to room temperature and stirred for an additional 12 hr. The solution is then washed with 100 mL each of water, saturated ammonium chloride solution, saturated sodium bicarbonate solution, and brine, and dried over sodium sulfate. After filtration and concentration, the residue is recrystallized from methanol-ether (1:1) to give 20.1 g (74%) of 2,3-bis(phenylsulfinyl)-1,3-butadiene as a 1.1:1 mixture of diastereomers (Note 3).

B. 2,3-Bis(phenylsulfonyl)-1,3-butadiene (**1**). A solution of 17.0 g of 2,3-bis(phenylsulfinyl)-1,3-butadiene (56.3 mmol) and 34 mL of hydrogen peroxide (30-35%) (Note 4) in 170 mL of glacial acetic acid is placed in a 500-mL, three-necked flask equipped with a reflux condenser and thermometer. The solution is warmed using an oil bath such that a reaction temperature of 90°C is maintained for 5 hr (Note 5). To the hot solution is added 40 mL of water and the mixture is allowed to stand at room temperature for 12 hr. The resulting precipitate is collected, washed with water (50 mL), cold methanol (30 mL), and ether (100 mL). The filter cake (13.9 g) is dissolved in 100 mL of hot dichloromethane and 100 mL of hexane is added slowly. After standing at 25°C for 1 day, the crystalline solid is collected and washed with 100 mL of ether to give 11.6 g (61%) of 2,3-bis(phenylsulfonyl)-1,3-butadiene (Note 6).[2] Concentration and recrystallization of the mother liquor affords an additional 1.7 g (9%) of the disulfone.

C. Dimethyl (E)-5-methoxycarbonyl-2-hexenedioate. Into a flame-dried, 500-mL, three-necked, round-bottomed flask equipped with a magnetic stirring bar, a 100-mL dropping funnel, and a nitrogen inlet, is placed 4.2 g of sodium hydride (60% in oil, 105 mmol) which has been washed twice with 50 mL of hexane and then suspended in 400 mL of anhydrous tetrahydrofuran (THF). The mixture is cooled to 0°C with an ice bath and a solution containing 13.7 mL (15.8 g, 120 mmol) of dimethyl malonate in 30 mL of anhydrous THF is added over a 30-min interval. After being stirred for an additional 20 min at 0°C, the mixture is transferred by cannula into an ice-cold solution containing 19.9 g (110 mmol) of methyl 4-bromocrotonate (Note 7) in 300 mL of THF in a 1-L, round-bottomed flask fitted with a nitrogen inlet. Stirring is continued for an additional 20 min and the reaction is quenched by the addition of 100 mL of a saturated ammonium chloride solution. The solvent is removed with a rotary evaporator at aspirator vacuum, and the aqueous layer is extracted three times with ether (200 mL). The organic layer is washed with 100 mL each of water and brine, and dried over sodium sulfate. After filtration and concentration, the resulting residue is distilled using a 5-cm Vigreux column under reduced pressure to give 8.8 g (38%) of dimethyl (E)-5-methoxycarbonyl-2-hexenedioate (Note 8).

D. trans-4,7,7-Tricarbomethoxy-2-phenylsulfonylbicyclo[3.3.0]oct-1-ene. In a flame-dried, 2-L, three-necked, round-bottomed flask equipped with a magnetic stirrer, dropping funnel, and nitrogen inlet, is placed 1.44 g of sodium hydride (60% in oil, 36.0 mmol) which has been washed twice with 50 mL of hexane and then suspended in 500 mL of anhydrous THF. To the above mixture is added 7.59 g (33.0 mmol) of dimethyl (E)-5-methoxycarbonyl-2-hexenedioate in 50 mL of THF. After being stirred at 0°C for 30 min, a solution containing 10.02 g (30.0 mmol) of 2,3-bis(phenylsulfonyl)-1,3-butadiene in 900 mL of anhydrous THF is added over 30 min. The solution is stirred for 10 min at 0°C and then quenched with 100 mL of a saturated aqueous ammonium chloride solution. The solvent is removed under reduced pressure, and the residue is extracted three times with 300 mL of dichloromethane. The organic phase is washed with 100 mL each of water, brine, and dried over sodium sulfate. The filtrate is concentrated under reduced pressure and the residue is chromatographed on 350 g of silica gel eluting with 4 L of a 2:1-hexane/ethyl acetate mixture to give 9.54 g (75%) of trans-4,7,7-tricarbomethoxy-2-phenylsulfonylbicyclo[3.3.0]oct-1-ene as a clear oil (Note 9).

2. Notes

1. 2-Butyne-1,4-diol, sodium hydride (60% in oil) and dimethyl malonate were purchased from Aldrich Chemical Company, Inc. and used without further purification. Triethylamine and dichloromethane were distilled from calcium hydride prior to use.

2. Phenylsulfenyl chloride was prepared according to the known literature procedure.[3]

3. This product is a 1.1:1 mixture of two diastereoisomers, mp 127-129°C, and was used in the next step without further purification. The product has the following spectral properties: ^{1}H NMR (200 MHz, $CDCl_3$) δ: 5.6, 5.85, 6.18, 6.22 (four s, 4 H), 7.1-7.5 (m, 10 H).

4. Hydrogen peroxide (30-35%) and glacial acetic acid were purchased from Fisher Scientific Company and were used without further purification.

5. The yield is considerably reduced if the reaction temperature is allowed to exceed 90°C. Yields as high as 80% have been obtained with careful monitoring of the temperature.

6. The product has the following physical properties: mp 181-183°C (lit.[2] 183-185°C); IR (KBr) cm^{-1}: 3120, 1580, 1441, 1302, 1140, 1067, 744, 685; ^{1}H NMR (300 MHz, $CDCl_3$) δ: 6.61 (s, 2 H), 6.74 (s, 2 H), 7.20-7.30 (m, 4 H), 7.42-7.51 (m, 6 H); ^{13}C NMR (75 MHz, $CDCl_3$) δ: 128.4 (d), 129.1 (d), 130.8 (t), 133.7 (d), 137.1 (s), 140.8 (s).

Anal. Calcd for $C_{16}H_{14}O_4S_2$: C, 57.46; H, 4.22; S, 19.18. Found: C, 57.39; H, 4.23; S, 19.08.

7. Methyl 4-bromocrotonate was purchased from Lancaster Synthesis Inc. and distilled before use.

8. Dimethyl (E)-5-methoxycarbonyl-2-hexenedioate[4] has the following physical properties: bp 144-146°C (2.0 mm); IR (neat) cm^{-1}: 3004, 2947, 1737, 1431, 1268, 1033, 727; ^{1}H NMR (300 MHz, $CDCl_3$) δ: 2.79 (t, 2 H, J = 7.2), 3.51 (t, 1 H, J = 7.2), 3.70 (s, 3 H), 3.74 (s, 6 H), 5.98 (d, 1 H, J = 15.6), 6.86 (dt, 1 H, J = 15.6, 7.2); ^{13}C NMR (75 MHz, $CDCl_3$) δ: 30.9, 50.1, 51.3, 52.6, 123.3, 143.7, 166.1, 168.5. This compound has also been prepared by the ozonolysis of dimethyl allylmalonate followed by reaction with methyl (triphenylphosphoranylidene)acetate.[5] While this procedure is reported to afford the triester in 60-65% overall yield, we found the above method to be simpler and more convenient for the preparation of larger quantities of material.

9. trans-4,7,7-Tricarbomethoxy-2-(phenylsulfonyl)bicyclo[3.3.0]oct-1-ene has the following spectral properties: IR (neat) cm^{-1}: 3004, 2947, 1730, 1438, 1147, 1075, 727; ^{1}H NMR (300 MHz, $CDCl_3$) δ: 1.86 (t, 1 H, J = 12.3), 2.56 (dd, 1 H, J = 13.2, 7.8), 2.79-2.99 (m, 3 H), 3.17 (brd, 1 H, J = 21.0), 3.34 (brd, 1 H, J = 21.0), 3.4-3.7 (m, 1 H), 3.56 (s, 3 H), 3.65 (s, 3 H), 3.67 (s, 3 H), 7.47 (t, 2 H, J = 7.5), 7.56 (t, 1 H, J = 7.2), 7.78 (d, 2 H, J = 7.5); ^{13}C NMR (75 MHz, $CDCl_3$) δ: 33.2, 37.4, 39.4, 48.9, 51.9, 52.8, 53.0, 54.7, 63.1, 127.3, 129.0, 129.1, 133.4, 139.7, 161.0, 170.7, 171.2, 172.4; HRMS Calcd for $C_{20}H_{22}O_8S$: m/e 422.1035. Found m/e 422.1030.

Waste Disposal Information

All toxic materials were disposed of in accordance with "Prudent Practices in the Laboratory"; National Academic Press; Washington, DC, 1996.

3. Discussion

The chemistry of phenylsulfonyl-substituted 1,3-butadienes is receiving increasing attention because of their synthetic versatility and the efficient π-bond activation by the sulfonyl group.[6-10] Some of the procedures currently available for their preparation include phenylsulfonyl-mercuration of 1,3-dienes,[11] condensation of allyl sulfones with aldehydes followed by acetylation and subsequent elimination,[12] thermal SO_2 extrusion from 2-(arylsulfonyl)sulfolenes,[13,14] cheletropic ring-opening of sulfolanes,[15] palladium(II)-catalyzed chloroacetoxylation of 1,3-dienes,[16] and the 2-tosylvinyl sulfone coupling with vinylstannanes.[17] Methods of preparing bis(phenylsulfonyl)-substituted 1,3-dienes, however, are limited to relatively few routes.[18] In spite of its simplicity and its obvious potential as an activated diene, 2,3-bis(phenylsulfonyl)-1,3-butadiene (**1**) has not been used extensively in organic synthesis. This reagent was prepared by a modification of the procedure of Jeganathan and Okamura in multigram quantities.[2] Treatment of 2-butyne-1,4-diol with benzenesulfenyl chloride produces the disulfenate ester as a transient species, which rapidly undergoes a series of 2,3-sigmatropic rearrangements to give 2,3-bis(phenylsulfinyl)-1,3-butadiene. This material is readily oxidized with hydrogen peroxide to the bis(phenylsulfonyl)diene **1** in 70-80% yield.

Over the past several years, the submitter has demonstrated the use of diene **1** as a versatile building block in organic synthesis, particularly for [4+2]-cycloaddition chemistry.[19] In an early report, the cycloaddition of this diene with various oximes as a method for piperidine formation was described.[20] In subsequent investigations, disulfone **1** was demonstrated to undergo cycloaddition to a variety of simple imines[21] and enamines[22] under mild conditions producing novel rearranged heterocycles. This diene also played an important role in the successful outcome of the submitter's [4+1]-annulation strategy for pyrrolidine formation, since it is highly activated toward nucleophilic addition.[23] More recently, disulfone **1** was found to undergo a [4+1]-annulation reaction with a variety of soft carbanions to give phenylsulfonyl-substituted cyclopentenes in good yield.[24] The pivotal step in this reaction involves addition of a stabilized carbanion onto the highly activated π-bond of diene **1** followed by $PhSO_2^-$ elimination to give a phenylsulfonyl-substituted allene as a key intermediate. Further reaction of the allene with benzenesulfinate anion eventually provides the five-membered ring. Several phenylsulfonyl alkenyl-substituted allenes were also prepared and found to undergo a highly chemo- and stereospecific intramolecular [2+2]-cycloaddition reaction to give bicyclo[4.2.0]oct-1-enes in high yield.[25]

The procedure described here provides a simple and general approach for the construction of bicyclo[3.3.0]oct-1-enes using 2,3-bis(phenylsulfonyl)-1,3-butadiene (**1**) as a key reagent. The [3+2]-cycloaddition of an allyl anion with an activated olefin to give cyclopentyl anion has been an active area of investigation for many years.[26] The synthetic potential of disulfone **1** for [3+2]-cyclization chemistry is demonstrated here by taking advantage of two properties of the molecule: (1) the ability of the phenylsulfonyl group to activate the double bond toward Michael addition with soft β-dicarbonyl anions and (2) the facility with which the phenylsulfonyl group can be eliminated. Thus, conjugate addition of the anion derived from dimethyl (E)-5-methoxycarbonyl-2-hexenedioate to diene **1** is followed by a series of consecutive tandem Michael additions to the activated π-bonds.[27] Eventual phenylsulfinate elimination produces the bicyclo[3.3.0]oct-1-ene skeleton. Since disulfone **1** can react with a variety of β-dicarbonyl anions, a wide assortment of bicyclo[3.3.0]octenes is easily available. This strategy can clearly be applied to more complex targets.

SO_2Ph

MeO_2C CO_2Me

Na CO_2Me CO_2Me

SO_2Ph CH_2 CH_2 SO_2Ph

RNH_2

SO_2Ph N R

R=CH_2Ph (98%)
R=$(CH_2)_2OCH_3$

$R_1(R_2)C{=}NOH$

O N R_1 H R_2 PhO_2S SO_2Ph

$R_1=R_2=CH_3$ (95%)
R_1=H; $R_2=CH_3$(95%)

MeO_2C Na H MeO_2C C H_2 R_1 R_2

$H_2C{=}C{=}$ SO_2Ph CH_2 H MeO_2C MeO_2C C H_2 R_1 R_2

$R_1=R_2=CH_3$ (80%)
$R_1=R_2$=H (52%)

Δ

SO_2Ph R_1 CO_2Me CO_2Me R_2 H

$R_1=R_2=CH_3$ (85%)
$R_1=R_2$=H (90%)

1. Department of Chemistry, Emory University, Atlanta, GA 30322.
2. Jeganathan, S.; Okamura, W. H. *Tetrahedron Lett.* **1982**, *23*, 4763.
3. Barrett, A. G. M.; Dhanak, D.; Graboski, G. G.; Taylor, S. J. *Org.* Synth., *Coll. Vol. VIII* **1993**, 550.
4. Colonge, J.; Cayrel, J. P. *Bull. Soc. Chim. Fr.* **1965**, 3596; Jackson, W. R.; Strauss, J. U. *Aust. J. Chem.* **1977**, *30*, 553.
5. Bunce, R. A.; Pierce, J. D. *Org. Prep. Proced. Int.* **1987**, *19*, 67.
6. Bäckvall, J.-E.; Juntunen, S. K. *J. Am. Chem. Soc.* **1987**, *109*, 6396; Bäckvall, J.-E.; Juntunen, S. K. *J. Org. Chem.* **1988**, *53*, 2398; Bäckvall, J.-E.; Plobeck, N. A. *J. Org. Chem.* **1990**, *55*, 4528; Plobeck, N. A.; Bäckvall, J.-E. *J. Org. Chem.* **1991**, *56*, 4508.
7. Masuyama, Y.; Yamazaki, H.; Toyoda, Y.; Kurusu, Y. *Synthesis* **1985**, 964.
8. Lee, S.-J.; Lee, J.-C.; Peng, M.-L.; Chou, T. S. *J. Chem. Soc., Chem. Commun.* **1989**, 1020.
9. Hardinger, S. A.; Fuchs, P. L. *J. Org. Chem.* **1987**, *52*, 2739.
10. Padwa, A.; Murphree, S. S. *Org. Prep. Proced. Int.* **1991**, *23*, 545.
11. Andell, O. S.; Bäckvall, J.-E. *Tetrahedron Lett.* **1985**, *26*, 4555.
12. Cuvigny, T.; Hervé du Penhoat, C.; Julia, M. *Tetrahedron* **1986**, *42*, 5329; Cuvigny, T.; Hervé du Penhoat, C.; Julia, M. *Tetrahedron Lett.* **1983**, *24*, 4315.
13. Inomata, K.; Kinoshita, H.; Takemoto, H.; Murata, Y. Kotake, H. *Bull. Chem. Soc. Jpn.* **1978**, *51*, 3341.
14. Chou, T. S.; Lee, S.-J.; Peng, M.-L.; Sun, D.-J.; Chou, S.-S. P. *J. Org. Chem.* **1988**, *53*, 3027.
15. Näf, F.; Decorzant, R.; Escher, S. D. *Tetrahedron Lett.* **1982**, *23*, 5043.
16. Akermark, B.; Nyström, J.-E.; Rein, T.; Bäckvall, J.-E.; Helquist, P.; Aslanian, R.*Tetrahedron Lett.* **1984**, *25*, 5719.
17. Marino, J. P.; Long, J. K. *J. Am. Chem. Soc.* **1988**, *110*, 7916.

18. Masuyama, Y.; Sato, H.; Kurusu, Y. *Tetrahedron Lett.* **1985**, *26*, 67; Padwa, A.; Harrison, B.; Murphree, S. S.; Yeske, P. E. *J. Org. Chem.* **1989**, *54*, 4232; Ni, Z.; Wang, X.; Rodriguez, A.; Padwa, A. *Tetrahedron Lett.* **1992**, *33,* 7303.

19. Padwa, A.; Murphree, S. S. *Rev. Heteroat. Chem.* **1992**, *6*, 241.

20. Padwa, A.; Norman, B. H. *Tetrahedron Lett.* **1988**, *29*, 2417; Norman, B. H.; Gareau, Y.; Padwa, A. *J. Org. Chem.* **1991**, *56*, 2154.

21. Padwa, A.; Harrison, B.; Norman, B. H. *Tetrahedron Lett.* **1989**, *30*, 3259; Padwa, A.; Gareau, Y.; Harrison, B.; Norman, B. H. *J. Org. Chem.* **1991**, *56*, 2713.

22. Padwa, A.; Gareau, Y.; Harrison, B.; Rodriguez, A. *J. Org. Chem.* **1992**, *57*, 3540.

23. Padwa, A.; Norman, B. H. *Tetrahedron Lett.* **1988**, *29*, 3041; Padwa, A.; Norman, B. H. *J. Org. Chem.* **1990**, *55*, 4801.

24. Padwa, A.; Filipkowski, M. A. *Tetrahedron Lett.* **1993**, *34*, 813.

25. Padwa, A.; Filipkowski, M. A.; Meske, M.; Watterson, S. H.; Ni, Z. *J. Am. Chem. Soc.* **1993**, *115*, 3776.

26. For some leading references, see Beak, P.; Burg, D. A. *J. Org. Chem.* **1989**, *54*, 1647.

27. Padwa, A.; Watterson, S. H.; Ni, Z. *J. Org. Chem.* **1994**, *59*, 3256.

Appendix

Chemical Abstracts Nomenclature (Collective Index Number); (Registry Number)

2,3-Bis(phenylsulfonyl)-1,3-butadiene: Benzene, 1,1'-[[1,2-bis(methylene)-1,2-ethanediyl]bis(sulfonyl)]bis- (11); (85540-20-1)

2,3-Bis(phenylsulfinyl)-1,3-butadiene: Benzene, 1,1'-[[1,2-bis(methylene)-1,2-ethanediyl]bis(sulfinyl)bis- (11); (85540-18-7)

2-Butyne-1,4-diol (8,9); (110-65-6)

Triethylamine (8); Ethanamine, N,N-diethyl- (9); (121-44-8)

Benzenesulfenyl chloride (8,9); (931-59-9)

Hydrogen peroxide (8,9); (7722-84-1)

Dimethyl (E)-5-methoxycarbonyl-2-hexenedioate (11); 3-Butene-1,1,4-tricarboxylic acid, trimethyl ester, (E)- (11); (93279-60-8)

Dimethyl malonate: Malonic acid, dimethyl ester (8); Propanedioic acid, dimethyl ester (9); (108-59-8)

Methyl 4-bromocrotonate: Crotonic acid, 4-bromo-, methyl ester (8); 2-Butenoic acid, 4-bromo-, methyl ester (9); (1117-71-1)

Sodium hydride (8,9); (7646-69-7)

PREPARATION OF BICYCLO[3.2.0]HEPT-3-EN-6-ONES: 1,4-DIMETHYLBICYCLO[3.2.0]HEPT-3-EN-6-ONE

(Bicyclo[3.2.0]hept-3-en-6-one, 1,4-dimethyl-, cis-(±)-)

A. $BrCH_2COOMe$, Zn, $B(OCH_3)_3$

B. 1. KOH in methanol, rt, 24 hr
2. H_3O^+

C. 1. Ac_2O, AcOK, rt, 2 hr
2. Δ, 3.5 hr
3. H_2O, 12 hr, rt

97 : 3

Submitted by Goffredo Rosini,[1] Giovanni Confalonieri,[2] Emanuela Marotta,[1] Franco Rama,[2] and Paolo Righi.[1]

Checked by Jian-Ping Chao and Robert K. Boeckman, Jr.

1. Procedure

A. Methyl 3,6-dimethyl-3-hydroxy-6-heptenoate (Note 1). An oven-dried, three-necked, 500-mL, round-bottomed flask, is fitted with an efficient mechanical stirrer, a Claisen adapter bearing a reflux condenser with a nitrogen inlet and rubber septum, and a thermometer. The flask is maintained under a static nitrogen pressure and charged with 5-methyl-5-hexen-2-one (24.20 g, 0.216 mol) (Note 2), trimethyl borate

(60 mL) (Note 3), tetrahydrofuran (60 mL) (THF) (Note 4), and freshly activated 20-mesh zinc granules (16.96 g, 0.259 g-atom) (Note 5). The reaction flask is immersed in an oil bath at 25°C. Stirring is initiated and methyl bromoacetate (39.65 g, 0.259 mol) (Note 6) is added in a single portion through the septum. After an induction time (Note 7), a white precipitate starts to form and an increase of the internal temperature, up to reflux, is observed (Note 8). The mixture is stirred for 3 hr, after which time starting material is completely consumed (Note 9). The reaction is quenched by the sequential addition of glycerol (60 mL) and saturated aqueous ammonium chloride (60 mL), then is transferred to a separatory funnel using 120 mL of diethyl ether. The aqueous layer is separated and extracted again with diethyl ether (3 x 60 mL). The combined ether extracts are washed with aqueous 30% ammonium hydroxide solution (3 x 30 mL) and then with saturated sodium chloride solution (2 x 30 mL). The organic extracts are dried over magnesium sulfate ($MgSO_4$) and concentrated at reduced pressure with a rotary evaporator to afford 37.32 - 38.92 g of crude β-hydroxy ester (93 - 97% yield) (Note 10).

B. 3,6-Dimethyl-3-hydroxy-6-heptenoic acid. In a 500-mL flask, the crude ester prepared in Part A (37.00 g, 0.199 mol) is dissolved in a 2 N solution of potassium hydroxide (KOH) in methanol (130 mL, 0.260 mol). The solution is stirred at 25°C and disappearance of the starting material is monitored by GLC (Note 9). After 5 hr saponification is complete and the methanol is evaporated at reduced pressure (Note 11). The residue is taken up with water (500 mL), extracted with diethyl ether (3 x 100 mL), and the organic phase is discarded. The aqueous phase is acidified (pH 2.5 on universal pH indicator paper) with ~ 6 N hydrochloric acid (about 60 mL) and extracted with diethyl ether (5 x 100 mL). These latter ethereal extracts are washed with water (2 x 30 mL) and then with saturated sodium chloride (2 x 30 mL). The organic phase is dried over sodium sulfate, and filtered. Evaporation at reduced pressure affords crude 3,6-dimethyl-3-hydroxy-6-heptenoic acid as a viscous yellow oil that can be used in

Part C without further purification. Evacuation at 0.5 mm at room temperature with internal magnetic stirring for 24 hr gives 28.07 - 29.40 g (82 - 86% yield) of a solvent-free material (Notes 12 and 13).

C. 1,4-Dimethylbicyclo[3.2.0]hept-3-en-6-one. A three-necked, 500-mL, round-bottomed flask, equipped with a condenser fitted with a calcium chloride tube, an efficient mechanical stirrer, and an immersion thermometer, is charged with crude 3,6-dimethyl-3-hydroxy-6-heptenoic acid (29.40 g, 0.171 mol), acetic anhydride (185 mL) (Note 14) and potassium acetate (40.28 g, 0.410 mol). The reaction mixture is stirred for 2 hr at room temperature. During this time, an exotherm is observed (up to 50°C) followed by a slow return to room temperature, and the suspension becomes thicker. The reaction is brought to reflux (in ca. 20 min) by means of a heating mantle and stirring is continued for another 3.5 hr. A two-necked, 2-L, round-bottomed flask fitted with a condenser and magnetic stirrer is charged with crushed ice (400 g) and water (100 g), and the hot mixture is added carefully with good stirring. Light petroleum ether (500 mL) is added and the mixture is stirred for 12 hr at room temperature. The reaction mixture is transferred into a 2-L separatory funnel and the aqueous layer is separated and extracted with light petroleum ether (4 x 100 mL). The combined organic layers are washed with saturated sodium bicarbonate solution (3 x 50 mL) and saturated sodium chloride solution (2 x 30 mL), dried over anhydrous sodium sulfate, filtered, and concentrated at ambient pressure to give the crude product as a dark oil. Final purification is achieved by distillation under reduced pressure (Note 15) affording 17.70 - 18.86 g (76 - 81% yield) of a 99% pure mixture (97:3) of the keto olefin isomers as a colorless oil, bp 84-85°C/26-28 mm (Note 16).

2. Notes

1. This procedure is essentially that of Rathke and Lindert.[3]

2. The submitters used freshly distilled ketone prepared according to the *Organic Syntheses* procedure.[4] 5-Methyl-5-hexen-2-one is also available from Aldrich Chemical Company, Inc., and can be used as purchased.

3. Trimethyl borate was distilled from calcium hydride and stored under nitrogen.

4. Tetrahydrofuran (reagent grade) was dried by distillation from sodium/benzophenone ketyl and freshly distilled before use.

5. Zinc granules (20-mesh, 99.8%, A.C.S. reagent grade, Pb $\leq$ 0.01%, Fe $\leq$ 0.01%) obtained from Aldrich Chemical Company, Inc., were activated according to the procedure of Newman and Evans.[5] About 25 g of 20-mesh zinc was covered with 5% hydrochloric acid (30 mL) and stirred vigorously for 3 min. The zinc was washed by decantation with distilled water (3 x 30 mL), with acetone (2 x 20 mL), with diethyl ether (2 x 20 mL), and finally dried and stored in a vacuum desiccator.

6. Methyl bromoacetate was distilled and stored under nitrogen. Ethyl bromoacetate can also be used: in this case a mixture of ethyl and methyl esters is obtained.

7. Induction times from 15 to 45 min have been observed. In one instance, the reaction did not start even after 1.5 hr, and required the mixture to be heated to 60°C for several minutes until a white precipitate began to form. The checkers observed induction periods of 25 - 35 min.

8. On some occasions, the exotherm induced a vigorous reflux that was difficult to control. In such cases, stirring had to be stopped occasionally to control the reflux rate. Although the use of 20-mesh zinc granules reduces the possibility of hard-to-control exotherms, a cold water bath was kept ready to provide cooling if needed. The

checkers did not observe an uncontrollable exotherm, but this type of reaction is well-known to be susceptible to such a problem. The checkers recommend that the mesh size not be reduced further without first testing the procedure on smaller scale. The checkers successfully performed this reaction at twice the reported scale without event and obtained the same yield (93%).

9. Progress of the reactions and analysis of the products was monitored by GLC using a Hewlett Packard 5890 Series II gas chromatograph equipped with an HP 3396 Series II Integrator and a Quadrex 007 Series bonded-phase fused silica capillary column (methyl 5% phenyl silicone; length 25 m; internal diameter 0.25 mm; film thickness 0.25 µm; carrier gas: nitrogen 75 kPa). Injector: 200°C; detector (FID): 250°C; Temperature program: 75°C (2 min); 15°/min, 250°C (3 min). Retention times (min): methyl bromoacetate, 2.24 (2.54); 5-methyl-5-hexen-2-one, 2.68 (2.94); 1,4-dimethylbicyclo[3.2.0]hept-3-en-6-one, 4.42 (4.77); 1-methyl-4-methylidenebicyclo-[3.2.0]heptan-6-one, 4.73 (5.08); methyl 3,6-dimethyl-3-hydroxy-6-heptenoate, 7.05 (7.44). The checkers employed a similar column 30 m in length with the same stationary phase and observed the retention times shown above in parentheses.

10. The crude ester so obtained is greater than 97% pure by GLC analysis and is used in the next step without further purification. The checkers determined that the remainder of the material (~3%) is unreacted methyl bromoacetate. Spectral data for the crude ester are as follows: IR (neat) cm^{-1}: 3516, 2951, 1735, 1649, 1439, 1216; ^{1}H NMR (200 MHz, $CDCl_3$) δ: 1.22 (s, 3 H), 1.70 (m, 2 H), 1.75 (s, 3 H), 2.11 (m, 2 H), 2.48 (d, 1 H, J = 17.2), 2.54 (d, 1 H, J = 17.2), 3.52 (s, 1 H, disappears after D_2O exchange), 3.73 (s, 3 H), 4.70 (s, 2 H); ^{13}C NMR (50.3 MHz, $CDCl_3$) δ: 23.07 (CH_3), 27.04 (CH_3), 32.43 (CH_2), 40.33 (CH_2), 45.16 (CH_2), 52.12 (CH_3), 71.26 (C), 110.2 (CH_2), 146.2 (C), 173.8 (C).

11. The submitters employed a 1 N solution of KOH in methanol and observed the saponification to be complete in 24 hr. The checkers observed that the saponification required ~48 hr for completion under these conditions. The checkers employed a 2 N solution of KOH in methanol and observed that saponification was complete in 5 hr.

12. The submitters reported 87-92% yields of the acid. In the hands of the checkers, the yield of this reaction appears to be scale dependent. At one half the reported scale, the yield was reduced by 10-12%. This variation appears to result from losses during the isolation procedure. The acid has a very disagreeable butyric acid-like odor. The checkers recommend that all manipulations of this material be performed in a good hood, and that all apparatus employed be kept in the hood until base washed.

13. Spectral data for the crude acid are as follows: IR (neat) cm^{-1}: 3380, 2971, 1707, 1646, 1222, 888. ^1H NMR (200 MHz, $CDCl_3$) δ: 1.30 (s, 3 H), 1.67 (m, 2 H), 1.72 (s, 3 H), 2.10 (m, 2 H), 2.55 (d, 1 H, J = 17.1), 2.60 (d, 1 H, J = 17.1), 4.70 (s, 2 H), 6.61 [s (broad), 2 H]; ^{13}C NMR (50.3 MHz, $CDCl_3$) δ: 23.11 (CH_3), 26.33 (CH_3), 32.40 (CH_2), 40.18 (CH_2), 45.13 (CH_2), 71.84 (C), 110.5 (CH_2), 146.0 (C), 177.9 (C).

14. Acetic anhydride was distilled from anhydrous sodium acetate and stored under nitrogen.

15. To control foaming, the submitters recommend the use of a distillation flask larger than usual. In this instance, they used a 250-mL flask fitted with a Claisen adapter.

16. The product is about 99% pure by GLC analysis (Note 9), containing about 1% of an unknown impurity having a retention time of 8.01 min. The ratio of the olefin isomers, 1,4-dimethylbicyclo[3.2.0]hept-3-en-6-one and 1-methyl-4-methylenebicyclo-[3.2.0]heptan-6-one, was determined to be 97:3. Spectral data for the major isomer are as follows: IR (neat) cm^{-1}: 1770; ^1H NMR (200 MHz, $CDCl_3$) δ: 1.35 (s, 3 H), 1.73

(s, 3 H), 2.55 (m, 2 H), 2.87 (dd, 1 H, J = 17.8, 4.7), 3.03 (dd, 1 H, J = 17.8, 2.9), 3.58 (m, 1 H), 5.45 (m, 1 H); ^{13}C NMR (50.3 MHz, $CDCl_3$) δ: 15.34 (CH_3), 24.03 (CH_3), 35.35 (C), 47.08 (CH_2), 58.81 (CH_2), 79.94 (CH), 126.8 (CH), 135.2 (C), 207.9 (C).

Waste Disposal Information

All toxic materials were disposed of in accordance with "Prudent Practices in the Laboratory"; National Academic Press; Washington, DC, 1996.

3. Discussion

Bicyclo[3.2.0]hept-3-en-6-ones are fused ring compounds that offer different functionalities in each ring, and have different ring size. They are suitable for chemo-, regio- and stereo-controlled manipulations and are useful for the assembly of more complex structures in a predictable fashion. The preparation reported here illustrates the simplicity of the procedure and the selectivity by which the thermodynamically more stable isomer can be prepared in high purity and good yield. Both yield and stereoselectivity are superior to those observed in the syntheses of 1,4-dimethylbicyclo[3.2.0]hept-3-en-6-one previously reported in the literature.[6,7] This process involves the generation of an α,β-unsaturated ketene intermediate that undergoes intramolecular [2+2] cyclization to give a bicyclo[3.2.0]hept-3-en-6-one. An equilibrium among the possible isomers of the α,β-unsaturated ketene intermediate could account for the good yield as well as the high selectivity in generating the thermodynamically more stable endo-ene isomer.[8]

This foregoing procedure is rather general as demonstrated by the preparation of a number of bicyclo[3.2.0]hept-3-en-6-ones by the bicyclization of a variety of secondary and tertiary 3-hydroxy-6-alkenoic acids (Table). The use of bicyclo[3.2.0]hept-3-en-6-ones as starting materials has been reported in the synthesis of racemic grandisol,[9] lineatin,[1,2,4] filifolone,[10] and several intermediates[5,11] in Curran's synthesis of linear condensed sesquiterpenes such as hirsutene,[12,13] Δ^2-capnellene,[14] hypnophilin[15] and coriolin.[15] The successful use of these intermediates in the previously mentioned applications, and their unusual reactivity,[16] suggest the broad usefulness of the bicyclo[3.2.0]hept-3-en-6-ones made readily available by the present procedure. Some bicyclo[3.2.0]hept-3-en-6-ones have been obtained enantiomerically pure[17] by the oxidation of the enantiomers of bicyclo[3.2.0]hept-3-en-6-endo-ols, resolved using (-)-(1S,4R)-camphanic acid chloride.[18]

1. Dipartimento di Chimica Organica "A. Mangini" dell'Università di Bologna, Viale Risorgimento n.4, I-40136 Bologna, Italy.
2. ISAGRO Ricerca s.r.l. – Via Fauser n.4, I-28100 Novara, Italy.
3. Rathke, M. W.; Lindert, A. *J. Org. Chem.* **1970**, *35*, 3966-3967.
4. Boatman, S.; Hauser, C. R. *Org. Synth., Coll. Vol. V* **1973**, 767.
5. Newman, M. S.; Evans, F. J., Jr. *J. Am. Chem. Soc.* **1955**, *77*, 946-947.
6. Baeckström, P.; Li, L.; Polec, I.; Unelius, C. R.; Wimalasiri, W. R. *J. Org. Chem.* **1991**, *56*, 3358-3362.
7. Snider, B. B.; Ron, E.; Burbaum, B. W. *J. Org. Chem.* **1987**, *52*, 5413-5419.
8. Marotta, E.; Medici, M.; Righi, P.; Rosini, G. *J. Org. Chem.* **1994**, *59*, 7529-7531.
9. Confalonieri, G.; Marotta, E.; Rama, F.; Righi, P.; Rosini, G.; Serra, R.; Venturelli, F. *Tetrahedron* **1994**, *50*, 3235-3250.
10. Marotta, E.; Pagani, I.; Righi, P.; Rosini, G. *Tetrahedron* **1994**, *50*, 7645-7656.
11. Marotta, E.; Righi, P.; Rosini, G. *Tetrahedron Lett.* **1994**, *35*, 2949-2950.
12. Curran, D. P.; Rakiewicsz, D. M. *Tetrahedron* **1985**, *41*, 3943-3958.
13. Curran, D. P.; Rakiewicz, D. *J. Am. Chem. Soc.* **1985**, *107*, 1448-1449
14. Curran, D. P.; Chen, M.-H. *Tetrahedron Lett.* **1985**, *26*, 4991-4994.

15. Fevig, T. L.; Elliott, R. L.; Curran, D. P. *J. Am. Chem. Soc.* **1988**, *110*, 5064-5067.

16. Marotta, E.; Piombi, B.; Righi, P.; Rosini, G. *J. Org. Chem.* **1994**, *59*, 7526-7528.

17. Marotta, E.; Pagani, I.; Righi, P.; Rosini, G.; Bertolasi, V.; Medici, A. *Tetrahedron: Asymmetry* **1995**, *6*, 2319-2328.

18. Gerlach, H.; Kappes, D.; Boeckman, R. K., Jr.; Maw, B. N. *Org. Synth.* **1993**, *71*, 48-55.

TABLE
PREPARATION OF BICYCLO[3.2.0]HEPT-3-EN-6-ONES

Starting Material	Product	Yield (%)
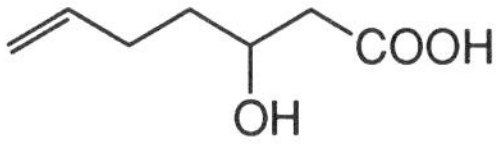 COOH, OH	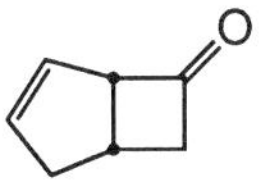 O	62
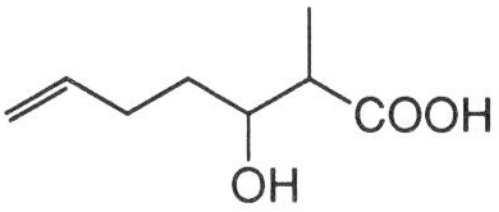 COOH, OH	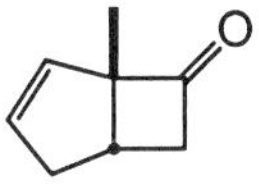 O	63
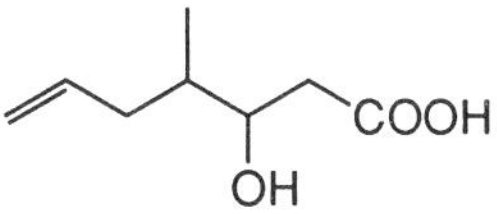 COOH, OH	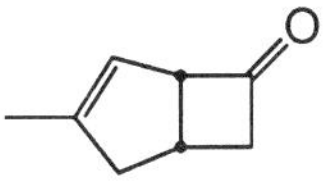 O	63
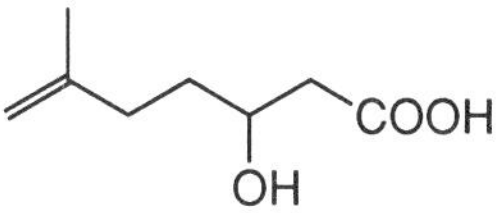 COOH, OH	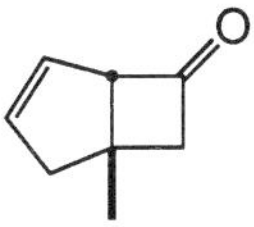 O	78
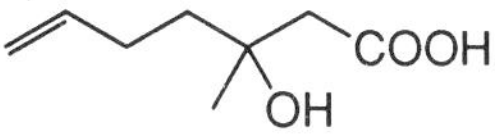 COOH, OH	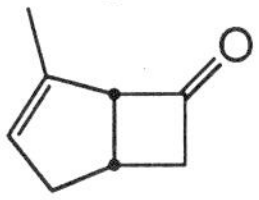 O	79
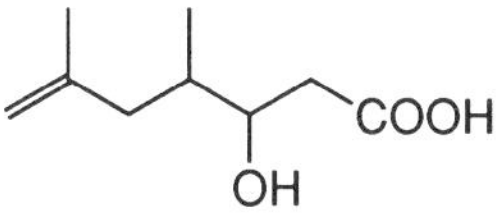 COOH, OH	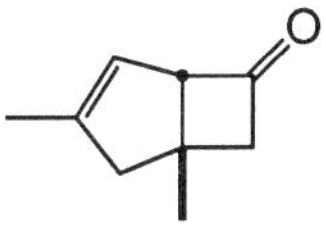 O	55
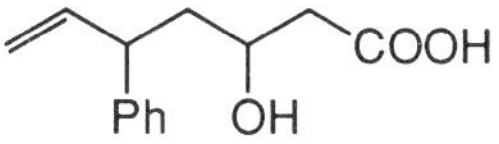 COOH, Ph, OH	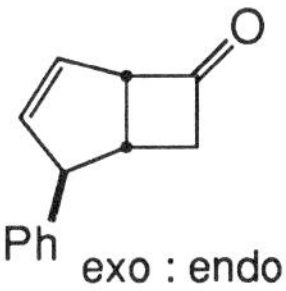 O, Ph, exo : endo = 3:1	60
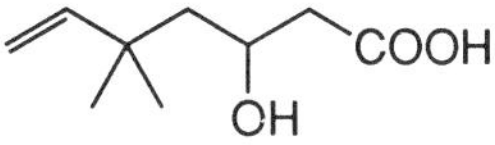 COOH, OH	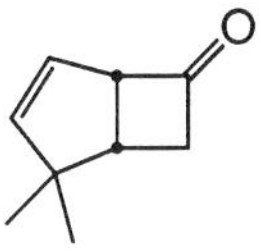 O	48
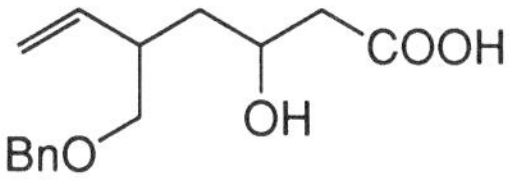 COOH, OH, BnO	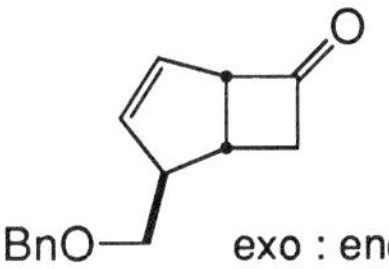 O, BnO, exo : endo = 3:1	70

Appendix

Chemical Abstracts Nomenclature (Collective Index Number); (Registry Number)

1,4-Dimethylbicyclo[3.2.0]hept-3-en-6-one: Bicyclo[3.2.0]hept-3-en-6-one, 1,4-dimethyl-, cis- (±)- (12); (133700-21-7)

5-Methyl-5-hexen-2-one: 5-Hexen-2-one, 5-methyl- (8,9); (3240-09-3)

Trimethyl borate: Boric acid, trimethyl ester (8,9); (121-43-7)

Zinc (8,9); (7440-66-6)

Methyl bromoacetate: Acetic acid, bromo-, methyl ester (8,9); (96-32-2)

Acetic anhydride (8); Acetic acid anhydride (9); (108-24-7)

PREPARATION OF POLYQUINANES BY DOUBLE ADDITION OF VINYL ANIONS TO SQUARATE ESTERS: 4,5,6,6a-TETRAHYDRO-3a-HYDROXY-2,3-DIISOPROPOXY-4,6a-DIMETHYL-1(3aH)-PENTALENONE

i-PrO, i-PrO, O, O; $H_2C{=}C(CH_3)Li$; THF, -78°C → 20°C; NH_4Cl, H_2O; (33 : 1)

Submitted by Tina Morwick and Leo A. Paquette.[1]
Checked by Lisa Frey and Ichiro Shinkai.

1. Procedure

A 1000-mL, two-necked, round-bottomed flask (Note 1) is equipped with a rubber septum, magnetic stirring bar, and gas inlet connected to a vacuum/argon line via a Firestone valve. The flask is flame-dried under vacuum, then filled with argon. Ten cycles of evacuation and argon fill are carried out. The flask, in which a positive flow of argon is maintained throughout the entire procedure, is charged with 6.66 mL (75 mmol) of 2-bromopropene (Note 2) dissolved in dry tetrahydrofuran (250 mL) (Note 3). The solution is cooled to -78°C by a dry ice/acetone bath, at which point 88 mL of 1.7 M tert-butyllithium in pentane (150 mmol) is introduced dropwise in three roughly equal portions from a 50-mL syringe (Note 4) during 45 min. The colorless reaction mixture is allowed to stir at -78°C for approximately 30 min (Note 5).

A 250 mL, two-necked, round-bottomed flask, equipped with a rubber septum and gas inlet, is flame-dried under vacuum, filled with argon, and charged with a solution of diisopropyl squarate (5.94 g, 30 mmol) (Note 6) in dry tetrahydrofuran (150 mL). The solution is cooled to -78°C and cannulated rapidly into the reaction flask (Note 7). After completion of the addition, the mixture is stirred for 2 hr at 0°C, then for an additional 15 hr at room temperature. Following recooling to 0°C, 180 mL of saturated ammonium chloride solution (Note 8) is added via syringe, the ice bath is removed, and stirring is continued for an additional 7 hr.

The reaction mixture is poured into a 2000-mL separatory funnel containing water (300 mL) and ether (300 mL). After thorough mixing, the aqueous layer is separated and extracted with ether (2 x 200 mL). The combined ethereal solutions are washed with water (300 mL) and brine (300 mL), dried over magnesium sulfate, and evaporated under reduced pressure to leave 8.5 g of a pale yellow oil. This material is subjected to flash chromatography on silica gel using 20% ethyl acetate in hexanes as the mobile phase (Note 9). The minor diquinane (230 mg, Note 10) is eluted first. A mixed fraction of the two isomers follows (570 mg, Note 11) in advance of the pure title compound (6.36 g, Notes 12, 13).

2. Notes

1. All apparatus was washed with base and oven-dried overnight.

2. 2-Bromopropene was purchased from the Aldrich Chemical Company, Inc., and used as received.

3. Tetrahydrofuran was freshly distilled from sodium benzophenone ketyl.

4. tert-Butyllithium was purchased from the Aldrich Chemical Company, Inc., and titrated before use with diphenylacetic acid according to an established procedure.[2] The syringe and needle were oven-dried overnight prior to use. The plunger is rotated slowly and continuously throughout the addition to avoid loss of mobility brought on by the adventitious formation of lithium hydroxide. This alkyllithium

is extremely pyrophoric and must be treated cautiously to avoid exposure to the atmosphere.

5. tert-Butyllithium forms a yellow complex with tetrahydrofuran at low temperatures. When all of this reagent is consumed, the yellow color disappears and a colorless solution of the vinyl anion is obtained. If the color continues the remaining tert-butyllithium can be consumed by the slow addition of a few drops of 2-bromopropene until a colorless solution results.

6. Diisopropyl squarate, available from the Aldrich Chemical Company, Inc., can be readily prepared from squaric acid according to the procedure reported by Liebeskind.[3]

7. An oven-dried, 18-gauge cannula wrapped with glass wool and aluminum foil was used.

8. This solution was deoxygenated by bubbling argon through it for a period of 15 min immediately before use.

9. The dimensions of the column were 6.5 cm by 36 cm. The relevant R_f values of the isomers are 0.34 and 0.26.

10. This product, which can be recrystallized from hexane (colorless crystals, mp 110-111°C), exhibits the following spectral properties: IR ($CHCl_3$) cm^{-1}: 3560, 2980, 1650, 1600; ^1H NMR (300 MHz, C_6D_6) δ: 0.92-1.12 (m, 1 H), 1.02 (d, 3 H, J = 6.0), 1.03 (d, 3 H, J = 6.0), 1.05 (d, 3 H, J = 6.0), 1.10 (d, 3 H, J = 7.0), 1.15 (d, 3 H, J = 6.0), 1.26 (s, 3 H), 1.25-1.35 (m, 1 H), 1.42-1.50 (m, 1 H), 1.92-2.07 (m, 2 H), 3.08 (s, 1 H), 5.20 (hep, 1 H, J = 6.0), 5.31 (hep, 1 H, J = 6.0); ^{13}C NMR (75 MHz, C_6D_6) δ: 15.3, 19.9, 22.3 (2C), 22.6, 22.8, 30.8, 34.3, 46.6, 52.8, 71.1, 73.4, 83.9, 131.4, 171.9, 199.9.

11. Assay of this fraction by VPC (SE-30, 70-250°C/min) showed its composition to consist of 8% of the less polar isomer and 92% of the more polar product. The overall yields are consequently 2.5% and 81% respectively.

12. The title compound was obtained as a colorless oil that slowly crystallized on standing, mp 52-53°C. Its spectral characteristics are as follows: IR ($CHCl_3$) cm^{-1}: 3589, 2979, 1697, 1616; ^1H NMR (300 MHz, C_6D_6) δ: 0.84-0.98 (m, 1 H), 1.01 (d, 3 H, J = 7.0), 1.07 (d, 3 H, J = 6.0), 1.09 (d, 3 H, J = 6.0), 1.10 (d, 3 H, J = 6.0), 1.11 (d, 3 H, J = 6.0), 1.28 (s, 3 H), 1.23-1.43 (m, 2 H), 1.86-1.94 (m, 1 H), 2.18-2.24 (m, 2 H), 5.24-5.38 (m, 2 H); ^{13}C NMR (75 MHz, C_6D_6) δ: 15.5, 19.6, 22.5, 22.6 (2 C), 22.9, 31.3, 35.0, 46.9, 56.7, 71.2, 73.6, 83.1, 132.5, 165.4, 202.7.

13. The checkers isolated an additional side product **A**, not observed by the submitters, in 2% yield; its structure is based upon NMR $^1H/^{13}C$ correlations and NOE data:

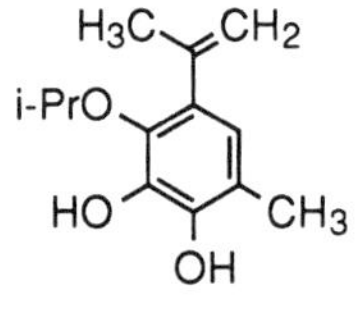

H_3C CH_2
i-PrO
HO CH_3
OH

A

^1H NMR (300 MHz, $CDCl_3$) δ: 1.2 [(d, 6 H) ($OCH(\underline{CH_3})_2$)], 2.08 [(m, 3 H) ($CH_2=C-\underline{CH_3}$)], 2.4 [(s, 3 H) (ar-CH_3)]; 4.3 [(m, 1 H) ($O\underline{CH}(CH_3)_2$)], 5.08 (m, 2 H, =CH_2), 5.2 (s, 1 H, 3-OH), 5.6 (s, 1 H, 2-OH), 6.5 (s, 1 H, ar-H).

Waste Disposal Information

All toxic materials were disposed of in accordance with "Prudent Practices in the Laboratory"; National Academic Press; Washington, DC, 1996.

3. Discussion

Twofold addition of the same or different vinyl anions to squarate esters leads to polyquinane products.[4] The principal pathway involves trans-1,2-addition of the organometallic reagent to generate a cyclobutene dialkoxide such as **1**. Ring strain and electrostatic factors promote the rapid conrotatory opening of **1** in that sense

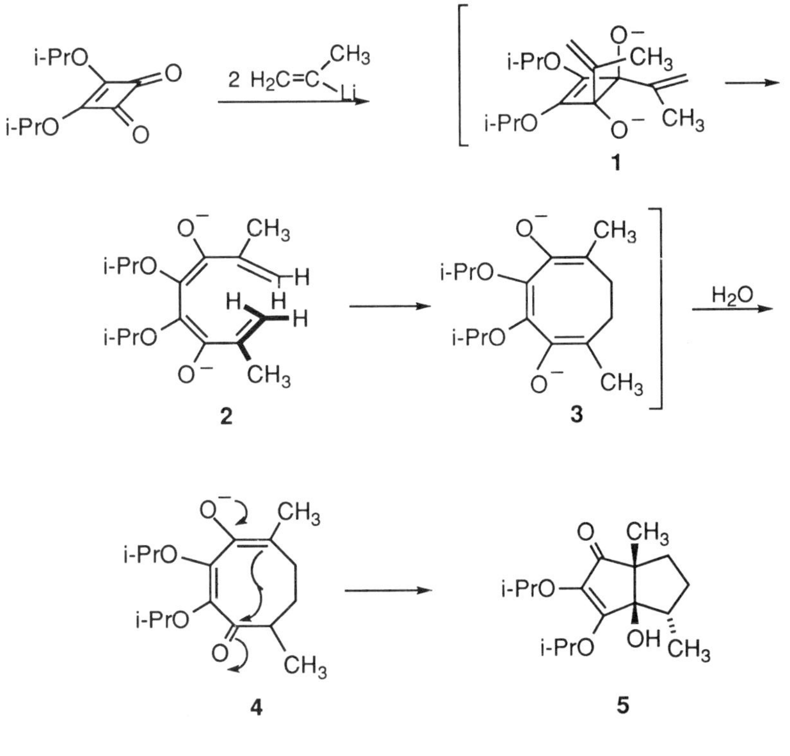

leading to outward splaying of the oxido functional groups.[5] The resultant doubly-charged 1,3,5,7-octatetraenes shown by **2** undergo symmetry-controlled 8π electrocyclization from a coiled conformation, thereby giving rise to cyclooctenyl dienolates **3**. In symmetrical examples such as that illustrated, protonation at either available reactive center delivers **4,** and sets the stage for intramolecular aldolization via transannular cyclization. In unsymmetrical cases, both aldols are sometimes produced, with steric discrimination occurring.[6] A strong interdependence of the efficiency in this cascade process and vinyl anion substitution has been noted.[6] 2-Propenyllithium is particularly conducive to product formation in good yields, presumably because the presence of methyl groups at C-2 and C-7 in **2** favors the adoption of the conformer shown over others less conducive to the conrotatory requirements for the conversion to **3**.

The cascade sequence associated with this remarkable series of chemical events is tolerant of structurally varied vinyl anions provided that steric bulk is not excessive. The end result is the potential for establishing many stereogenic centers from a triad of achiral reactants in a single laboratory operation. The very substantial increase in complexity attainable from these tandem stereoregulated chemical events is shown in the Table.

Two minor processes sometimes operate competitively with that illustrated in the scheme. One of these involves 1,4-addition of the second vinyl anion to give a reactive intermediate that differs structurally from **1**, but is capable of setting into motion a closely related sequence of chemical events leading to an isomeric diquinane.[4] This is the route followed to produce the minor product characterized here. The other option consists of cis-1,2-addition, an event that is followed by a dianionic oxy-Cope rearrangement via a boat-like transition state.[4] When sufficient substitution is present to allow the installation of multiple stereogenic centers, the adoption of this pathway is easily distinguished from the electrocyclic alternative since a cis relationship between relevant substituents is in place, instead of the trans arrangement required by the electrocyclization cascade.

1. Evans Chemical Laboratories, The Ohio State University, Columbus, Ohio 43210.
2. Kofron, W. G.; Baclawski, L. M. *J. Org. Chem.* **1976**, *41*, 1879.
3. Liebeskind, L. S.; Fengl, R. W.; Wirtz, K. R.; Shawe, T. T. *J. Org. Chem.* **1988**, *53*, 2482.
4. Negri, J. T.; Morwick, T.; Doyon, J.; Wilson, P. D.; Hickey, E. R.; Paquette, L. A. *J. Am. Chem. Soc.* **1993**, *115*, 12189.
5. Rondan, N. G.; Houk, K. N. *J. Am. Chem. Soc.* **1985**, *107*, 2099.
6. Paquette, L. A.; Morwick, T. submitted for publication.

Appendix

Chemical Abstracts Nomenclature (Collective Index Number); (Registry Number)

2-Bromopropene: Propene, 2-bromo- (8); 1-Propene, bromo- (9); (557-93-7)

tert-Butyllithium: Lithium, tert-butyl- (8); Lithium, (1,1-dimethylethyl)- (9); (594-19-4)

Diisopropyl squarate: 3-Cyclobutene-1,2-dione, 3,4-bis(1-methylethoxy)- (10); (61999-62-5)

TABLE
POLYQUINANES PRODUCED FROM DIISOPROPYL SQUARATE VIA AN ELECTROCYCLIC CASCADE

First Anion	Second Anion	Products (isolated yield)
	CH_2=CHLi	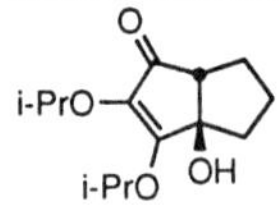 (45%)
	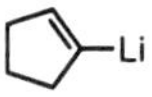 Li	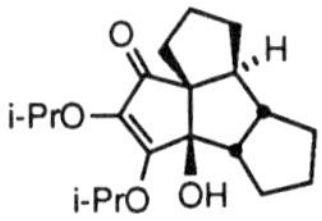 (40%)
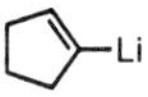 Li	 CH_3 Li	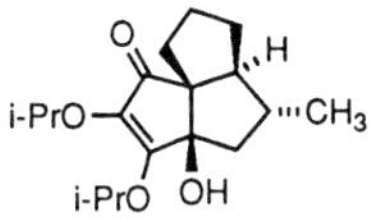 (62%)
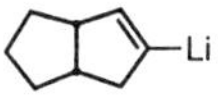 Li	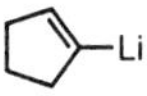 Li	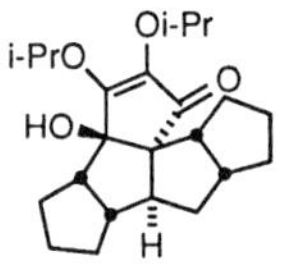 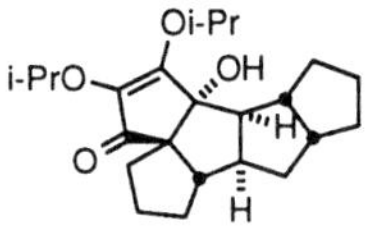 (38%) + (27%)
	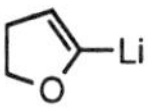 Li	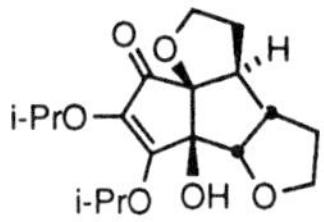 (38%)
CH_2=CHLi	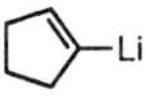 Li	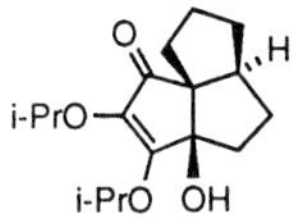 (67%)
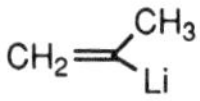 CH_2=C(CH_3)Li	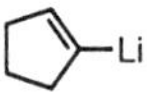 Li	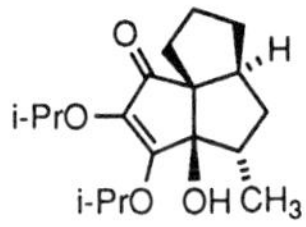 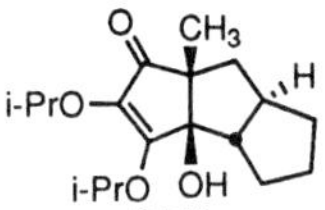 (27%) + (61%)

TABLE contd.

POLYQUINANES PRODUCED FROM DIISOPROPYL SQUARATE VIA AN ELECTROCYCLIC CASCADE

First Anion	Second Anion	Products (isolated yield)
	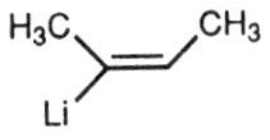 H_3C, CH_3, Li	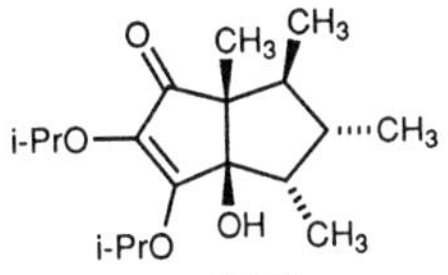 O, CH_3, CH_3, i-PrO, CH_3, i-PrO, OH, CH_3 (73%)
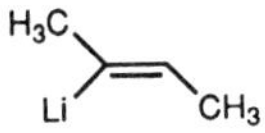 H_3C, Li, CH_3	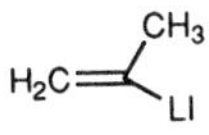 CH_3, H_2C=, LI	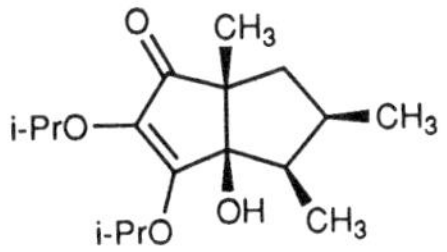 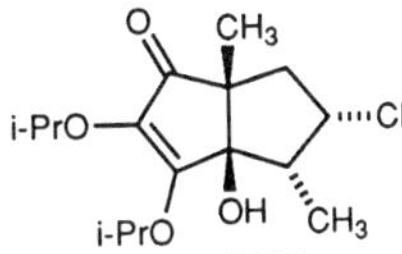 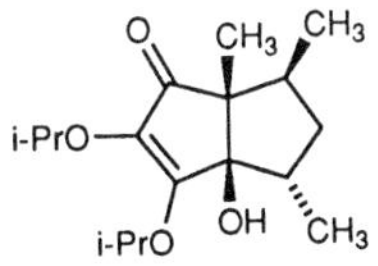 O, CH_3, i-PrO, CH_3, i-PrO, OH, CH_3 (30%) + O, CH_3, i-PrO, CH_3, i-PrO, OH, CH_3 (28%) + O, CH_3, CH_3, i-PrO, i-PrO, OH, CH_3 (22%)

REGIO- AND STEREOSELECTIVE CARBOXYLATION OF ALLYLIC BARIUM REAGENTS: (E)-4,8-DIMETHYL-3,7-NONADIENOIC ACID

(3,7-Nonadienoic acid, 4,8-dimethyl-, (E)-)

A. Li + biphenyl $\xrightarrow[\text{rt, 2 hr}]{\text{THF}}$ Li^+[biphenyl]$^{\cdot -}$

BaI_2 + 2 Li^+[biphenyl]$^{\cdot -}$ $\xrightarrow[\text{rt, 30 min}]{\text{THF}}$ Ba*

B. (E)-geranyl chloride $\xrightarrow[\text{THF, -78°C}]{\text{Ba*}}$ $\xrightarrow[\text{-78°C}]{CO_2}$ (E)-4,8-dimethyl-3,7-nonadienoic acid (CO_2H)

Submitted by Akira Yanagisawa, Katsutaka Yasue, and Hisashi Yamamoto.[1]
Checked by D. Scott Coffey and William R. Roush.

1. Procedure

A. Active barium. An oven-dried, three-necked, round-bottomed, 300-mL flask, equipped with a Teflon-coated magnetic stirring bar, is flushed with argon (Note 1). Freshly cut lithium (210 mg, 30.3 mmol) and biphenyl (4.7 g, 30.5 mmol) are placed in the apparatus and covered with dry tetrahydrofuran (THF) (80 mL) (Note 2), and the mixture is stirred for 2 hr at room temperature (Note 3). In a separate, oven-dried, three-necked, round-bottomed, 500-mL flask, equipped with a Teflon-coated magnetic stirring bar and a 100-mL addition funnel, is placed anhydrous barium iodide (BaI_2) (6.0 g, 15.3 mmol) (Notes 4, 5) under an argon atmosphere; this is covered with dry THF (80 mL), and stirred for 5 min at room temperature. To the resulting yellowish solution of BaI_2 in THF is added at room temperature a solution of the lithium biphenylide through a stainless steel cannula under an argon stream. The reaction mixture is stirred for 1 hr at room temperature, and the resulting dark brown suspension of active barium thus prepared is ready for further use.

B. (E)-4,8-Dimethyl-3,7-nonadienoic acid. To the suspension of active barium in THF is added dropwise over 20 min a solution of geranyl chloride (1.19 g, 6.89 mmol) (Note 6) in THF (40 mL) from the 100-mL dropping funnel at -78°C (Note 7), and the mixture is stirred at this temperature for 30 min (Note 8). An excess of dry ice (ca. 10 g) (Note 9) is added at -78°C and stirring continued for 10 min. The reaction mixture is quenched with 1 N hydrochloric acid (HCl) (40 mL) at -78°C, warmed to room temperature, and poured into a mixture of water (H_2O) (200 mL) and ethyl acetate (EtOAc) (200 mL). After the organic layer is shaken vigorously, it is separated and washed with dilute sodium thiosulfate solution (200 mL). The two aqueous layers are combined, acidified (pH <3) with concd HCl, and extracted twice with EtOAc (2 x 100 mL). The combined organic extracts are washed with H_2O (200 mL), dried over anhydrous magnesium sulfate, and concentrated under reduced pressure. The residue is then dissolved in methanol (MeOH) (10 mL) by gentle heating and placed in a freezer (0°C) for over 1 hr to crystallize the biphenyl. The white solid is filtered off (Note 10), washed with cold MeOH (0°C, 40 mL), and the filtrate is concentrated under reduced pressure. The residual oil is purified by flash-column chromatography on silica gel (70 g, Note 11) using 2% EtOAc/hexane (500 mL), 20% EtOAc/hexane (600 mL), and then 30% EtOAc/hexane (1 L) as eluant, to afford the crude β,γ-unsaturated carboxylic acid (1.43 g). An additional vacuum distillation (106°C/0.4 mm, Note 12) provides pure (E)-4,8-dimethyl-3,7-nonadienoic acid (0.91–0.96 g, 72–76% yield, Notes 13, 14, 15) as a colorless oil. The isomeric purity is determined to be ≥97 : 3 [(E/Z) of 98 : 2 and α:γ of >99 : 1] by GC analysis after conversion to the corresponding methyl ester (Note 16).

2. Notes

1. The submitters used standard grade argon gas (oxygen <10 ppm) which was further purified by passing through a GAS CLEAN column (GC-RX, NIKKA SEIKO Co.) to remove traces of oxygen. The checkers used UHP/Zero grade argon that was passed through a tube of Dri-Rite before use. However, no special precautions were made to remove oxygen.

2. Lithium (wire, 99.9%) was purchased from Aldrich Chemical Company, Inc., (submitters) or EM Science (checkers). The wire was cut into 20–30-mg pieces that were rinsed with dry hexane before use. Biphenyl (guaranteed reagent) was used as purchased from Nacalai Tesque (submitters) or EM Science (checkers). The submitters used dry THF as purchased from Aldrich Chemical Company, Inc. (anhydrous, 99.9%). The checkers used THF (99.5%, EM Science) that was distilled from benzophenone ketyl.

3. The submitters reported that lithium was completely consumed within 2 hr at room temperature (20~25°C). However, the checkers found that a small amount of Li (ca. 20-40 mg) remained at the end of the reaction.

4. The submitters report that $BaI_2 \cdot 2H_2O$ purchased from Nacalai Tesque (extra pure reagent), Aldrich Chemical Company, Inc., Fluka Chemical Corp., Kishida Chemical, or Wako Pure Chemical can be used with equal efficiency. The checkers used $BaI_2 \cdot 2H_2O$ purchased from Aldrich Chemical Company, Inc.

5. The submitters prepared anhydrous BaI_2 by drying $BaI_2 \cdot 2H_2O$ at 150°C for 2 hr under reduced pressure (<10 mm). However, the checkers were unsuccessful in attempts to generate active barium from BaI_2 that was dried according to these specifications. The checkers obtained good results when finely ground $BaI_2 \cdot 2H_2O$ was dried at 150°C (1-2 mm) for 12-24 hr while being stirred (see Note 13). The color

of the $BaI_2 \cdot 2H_2O$ changes from light yellow to white during the first 1-2 hr, with no subsequent color changes observed.

6. Geranyl chloride (95%, Aldrich Chemical Company, Inc.) was purified by distillation immediately before use.

7. A 5-L Dewar flask (I.D. 200 mm) was employed for the -78°C cooling bath (dry ice/methanol).

8. A dark red suspension or wine-red solution is obtained.

9. Dry ice was cut into appropriate size pieces and added from the middle inlet of the three-necked, round-bottomed flask. The checkers obtained good results (65-74%) by bubbling carbon dioxide (CO_2) gas (Air Products and Chemicals, Inc.) vigorously into the reaction mixture through an 18-gauge needle for 20 min. The CO_2 was dried by passing through sulfuric acid and then through a drying tube packed with Dri-Rite.

10. After recrystallization (methanol) ca. 2.8 g of biphenyl was recovered (60%).

11. Silica gel 60 (E. Merck 9385, 230-400 mesh) was used.

12. A bulb to bulb distillation apparatus was used. The checkers found the bp to be 104-110°C/0.4 mm.

13. The checkers obtained 0.93 g of product (74% yield) with isomeric purity of 97 : 3 from an experiment run with 210 mg of lithium (Li), BaI_2 that was dried for 24 hr (see Note 5), and CO_2 gas. The yield was 62-65% (isomeric purity 97 : 3) from experiments run with 230 mg of Li and BaI_2 that was dried for only 12-14 hr; one experiment was quenched with dry ice and the other with CO_2 gas.

14. The physical properties of (E)-4,8-dimethyl-3,7-nonadienoic acid are as follows: TLC R_f = 0.50 (1:1 ethyl acetate/hexane); bp 106°C/0.4 mm; IR (neat) cm^{-1}: 2969, 2919, 1713, 1416, 1300, 1225, 1156, 1109, 941, 831; ^1H NMR (200 MHz, $CDCl_3$) δ: 1.60 (s, 3 H, CH_3), 1.65 (s, 3 H, CH_3), 1.68 (s, 3 H, CH_3), 2.07 (m, 4 H, 2 CH_2), 3.10 (d, 2 H, J = 7.0, CH_2), 5.05-5.13 (m, 1 H, vinyl), 5.31 (t, 1 H, J = 7.0, vinyl),

10.2-11.4 (br, 1 H, CO_2H); ^{13}C NMR (125 MHz, $CDCl_3$) δ: 16.4, 17.7, 25.7, 26.4, 33.5, 39.5, 114.9, 123.9, 131.7, 139.8, 178.8; MS (EI) m/e (rel intensity): 170 (5.86, M -12), 149 (7.45), 139 (17.87), 122 (9.11), 69 (62.75); MS (FAB) m/e 183 ($M^{+}+1$). Anal. Calcd for $C_{11}H_{18}O_2$: C, 72.49; H, 9.95. Found: C, 72.51; H, 10.20.

15. The checkers performed this procedure on five times the reported scale and obtained 2.6 g (41%) of crude (impure) (E)-4,8-dimethyl-3,7-nonadienoic acid following chromatographic purification.

16. GC analysis was performed on a Shimadzu GC-8A instrument equipped with a flame ionization detector and a capillary column of PEG-HT (0.25 x 25000 mm) using nitrogen as carrier gas.

Waste Disposal Information

All toxic materials were disposed of in accordance with "Prudent Practices in the Laboratory"; National Academic Press; Washington, DC, 1996.

3. Discussion

β,γ-Unsaturated carboxylic acids and their derivatives are valuable synthetic intermediates for various natural products. Two typical multi-step processes for the synthesis of β,γ-unsaturated acids, Knoevenagel reaction/isomerization with base[2] and allylic cyanide/hydrolysis,[3] are those most commonly used. Other new methods have been developed;[4-7] however, there is a problem with E/Z stereoselectivity. One straightforward way to obtain β,γ-unsaturated acids is by the carboxylation of an allyl metal intermediate. In the substituted allylic series, the reaction usually occurs at the more sterically hindered terminus.[8] A stereospecific route for the synthesis of homogeranic acid and homoneric acid by carboxylation of the lithiated allylic sulfone has also been reported.[9] In contrast, we have found that allylic barium reagents are prepared directly by reaction of in situ generated barium metal with various allylic chlorides, and react with carbonyl compounds or allylic halides in a highly α-selective manner without loss of the double bond geometry.[10] As illustrated in the present procedure, treatment of the allylic barium reagent with excess carbon dioxide results in α-carboxylation, whereas γ-carboxylation occurred with allylic magnesium reagent.[8] Results of carboxylation of allylic barium reagents are summarized in the Table.[11] The characteristic features of the reaction are as follows: (1) Allylic barium reagents generated from a variety of γ-mono- and γ-disubstituted allyl chlorides showed high α-selectivities without exception. (2) The double-bond geometry of the allyl chloride precursor was completely retained in each case. (3) The alkyl substituent at the β-position of an allylic barium reagent had no effect on the regioselectivity.

In conclusion, this is one of the most straightforward and practical methods available for the regioselective and stereospecific synthesis of β,γ-unsaturated carboxylic acids.

1. School of Engineering, Nagoya University, Chikusa, Nagoya 464-01, Japan.
2. (a) Caspi, E.; Varma, K. R. *J. Org. Chem.* **1968**, *33*, 2181; (b) Maercker, A.; Streit, W. *Chem. Ber.* **1976**, *109*, 2064; (c) Mikolajczak, K. L.; Smith, C. R. Jr. *J. Org. Chem.* **1978**, *43*, 4762; (d) Grob, C. A.; Waldner, A. *Helv. Chim. Acta* **1979**, *62*, 1854; (e) Ragoussis, N. *Tetrahedron Lett.* **1987**, *28*, 93.
3. (a) Katagiri, T.; Agata, A.; Takabe, K.; Tanaka, J. *Bull. Chem. Soc. Jpn.* **1976**, *49*, 3715; (b) Hirai, H.; Matsui, M. *Agric. Biol. Chem.* **1976**, *40*, 169; (c) Garbers, C. F.; Beukes, M. S.; Ehlers, C.; McKenzie, M. J. *Tetrahedron Lett.* **1978**, 77; (d) Hoye, T. R.; Kurth, M. J. *J. Org. Chem.* **1978**, *43*, 3693; (e) Gosselin, P.; Rouessac, F. *C. R. Séances Acad. Sci., Ser. 2* **1982**, *295*, 469; (f) Mori, K.; Funaki, Y. *Tetrahedron* **1985**, *41*, 2369.

4. Ene reaction of diethyl oxomalonate: Salomon, M. F.; Pardo, S. N.; Salomon, R. G. *J. Am. Chem. Soc.* **1980**, *102*, 2473.
5. β-Vinyl-β-propiolactone/organocopper reagent: Kawashima, M.; Sato, T.; Fujisawa, T. *Bull. Chem. Soc. Jpn.* **1988**, *61*, 3255.
6. Transition-metal-catalyzed carbonylation: (a) Alper, H.; Amer, I. *J. Mol. Catal.* **1989**, *54*, L33; (b) Garlaschelli, L.; Marchionna, M.; Iapalucci, M. C.; Longoni, G. *J. Organomet. Chem.* **1989**, *378*, 457; (c) Satyanarayana, N.; Alper, H.; Amer, I. *Organometallics* **1990**, *9*, 284. For synthesis of β,γ-unsaturated ester: (d) [Review] Murahashi, S.-I.; Imada, Y. *Yuki Gosei Kagaku Kyokaishi* **1991**, *49*, 919 and references cited therein; (e) Murahashi, S.-I.; Imada, Y.; Taniguchi, Y.; Higashiura, S. *J. Org. Chem.* **1993**, *58*, 1538; (f) Okano, T.; Okabe, N.; Kiji, J. *Bull. Chem. Soc. Jpn.* **1992**, *65*, 2589.
7. Review: Colvin, E. W. In "Comprehensive Organic Chemistry"; Barton, D. H. R., Ollis, W. D., Eds.; Pergamon: Oxford, 1979; Vol. 2, p 593.
8. Review: Courtois, G.; Miginiac, L. *J. Organometal. Chem.* **1974**, *69*, 1.
9. Gosselin, P.; Maignan, C.; Rouessac, F. *Synthesis*, **1984**, 876.
10. (a) Yanagisawa, A.; Habaue, S.; Yamamoto, H. *J. Am. Chem. Soc.* **1991**, *113*, 8955; (b) Yanagisawa, A.; Hibino, H.; Habaue, S.; Hisada, Y.; Yamamoto, H. *J. Org. Chem.* **1992**, *57*, 6386.
11. (a) Yanagisawa, A.; Yasue, K.; Yamamoto, H. *Synlett* **1992**, 593; (b) Yanagisawa, A.; Habaue, S.; Yasue, K.; Yamamoto, H. *J. Am. Chem. Soc.* **1994**, *116*, 6130.

TABLE

REGIO- AND STEREOSELECTIVE CARBOXYLATION OF ALLYLIC BARIUM REAGENTS

Allylic chloride[a]	Product	Yield, %[b]	$\alpha:\gamma$[c]	E : Z[c]
(E)-$C_7H_{15}\overset{\gamma}{C}H{=}CH\overset{\alpha}{C}H_2Cl$	(E)-$C_7H_{15}CH{=}CHCH_2CO_2H$	82	98 : 2	99 : 1
(Z)-$C_7H_{15}\overset{\gamma}{C}H{=}CH\overset{\alpha}{C}H_2Cl$	(Z)-$C_7H_{15}CH{=}CHCH_2CO_2H$	65	82 : 18	1 : 99
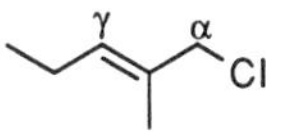 Cl	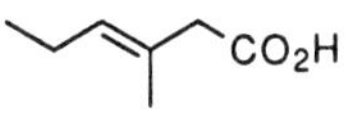 CO_2H	58	95 : 5	99 : 1
$(CH_3)_2\overset{\gamma}{C}{=}CH\overset{\alpha}{C}H_2Cl$	$(CH_3)_2C{=}CHCH_2CO_2H$	59	> 99 : 1	
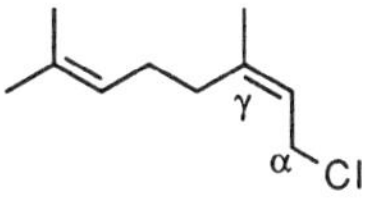 Cl	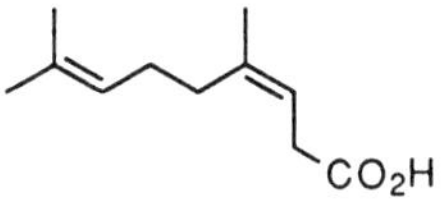 CO_2H	51	> 99 : 1	< 1 : 99
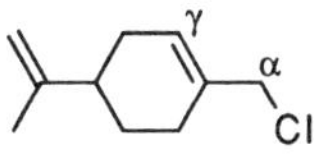 Cl	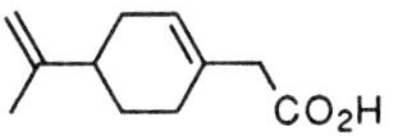 CO_2H	79	98 : 2	

[a] Stereochemically pure (> 99%) allylic chloride was used. [b] Isolated yield.
[c] Determined by GC analysis after conversion to methyl ester.

Appendix

Chemical Abstracts Nomenclature (Collective Index Number); (Registry Number)

(E)-4,8-Dimethyl-3,7-nonadienoic acid: 3,7-Nonadienoic acid, 4,8-dimethyl-, (E)- (8,9); (459-85-8)

Lithium (8,9); (7439-93-2)

Biphenyl (8); 1,1'-Biphenyl (9): (92-52-4)

Barium iodide dihydrate (8); Barium iodide, dihydrate (9); (7787-33-9)

Geranyl chloride: 2,6-Octadiene, 1-chloro-3,7-dimethyl-, (E)- (8,9); (5389-87-7)

NICKEL-CATALYZED, GEMINAL DIMETHYLATION OF ALLYLIC DITHIOACETALS: (E)-1-PHENYL-3,3-DIMETHYL-1-BUTENE

(Benzene, (3,3-dimethyl-1-butenyl)- (E)-)

A. Ph–CH=CH–C(=O)Me + $HSCH_2CH_2SH$ $\xrightarrow{BF_3 \cdot OEt_2}$ (E)-2-methyl-2-(2-phenylethenyl)-1,3-dithiolane

B. MeI + Mg $\longrightarrow$ MeMgI

C. (E)-2-methyl-2-(2-phenylethenyl)-1,3-dithiolane + MeMgI $\xrightarrow{NiCl_2(dppp)}$ Ph–CH=CH–CMe$_3$

Submitted by Tien-Min Yuan and Tien-Yau Luh.[1]

Checked by Yugang Liu and Robert K. Boeckman, Jr.

1. Procedure

Caution! 1,2-Ethanedithiol has a powerful stench. Steps A and C should be performed in a well-ventilated hood.

A. (E)-2-Methyl-2-(2-phenylethenyl)-1,3-dithiolane. In a 500-mL, round-bottomed flask equipped with a magnetic stirring bar are placed 29.2 g (0.2 mol) of (E)-4-phenyl-3-buten-2-one (Note 1) and 20.7 g (0.22 mol) of 1,2-ethanedithiol in 200 mL of methanol. To the stirred solution is added 11 mL (12.4 g, 0.087 mol) of boron trifluoride etherate. The mixture is stirred at room temperature for 45 min. To the mixture is added 50 mL of aqueous 10% sodium hydroxide, and the methanol is removed under reduced pressure. Dichloromethane (300 mL) is introduced and the solution is washed with aqueous 10% sodium hydroxide solution (2 x 100 mL). The aqueous layer is extracted with dichloromethane (3 x 200 mL). The combined organic layers are washed with 200 mL of water, dried over anhydrous magnesium sulfate, and filtered. The solvent is removed under reduced pressure and the residue is fractionally distilled to give 24.9-27.1 g (56-61%) of (E)-2-methyl-2-(2-phenylethenyl)-1,3-dithiolane (Notes 2 and 3) as a yellowish liquid, bp 109-112°C (0.16 mm), that solidifies on standing, mp 52-53°C.

B. Methylmagnesium iodide. A 500-mL, three-necked, round-bottomed flask containing 9.7 g (0.4 g-atom) of magnesium turnings is equipped with a rubber septum, reflux condenser, an addition funnel, and a magnetic stirring bar. The system is flame-dried and flushed with nitrogen. Anhydrous ether (20 mL, Note 4) is introduced to cover the magnesium. As the contents of the flask are stirred, iodomethane (56.7 g, 0.4 mol) in 180 mL of anhydrous ether is added dropwise through the addition funnel. The addition requires about 2 hr; the mixture is then stirred for an additional 30 min.

C. (E)-3,3-Dimethyl-1-phenyl-1-butene. In a 1-L, two-necked, round-bottomed flask fitted with a reflux condenser, rubber septum, and a magnetic stirring bar are placed 17.8 g (0.08 mol) of 2-methyl-2-(2-phenylethenyl)-1,3-dithiolane and 2.17 g (0.004 mol) of [1,3-bis(diphenylphosphino)propane]nickel(II) chloride [$NiCl_2$(dppp)] (Note 5). The flask is evacuated and flushed with nitrogen three times. To the above mixture is added 300 mL of anhydrous tetrahydrofuran (Note 6). The ether solution of methylmagnesium iodide prepared in Step B is introduced with a double-ended needle in one portion (Note 7). The mixture is heated under reflux for 24 hr, cooled to room temperature, and treated with 200 mL of saturated ammonium chloride solution.

The organic layer is separated and the aqueous layer is extracted with ether (3 x 200-mL). The combined organic layers are washed twice with aqueous 10% sodium hydroxide solution (100 mL) and with brine (100 mL). The organic solution is dried over anhydrous magnesium sulfate. The solvent is removed under reduced pressure and the residue is filtered through a short column packed with 30 g of silica gel (Note 8) and flushed under a positive nitrogen pressure with 350 mL of hexane. After the solvent is evaporated under reduced pressure, the yellowish residue is distilled through a spinning band (Note 9) to give 10.4 g (81%) of (E)-3,3-dimethyl-1-phenyl-1-butene (Note 10) as a colorless liquid, bp 91-93°C/30 mm.

2. Notes

1. (E)-4-Phenyl-3-buten-2-one was purchased from Division of Janssen Pharmaceutica and used directly.

2. A 6-cm Vigreux column was employed for fractional distillation.

3. The spectral properties of the (E)-2-methyl-2-(2-phenylethenyl)-1,3-dithiolane are as follows: IR (neat) cm^{-1}: 3025, 2922, 1637, 1600, 1494, 1446, 1275, 1065, 964, 754, 692; ^{1}H NMR (300 MHz, $CDCl_3$) δ: 2.01 (s, 3 H), 3.33-3.46 (m, 4 H), 6.47 (d, 1 H, J = 15.5), 6.63 (d, 1 H, J = 15.5), 7.21-7.43 (m, 5 H); ^{13}C NMR (75 MHz, $CDCl_3$) δ: 29.6, 40.1, 65.4, 126.6, 127.0, 127.5, 128.4, 135.0, 136.4; MS m/e (rel) 222 (M^+, 100), 207 (17), 194 (59), 161 (44), 129 (45), 117 (39); HRMS calcd for $C_{12}H_{14}S_2$: 222.0537. Found 222.0546.

4. Ethyl ether is distilled from sodium-benzophenone ketyl before use.

5. [1,3-Bis(diphenylphosphino)propane]nickel(II) chloride was purchased from Aldrich Chemical Company, Inc., and used without further purification. The catalyst can also be prepared according to literature procedures.[2]

6. Tetrahydrofuran is distilled from sodium-benzophenone ketyl before use.

7. An excess of the Grignard reagent is required to maximize the yield; otherwise the reaction is incomplete.

8. Silica gel (230-400 mesh) was purchased from E. Merck Co.

9. A spinning band distillation setup is employed (800 Mirco Still, B/R Instrument Corporation, 3000 rpm, theoretical plates 23-26). The mixture was heated under reflux for 3 hr to reach equilibrium before distillation (reflux/distillation ratio = 5-10/1).

10. The spectral properties of the product are as follows: IR (neat) cm^{-1}: 3027, 2960, 1648, 1598, 1493, 1362, 1266, 968, 746, 693; ^{1}H NMR (300 MHz, $CDCl_3$) δ: 1.16 (s, 9 H), 6.30 (d, 1 H, J = 16.2), 6.34 (d, 1 H, J = 16.2), 7.18-7.41 (m, 5 H); ^{13}C NMR (75 MHz, $CDCl_3$) δ: 29.6, 33.3, 124.6, 126.0, 126.7, 128.4, 138.1, 141.8; MS m/e (rel) 160 (M^+, 35), 145 (100), 117 (21), 105 (11), 103 (11), 91 (25), 77 (12); HRMS calcd for $C_{12}H_{16}$: 160.1252, Found 160.1250.

Waste Disposal Information

All toxic materials were disposed of in accordance with "Prudent Practices in the Laboratory"; National Academic Press; Washington, DC, 1996.

3. Discussion

Geminal dimethylation at a carbon center is useful in organic synthesis. Much effort has been devoted to the attachment of a tert-butyl group or a quarternary carbon center to olefinic carbon atom(s) in order to synthesize crowded olefins.[3] Although Tebbe-like reagents are effective for converting a carbonyl group into a gem-dimethyl substituent, their application to an allylic carbonyl substrate is limited by poor regioselectivity.[4-9] The present procedure is based on a series of reports on the nickel-catalyzed cross coupling reactions of dithioacetals with Grignard reagents.[10-16] α,β-Unsaturated aldehydes or ketones react as the dithioacetal in a polar solvent (such as tetrahydrofuran) or non-polar aromatic solvent (such as benzene or toluene) with methylmagnesium iodide in the presence of a catalytic amount $NiCl_2$(dppp) or [1,2-bis(diphenylphosphino)ethane]nickel(II) chloride [$NiCl_2$(dppe)], under a nitrogen atmosphere to give the corresponding geminal dimethylation products.[11-13] Substrates with structural variety react smoothly.[11-13] Allylic orthothioesters, on the other hand, give the corresponding trimethylation products with different regioselectivity.[12]

1. Department of Chemistry, National Taiwan University, Taipei, Taiwan 106, Republic of China.
2. Van Hecke, G. R.; Horrocks, W. DeW., Jr. *Inorg. Chem.* **1966**, *5*, 1960.
3. For examples: (a) Mulzer, J.; Lammer, O. *Angew. Chem., Int. Ed. Engl.* **1983**, *22*, 628; (b) Russell, G. A.; Tashtoush, H.; Ngoviwatchai, P. *J. Am. Chem. Soc.* **1984**, *106*, 4622; (c) Ager, D. J. *J. Chem. Soc., Perkin Trans. I* **1986**, 183; (d) Smegal, J. A.; Meier, I. K.; Schwartz, J. *J. Am. Chem. Soc.* **1986**, *108*, 1322; (e) Palomo, C.; Aizpurua, J. M.; Garcia, J. M.; Ganboa, I.; Cossio, F. P.; Lecea, B.; López, C. *J. Org. Chem.* **1990**, *55*, 2498.
4. Posner, G. H.; Brunelle, D. J. *Tetrahedron Lett.* **1972**, 293.
5. Meisters, A.; Mole, T. *J. Chem. Soc., Chem. Commun.* **1972**, 595.
6. Reetz, M. T.; Westermann, J.; Steinbach, R. *Angew. Chem., Int. Ed. Engl.* **1980**, *19*, 900.
7. Brown-Wensley, K. A.; Buchwald, S. L.; Cannizzo, L.; Clawson, L.; Ho, C. S.; Meinhardt, D.; Stille, J. R.; Straus, D.; Grubbs, R. H. *Pure Appl. Chem.* **1983**, *55*, 1733.

8. Pine, S. H.; Pettit, R. J.; Geib, G. D.; Cruz, S. G.; Gallego, C. H.; Tijerina, T.; Pine, R. D. *J. Org. Chem.* **1985**, *50*, 1212.
9. Reetz, M. T.; Westermann, J.; Kyung, S.-H. *Chem. Ber.* **1985**, *118*, 1150.
10. Luh, T.-Y. *Acc. Chem. Res.* **1991**, *24*, 257.
11. Yang, P.-F.; Ni, Z.-J.; Luh, T.-Y. *J. Org. Chem.* **1989**, *54*, 2261.
12. Tzeng, Y.-L.; Yang, P.-F.; Mei, N.-W.; Yuan, T.-M.; Yu, C.-C.; Luh, T.-Y. *J. Org. Chem.* **1991**, *56*, 5289.
13. Yuan, T.-M.; Luh, T.-Y. *J. Org. Chem.* **1992**, *57*, 4550.
14. Ni, Z.-J.; Mei, N.-W.; Shi, X.; Tzeng, Y.-L.; Wang, M. C.; Luh, T.-Y. *J. Org. Chem.* **1991**, *56*, 4035 and references therein.
15. Ni, Z.-J.; Yang, P.-F.; Ng, D. K. P.; Tzeng, Y.-L.; Luh, T.-Y. *J. Am. Chem. Soc.* **1990**, *112*, 9356.
16. Yu, C.-C.; Ng, D. K. P.; Chen, B.-L.; Luh, T.-Y. *Organometallics* **1994**, 1487 and references therein.

Appendix

Chemical Abstracts Nomenclature (Collective Index Number); (Registry Number)

(E)-1-Phenyl-3,3-dimethyl-1-butene: 1-Butene, 3,3-dimethyl-1-phenyl-, (E)- (8); Benzene, (3,3-dimethyl-1-butenyl)-, (E)- (9); (3846-66-0)

1,2-Ethanedithiol (8,9); (540-63-6)

(E)-2-Methyl-2-(2-phenylethenyl)-1,3-dithiolane: 1,3-Dithiolane, 2-methyl-2-(2-phenylethenyl)-, (E)- (12); (107389-59-3)

(E)-4-Phenyl-3-buten-2-one: 3-Buten-2-one, 4-phenyl-, (E)- (8,9); (1896-62-4)

Boron trifluoride etherate: Ethyl ether, compd. with boron fluoride (1:1) (8); Ethane, 1,1'-oxybis-, compd. with trifluoroborane (1:1) (9); (109-63-7)

Methylmagnesium iodide: Magnesium, iodomethyl- (8,9); (917-64-6)

Magnesium (8,9); (7439-95-4)

Iodomethane: HIGHLY TOXIC. CANCER SUSPECT AGENT: Methane, iodo- (8,9); (74-88-4)

[1,3-Bis(diphenylphosphino)propane]nickel(II) chloride: $NiCl_2$(dppp): CANCER SUSPECT AGENT: Nickel, dichloro[trimethylenebis[diphenylphosphine]]- (8); Nickel, dichloro[1,3-propanediylbis[diphenylphosphine]-P,P']- (9); (15629-92-2)

[1,2-Bis(diphenylphosphino)ethane]nickel(II) chloride: $NiCl_2$(dppe): CANCER SUSPECT AGENT: Nickel, dichloro[ethylenebis[diphenylphosphine]]- (8); Nickel, dichloro[1,2-ethanediylbis[diphenylphosphine]-P,P']-, (SP-4-2)- (9); (14647-23-5)

A SIMPLE AND CONVENIENT METHOD FOR THE PREPARATION OF (Z)-β-IODOACROLEIN AND OF (Z)- OR (E)-γ-IODO ALLYLIC ALCOHOLS: (Z)- AND (E)-1-IODOHEPT-1-EN-3-OL

(2-Propenal, 3-iodo-, (Z)- and 1-Hepten-3-ol, 1-iodo-, (Z)- and (E)-)

A. H—≡—COOEt (**1**) + NaI —(AcOH, 70°C, 12 hr)→ (Z)-I–CH=CH–COOEt (**2**)

B. **2** —(1) DIBAL-H / -78°C; 2) BuMgX)→ **3Z**

C. **2** —(1) DIBAL-H / -78°C; 2) 0°C, 15 min; 3) BuMgBr)→ **3E**

D. **2** —(1) DIBAL-H / -78°C, CH_2Cl_2 / 30 min; 2) MeOH / -80°C; 3) NaO_2C-$(CHOH)_2$-CO_2K)→ (Z)-I–CH=CH–CHO (**4**)

Submitted by Ilane Marek, Christophe Meyer, and Jean-F. Normant.[1]

Checked by David A. Favor and Amos B. Smith, III.

1. Procedure

A. Ethyl (Z)-β-iodoacrylate (Note 1). A 250-mL, round-bottomed flask equipped with a magnetic stirring bar and nitrogen gas inlet is charged with 30.6 g (204 mmol) of dry sodium iodide (Note 2) and 100 mL of glacial acetic acid. To the stirred solution is added in one portion 20.6 mL (204 mmol) of ethyl propiolate (Note 2) and the resulting mixture is heated with an oil bath (bath temperature 70°C) for 12 hr. The brown solution is cooled to room temperature, and 100 mL of water and 100 mL of ether are added. The organic layer is separated and the aqueous layer is extracted with ether (2 x 20 mL). The combined organic layers are treated with 3 M aqueous potassium hydroxide (ca. 150 mL in 50-mL portions) until the aqueous phase becomes neutral (pH = 7), washed with 50 mL of brine, and dried over anhydrous magnesium sulfate. After evaporation of the solvent, the residual brown oil is distilled to give 40.6 g (88% yield) of ethyl (Z)-β-iodoacrylate (Note 3) as a pale yellow liquid, bp 57°C/0.1 mm.

B. (Z)-1-Iodohept-1-en-3-ol. A 100-mL, dry (Note 4), four-necked, round-bottomed flask equipped with a mechanical stirrer, an internal thermometer, a rubber septum, and a nitrogen gas inlet, is charged with 2.26 g (10 mmol) of ethyl (Z)-β-iodoacrylate, **2**, and 20 mL of anhydrous dichloromethane (Note 5). The stirred solution is cooled to -78°C by means of a liquid nitrogen bath and 10 mL (10 mmol) of a 1 M solution of diisobutylaluminum hydride in hexane (Note 6) is added dropwise with a syringe at such a rate that the temperature does not exceed -75°C. After stirring for 30 min at -78°C (Note 7) 11 mL (11 mmol, 1 M solution in Et_2O) of a butylmagnesium bromide solution in ether (Note 8) is added dropwise at -70°C with a syringe through the septum. The cooling bath is removed and the reaction mixture is allowed to warm to room temperature. Hydrolysis is carried out at -20°C by dropwise addition of 20 mL of a 1 M aqueous solution of hydrochloric acid, followed by addition of 30 mL of ether. The organic layer is separated, the aqueous layer is extracted with ether (2 x 20 mL), and the combined extracts are dried over magnesium sulfate. After rotary evaporation of the solvents, the residual pale yellow oil is purified by chromatography through 1.85 g of silica (Note 9) packed in a 4-cm diameter column and eluted with 15% ethyl acetate in hexane, to give 1.7 g of (Z)-1-iodohept-1-en-3-ol, **2**, (71% yield) as a pale yellow liquid (Note 10).

C. (E)-1-Iodohept-1-en-3-ol. A 100-mL, dry (Note 4) four-necked, round-bottomed flask equipped with a mechanical stirrer, an internal thermometer, a rubber septum, and a nitrogen gas inlet, is charged with 2.26 g (10 mmol) of ethyl Z-β-iodoacrylate, **2**, and 20 mL of anhydrous dichloromethane (Note 5). The stirred solution is cooled to -78°C by means of a liquid nitrogen bath and 10 mL (10 mmol) of a 1 M solution of diisobutylaluminum hydride in hexane (Note 6) is added dropwise with a syringe at such a rate that the temperature does not exceed -75°C. After the mixture is stirred for 30 min at -78°C (Note 7), it is allowed to warm slowly to 0°C in 45 min, stirred for 15 min at this temperature, and then cooled to -20°C; 11 mL (11 mmol, 1 M solution in Et_2O) of a butylmagnesium bromide solution in ether (Note 8) is added dropwise at -20°C with a syringe through the septum. The reaction mixture is allowed to warm to room temperature and worked up, as described for the Z isomer. (E)-1-Iodohept-1-en-3-ol (1.9 g) is obtained as a colorless liquid (79% yield) (Note 11) with an E/Z ratio of 96/4 .

D. (Z)-β-Iodoacrolein, **4**. A 100-mL, dry (Note 4), four-necked, round-bottomed flask equipped with a mechanical stirrer, an internal thermometer, a rubber septum, and a nitrogen gas inlet, is charged with 2.26 g (10 mmol) of ethyl (Z)-β–iodoacrylate, **2**, and 20 mL of anhydrous dichloromethane (Note 5). The stirred solution is cooled to -78°C in a liquid nitrogen bath and 10 mL (10 mmol) of a 1 M solution of diisobutylaluminum hydride in hexane (Note 6) is added dropwise with a syringe at such a rate that the temperature does not exceed -75°C. After the solution is stirred at -78°C (Note 7), 5 mL of cold methanol (Note 12) is added dropwise through the septum with a syringe at -80°C. Immediately after this addition, 25 mL of aqueous 20% potassium sodium tartrate is added to the cold reaction mixture in one portion, which causes the temperature to reach 0°C within a few seconds. The cooling bath is removed and 20 mL of ether is added to the hydrolyzed reaction mixture, which is further stirred for 20 min at room temperature (Note 13). It is then filtered through a pad of Celite and extracted with 30 mL of ether. The organic layer is washed with brine and dried over potassium carbonate. After careful rotary evaporation of the solvents, 1.6 g of (Z)-β-iodoacrolein is obtained as a yellow liquid (88%) (Note 14), which can be handled easily as ether or dichloromethane solutions, and can be stored in the refrigerator for a few weeks without decomposition or isomerization.

2. Notes

1. Part A of the procedure should be carried out in an efficient fume hood to avoid exposure to noxious vapor (acetic acid) and to lachrymatory ethyl propiolate.

2. Sodium iodide, 99%, was purchased from Janssen Chimica and used as received. Glacial acetic acid, 99%, was purchased from Prolabo. Ethyl propiolate was obtained from Janssen Chimica and used as received.

3. The product exhibits the following physical and spectral properties: IR (film) cm^{-1}: 3080, 2940, 1735, 1640, 1425, 1365, 1230, 1170, 1020, 925, 730; ^{1}H NMR (400 MHz, $CDCl_3$) δ: 1.32 (t, 3 H, J = 7.15, $\mathbf{CH_3}CH_2$); 4.25 (q, 2 H, J = 7.14, $O\mathbf{CH_2}$), 6.89 (d, 1 H, J = 9.34, =**CH**-CO_2Et), 7.44 (d, 1 H, J = 9.34, =**CH**I); ^{13}C NMR (100 MHz, $CDCl_3$) δ: 14.2, 60.8, 94.6, 129.9, 164.6.

4. All glassware is oven dried at 140°C overnight and assembled while hot under a nitrogen atmosphere.

5. Dichloromethane was distilled from calcium hydride and stored over 4 Å molecular sieves.

6. Diisobutylaluminum hydride was purchased from Aldrich Chemical Company, Inc.

7. At this point, TLC analysis of a hydrolyzed (1 M aqueous hydrochloric acid) aliquot, eluted with 10% ethyl acetate in hexane, indicated that the reaction is complete.

8. Butylmagnesium bromide was prepared from the corresponding butyl bromide and magnesium turnings in anhydrous ether.[2]

9. Silica gel was purchased from Merck: Geduran SI 60.

10. The product exhibits the following spectral properties: ^{1}H NMR (400 MHz, $CDCl_3$) δ: 0.90 (t, 3 H, J = 7.15, **C*H*$_3$**), 1.31-1.36 (m, 4 H), 1.51-1.63 (m, 2 H), 1.7 (s, 1 H, O**H**), 4.4 (m, 1 H, C**H**OH), 6.20 (t, 1 H, J = 7.69, =C**H**), 6.33 (d, 1 H, J = 7.69, =C**H**I); ^{13}C NMR (100 MHz, $CDCl_3$) δ: 14.0, 22.6, 27.1, 35.6, 74.4 (CHOH), 82.2 (=CI), 143.4 (=CH).

11. The product exhibits the following spectral properties: IR (film) cm^{-1}: 3350, 3060, 2950, 2920, 2850, 1605, 1450, 1265, 1165, 1015, 930; ^{1}H NMR (400 MHz, $CDCl_3$) δ: 0.90 (t, 3 H, J = 7.15, C**H**$_3$), 1.31-1.36 (m, 4 H), 1.51-1.63 (m, 2 H), 1.71 (s, 1 H), 4.1 (m, 1 H, C**H**OH), 6.34 (dd, 1 H, J = 14.3, 1.1, =C**H**I), 6.58 (dd, 1 H, J = 14.3, 6.6, =C**H**); ^{13}C NMR (100 MHz, $CDCl_3$) δ: 13.9, 22.5, 27.2, 36.2, 74.6 (CHOH), 77.2 (=CI), 148.6 (=CH).

12. Methanol was obtained from Merck & Company, Inc., and used as received.

13. The workup should be carried out under an efficient fume hood since the title compound is a lachrymator.

14. The (Z)-β-iodoacrolein exhibits the following spectral properties: IR (CCl_4) cm^{-1}: 3050, 2820, 2730, 1675, 1610, 690; ^{1}H NMR (400 MHz, $CDCl_3$) δ: 6.77 (dd, 1 H, J = 8.25, 6.6, =C**H**-CHO), 7.79 (d, 1 H, J = 8.25, =C**H**I), 9.67 (d, 1 H, J = 6.6, C**H**O); ^{13}C NMR (100 MHz, $CDCl_3$) δ: 103.0 (=CHI), 136.4 (=CH), 195.4 (C=O).

Waste Disposal Information

All toxic materials were disposed of in accordance with "Prudent Practices in the Laboratory"; National Academic Press; Washington, DC, 1996.

3. Discussion

The present procedure illustrates the simplest convenient method for the preparation of (Z)- or (E)-γ-iodo allylic alcohols and (Z)-β-iodoacrolein. It uses ethyl propiolate as a unique starting material: ethyl propiolate and ethyl tetrolate can be hydroiodinated regio- and stereospecifically (Z > 99%) by reaction with inexpensive sodium iodide in acetic acid[3] (Scheme 1). The same methodology has been applied

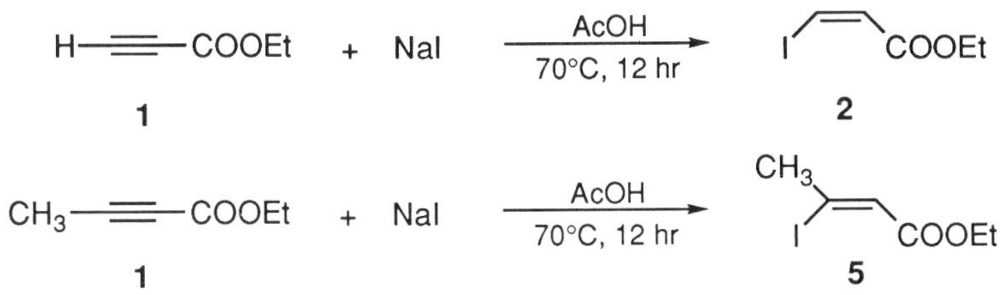

$$\text{H—≡—COOEt} \;\; + \;\; \text{NaI} \xrightarrow[70^\circ\text{C, 12 hr}]{\text{AcOH}} \text{I—CH=CH—COOEt}$$

1 → 2

$$\text{CH}_3\text{—≡—COOEt} \;\; + \;\; \text{NaI} \xrightarrow[70^\circ\text{C, 12 hr}]{\text{AcOH}} \text{(CH}_3\text{)(I)C=CH—COOEt}$$

1 → 5

Scheme 1

recently to the synthesis of enantiomerically pure (Z)-2-haloalkenyl sulfoxides.[4] By reaction of **2** with diisobutylaluminum hydride at low temperature, and subsequent reaction with a Grignard reagent, one can obtain, very easily, the secondary (Z)-γ-iodo allylic alcohols with good chemical yield and exclusive Z stereochemistry of the double bonds (Table I). This one-pot "reduction-C-alkylation" sequence of the ester group allows the preparation of very sensitive derivatives (entry 5 or 6), which are difficult to prepare by usual methods.[5]

However, reaction of ethyl (Z)-β-iodoacrylate with DIBAL-H at low temperature (-78°C, 15 min) followed by warming to 0°C and addition of a Grignard reagent (-20°C to 0°C) now leads to the E isomers of the secondary γ-iodo allylic alcohols with an E/Z ratio of 96/4 (Table II). This isomerization can be rationalized as occurring by the influence of the weak Lewis acid, $Al(iBu)_2OEt$, generated by α-elimination of the thermally labile aluminooxyacetal.[6] The (Z)-β-iodoacrolein **4** and crotonaldehyde **6**

(Scheme 2) are also easily obtained by dropwise addition of an excess of methanol to the aluminooxyacetal at low temperature, immediately followed by alkaline hydrolysis. (3Z)-Iodopropenals are valuable intermediates in organic synthesis because of the presence of three functional groups on three carbon atoms.[6,7]

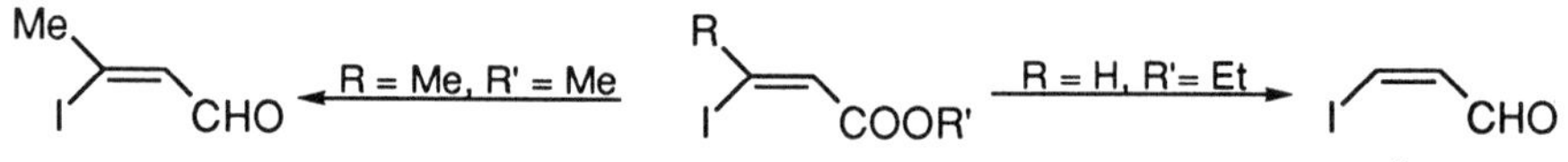

Me / I — CHO (6) ← R = Me, R' = Me — R / I — COOR' — R = H, R'= Et → I — CHO (4)

Scheme 2

Starting from ethyl (Z)-β-iodoacrylate, (E)- or (Z)-γ-iodo allylic alcohols are easily obtained in a highly stereoselective manner by a "reduction-C-alkylation" of the ester group without alteration of the nature or geometry of the iodo vinylic part. (Z)-β-Iodoacrolein can be easily isolated in good yields, and gives rise to various derivatives.

1. Laboratoire de Chimie des Organoéléments, Tour 44-45 E.2. 4, Place Jussieu. Boite 183. 75252 Paris Cedex 05, France.

2. Coleman, G. H.; Craig, D. *Org. Synth., Coll. Vol. II* **1943**, 179.
3. (a) Marek, I.; Alexakis, A.; Normant, J.-F. *Tetrahedron Lett.* **1991**, *32,* 5329. The same hydrohalogenation of a similar substrate with LiI has been described: (b) Ma, S.; Lu, X. *Org. Synth.* **1995**, *72*, 112; (c) Ma, S.; Lu, X. *J. Chem. Soc., Chem. Commun.* **1990**, 1643; (d) Ma, S.; Lu, X. *Tetrahedron Lett.* **1990**, *31*, 7653; (e) Ma, S.; Lu, X.; Li, Z. *J. Org. Chem.* **1992**, *57*, 709; (f) Lu, X.; Zhu, G.; Ma, S. *Chin. J. Chem.* **1993**, *11*, 267; *Chem. Abstr.* **1994**, *120*, 216690u.
4. de la Pradilla, R. F.; Morente, M.; Paley, R. S. *Tetrahedron Lett.* **1992**, *33*, 6101.

5. (a) Sato, F.; Kobayashi, Y. *Org. Synth., Coll. Vol. VIII* **1993**, 507; (b) Jones, T. K.; Denmark, S. E. *Org. Synth., Coll. Vol. VII* **1990**, 524; (c) Pasto, D. J.; Taylor, R. T. *Org. React.* **1991**, *40*, 91; (d) Corey, E. J.; Katzenellenbogen, J. A.; Posner, G. H. *J. Am. Chem. Soc.* **1967**, *89*, 4245.
6. Meyer, C.; Marek, I.; Normant, J. F. *Synlett* **1993**, 386.
7. The (Z)-β-iodoacrolein seems to be much more stable than the related β-bromo- or β-chloroacrolein: Meyers, A. I.; Babiak, K. A.; Campbell, A. L.; Comins, D. L.; Fleming, M. P.; Henning, R.; Heuschmann, M.; Hudspeth, J. P.; Kane, J. M.; Reider, P. J.; Roland, D. M.; Shimizu, K.; Tomioka, K.; Walkup, R. D. *J. Am. Chem. Soc.* **1983**, *105*, 5015.

Appendix
Chemical Abstracts Nomenclature (Collective Index Number); (Registry Number)

(Z)-β-Iodoacrolein: 2-Propenal, 3-iodo-, (Z)- (12); (138102-13-3)

Ethyl (Z)-β-iodoacrylate: Acrylic acid, 3-iodo-, ethyl ester, (Z)- (8);

2-Propenoic acid, 3-iodo-, ethyl ester, (Z)- (9); (3190-36-6)

Sodium iodide (8,9); (7681-82-5)

Ethyl propiolate: Propiolic acid, ethyl ester (8); 2-Propynoic acid, ethyl ester (9); (623-47-2)

(Z)-1-Iodohept-1-en-3-ol: 1-Hepten-3-ol, 1-iodo-, (Z)- (12); (138102-06-4)

Diisobutylaluminum hydride: Aluminum, hydrodiisobutyl- (8); Aluminum, hydrobis(2-methylpropyl)- (9); (1191-15-7)

Butylmagnesium bromide: Magnesium, bromobutyl- (8,9); (693-03-8)

(E)-1-Iodohept-1-en-3-ol: 1-Hepten-3-ol, 1-iodo, (E)- (13); (151160-08-6)

Potassium sodium tartrate: Tartaric acid, monopotassium monosodium salt, tetrahydrate, L-(+)- (8); Butanedioic acid, 2,3-dihydroxy-, [R-(R*,R*)]-, monopotassium monosodium salt, tetrahydrate (9); (6381-59-5)

TABLE I

Starting Ester	Grignard Reagents	Products	Yield
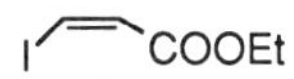 I COOEt **2**	EtMgBr	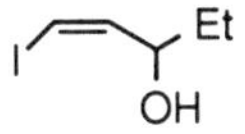 I Et OH	88%
2	BuMgBr	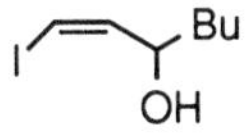 I Bu OH	79%
2	PhMgBr	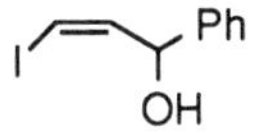 I Ph OH	85%
2	 MgBr	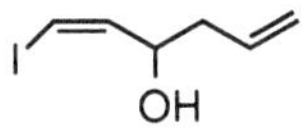 I OH	72%
2	nBu—≡—MgBr	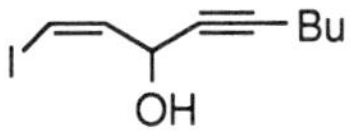 I Bu OH	81%
2	Me_3Si—≡—MgBr	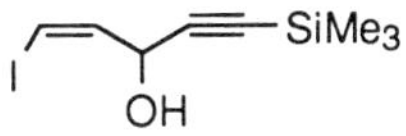 I $SiMe_3$ OH	63%
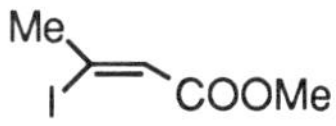 Me I COOMe **5**	 MgBr	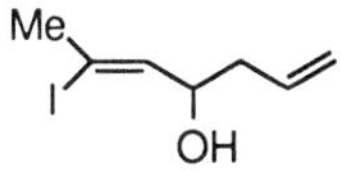 Me I OH	65%
5	BuMgBr	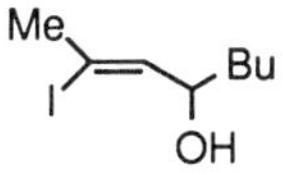 Me I Bu OH	81%

TABLE II

Starting Ester	Grignard Reagents	Products	Yield
I–CH=CH–COOEt **2**	CH₂=CH–CH₂–MgBr	I–CH=CH–CH(OH)–CH₂–CH=CH₂	80%
2	BuMgBr	I–CH=CH–CH(OH)–Bu	86%
2	Me_3Si–C≡C–MgBr	I–CH=CH–CH(OH)–C≡C–$SiMe_3$	72%

ALLYLIC ALCOHOLS BY ALKENE TRANSFER FROM ZIRCONIUM TO ZINC: 1-[(tert-BUTYL-DIPHENYLSILYL)OXY]-DEC-3-EN-5-OL

A. HO~~≡ → (t-BuPh$_2$SiCl, imidazole, DMAP, CH_2Cl_2) → t-BuPh$_2$SiO~~≡

B. t-BuPh$_2$SiO~~≡ → (Cp_2ZrHCl, CH_2Cl_2, 0 → 22°C) → t-BuPh$_2$SiO~~=~Zr(Cl)Cp_2 → (1. Me_2Zn, -60°C; 2. $H_3C(CH_2)_4CHO$, 0°C) → t-BuPh$_2$SiO~~=~CH(OH)~~~~

Submitted by Peter Wipf and Wenjing Xu.[1]

Checked by Thomas Wynn and Louis S. Hegedus.

1. Procedure

A. tert-Butyl(but-3-ynyloxy)diphenylsilane. A dry, 250-mL, round-bottomed flask is fitted with a rubber septum and a magnetic stirring bar. The flask is charged with 3-

butyn-1-ol (2.1 g, 30 mmol) (Note 1) and dry dichloromethane (60 mL) (Note 2). After addition of 8.5 g (31 mmol) of tert-butyldiphenylchlorosilane (Note 3), the reaction flask is immersed in a water bath and imidazole (2.86 g, 42 mmol) is added in one portion followed by 4-dimethylaminopyridine (0.37 g, 3 mmol). The water bath is removed and the reaction mixture is stirred at room temperature overnight. The white precipitate is filtered through a sintered glass funnel ('M' frit). The precipitate is washed with cold dichloromethane (50 mL). The combined filtrates are transferred to a 500-mL separatory funnel and washed with 1 M aqueous hydrochloric acid solution (50 mL) and water (100 mL), successively. The organic layer is dried over magnesium sulfate and concentrated by rotary evaporator. Bulb-to-bulb distillation of the residue (bp 150-152°C/1 mm) gives 8.90 g (96%) of tert-butyl(but-3-ynyloxy)diphenylsilane as a colorless oil (Note 4).

B. 1-[(tert-Butyldiphenylsilyloxy)]-dec-3-en-5-ol. A flame-dried, 250-mL, round-bottomed flask equipped with a gas inlet stopcock (Figure 1) is fitted with a rubber septum and a magnetic stirring bar, and flushed with nitrogen. The flask is charged with tert-butyl(but-3-ynyloxy)diphenylsilane (5.55 g, 18 mmol) and dry dichloromethane (60 mL), immersed in a cold water bath and stirred. Within 20 min, 5.10 g (19.8 mmol) of zirconocene hydrochloride (Note 5) is added in five portions. The water bath is removed and the reaction mixture is stirred at room temperature until a homogenous solution forms. The resulting golden-yellow solution is stirred for another 20 min (Note 6) and then cooled to -60°C. By syringe, dimethylzinc (2.0 M solution in toluene, 10.4 mL, 20.8 mmol) (Note 1) is added dropwise over 45 min while the bath temperature is kept at -60°C. The resulting orange-yellow solution is stirred for an additional 10 min at -60°C after the addition is completed. The reaction flask is immersed in an ice bath, and a solution of 2.16 g (21.6 mmol) of hexanal (Note 1) in dry dichloromethane (10 mL) is added via syringe over 45 min. The reaction mixture is stirred at 0°C for another 6 hr. The yellow solution is poured slowly into a beaker containing ice-cold aqueous 5% sodium bicarbonate solution (200 mL) and vigorous stirring is continued at room temperature until gas evolution subsides. The mixture is transferred to a 1-L separatory funnel and is extracted with diethyl ether (3 x 200 mL) (Note 7). The combined extracts are washed with a saturated aqueous sodium chloride solution (300 mL) and dried over sodium sulfate. The cloudy solution is filtered through a pad of Florisil (Note 8) loaded on a sintered glass funnel ('M' frit). The clear filtrate solution is concentrated by rotary evaporator. The residue is layered on a column of silica gel (150 g, column diameter: 6.0 cm) and eluted (ethyl acetate/hexane, 1:30 to 1:15 as eluents) to give 4.90 g (66%) of 1-[(tert-butyldiphenylsilyl)oxy]-dec-3-en-5-ol as a colorless oil (Note 9).

Figure 1

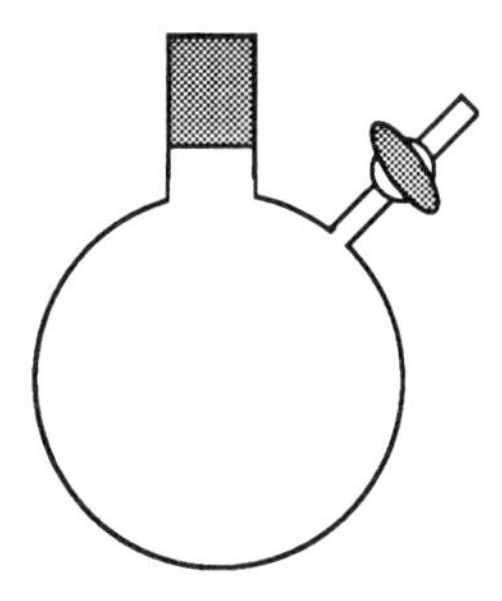

2. Notes

1. 3-Butyn-1-ol, dimethylzinc and hexanal were obtained from Aldrich Chemical Company, Inc., and used without purification.

2. Solvent grade dichloromethane was dried over calcium hydride, refluxed and distilled freshly before use.

3. tert-Butyldiphenylchlorosilane was obtained from United Chemical Technologies, Inc. or Aldrich Chemical Company, Inc., and used without purification.

4. The product has the following spectral properties: ^{1}H NMR (300 MHz, $CDCl_3$) δ: 1.06 (s, 9 H), 1.95 (t, 1 H, J = 2.6), 2.45 (dt, 2 H, J = 7.1, 2.7), 3.79 (t, 2 H, J = 7.1), 7.36-7.46 (m, 6 H), 7.67-7.70 (m, 4 H).

5. Zirconocene hydrochloride was prepared according to Buchwald's procedure[2] and stored in a Schlenk filter flask under argon in a freezer at -20°C or obtained from Aldrich Chemical Company, Inc., and used without further purification.

6. The reaction is conveniently monitored by TLC (silica gel, 4:1 hexane-ethyl acetate) with anisaldehyde stain (alkyne stains as a red spot, while the hydrozirconation product shows a blue color).

7. Significant amounts of white foam form between the ether and aqueous layers during extraction. Separation of these layers can be improved by the addition of 1-2 spoonfuls of solid sodium chloride.

8. Florisil (100-200 mesh) was obtained from Fisher Scientific Company.

9. The product has the following spectral data: IR (neat) cm^{-1}: 3350, 2912, 1454, 1417, 1099, 733, 698, 609; ^{1}H NMR (300 MHz, $CDCl_3$) δ: 0.88 (t, 3 H, J = 6.7), 1.05 (s, 9 H), 1.34-1.51 (m, 9 H), 2.26-2.33 (m, 2 H), 3.71 (t, 2 H, J = 6.6), 4.00-4.03 (m, 1 H), 5.49 (dd, 1 H, J = 15.5, 6.8), 5.64 (dt, 1 H, J = 15.5, 6.7), 7.35-7.46 (m, 6 H), 7.66-7.69 (m, 4 H); ^{13}C NMR ($CDCl_3$) δ: 14.1, 19.2, 22.6, 25.2, 26.9, 31.8, 35.6, 37.2, 63.6, 73.1, 127.7, 128.2, 129.6, 133.9, 135.2, 135.6; MS (EI) m/e (relative intensity) 353 ([M-

tert-butyl]+, 4), 335 (20), 229 (25), 199 (100), 135 (20), 91 (15), 57 (10); HRMS (EI) m/e calcd. for $C_{22}H_{29}O_2Si$ ($M-C_4H_9$): 353.1937, found: 353.1833.

Waste Disposal Information

All toxic materials were disposed of in accordance with "Prudent Practices in the Laboratory"; National Academic Press; Washington, DC, 1996.

3. Discussion

Organometallic derivatives of zirconium(IV) are readily obtained by hydrozirconation of alkenes and alkynes with Cp_2ZrHCl (Schwartz reagent). Because of the relatively low reactivity of the resulting organozirconocenes, transmetalation protocols are frequently applied for further carbon-carbon formation.[3] Transmetalation of alkenylzirconocenes to the corresponding organozinc compounds occurs rapidly at low temperature in the presence of stoichiometric amounts of commercially available dimethyl- or diethylzinc. Subsequent addition of aldehydes provides an in situ protocol for the conversion of alkynes into allylic alcohols in good to excellent yields.[4] Compared to standard organometallic methods applying lithium or Grignard reagents for this transformation, the use zirconocenes tolerates the presence of a wide range of functional groups in the substrate (Table). The zirconium→zinc transmetalation is related in scope to the boron→zinc transmetalation, which can also be applied for the conversion of alkynes to allylic alcohols.[5]

1. Department of Chemistry, University of Pittsburgh, Pittsburgh, PA 15260.
2. Buchwald, S. L.; LaMaire, S. J.; Nielsen, R. B.; Watson, B. T.; King, S. M. *Org. Synth.* **1993**, *71*, 77.

3. (a) Labinger, J. A. In "Comprehensive Organic Synthesis"; Trost, B. M.; Fleming, I., Eds.; Pergamon: Oxford, 1991; Vol. 8; pp 667-702; (b) Wipf, P. *Synthesis* **1993**, 537.

4. Wipf, P.; Xu, W. *Tetrahedron Lett.* **1994**, *35*, 5197.

5. (a) Srebnik, M. *Tetrahedron Lett.* **1991**, *32*, 2449; (b) Oppolzer, W.; Radinov, R. N. *J. Am. Chem. Soc.* **1993**, *115*, 1593.

Appendix

Chemical Abstracts Nomenclature (Collective Index Number); (Registry Number)

tert-Butyl(but-3-ynyloxy)diphenylsilane: Silane, (3-butynyloxy)(1,1-dimethylethyl)diphenyl- (11); (88138-68-3)

3-Butyn-1-ol (8,9); (927-74-2)

tert-Butylchlorodiphenylsilane: Silane, chloro(1,1-dimethylethyl)diphenyl- (9); (58479-61-1)

Imidazole (8); 1H-Imidazole (9); (288-32-4)

4-Dimethylaminopyridine: HIGHLY TOXIC: Pyridine, 4-(dimethylamino)- (8); 4-Pyridinamine, N,N-dimethyl- (9); (1122-58-3)

Zirconocene hydrochloride: ALDRICH: Zirconocene chloride hydride: Zirconium, chlorodi-π-cyclopentadienylhydro- (8); Zirconium, chlorobis(η^5-2,4-cyclopentadien-1-yl)hydro- (9); (37342-97-5)

Dimethylzinc: Zinc, dimethyl- (8,9); (544-97-8)

Hexanal (8,9); (66-25-1)

Table. Reaction of in situ generated alkenylalkylzinc reagents with aldehydes

Aldehyde	Alkyne	Time (hr)	Product	Yield(%)
PhCHO	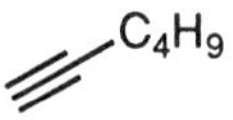 C_4H_9	1	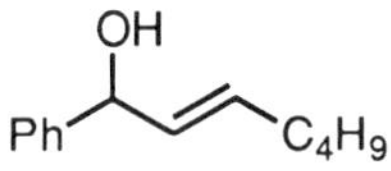 OH, Ph, C_4H_9	93
PhCHO	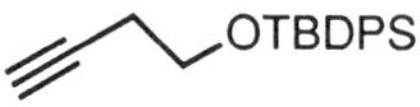 OTBDPS	2	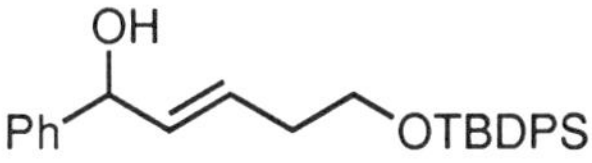 OH, Ph, OTBDPS	89
PhCHO		5	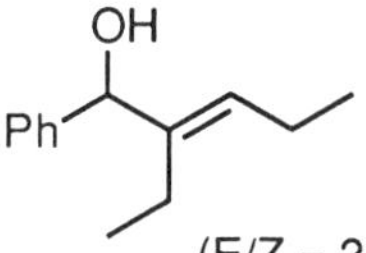 OH, Ph (E/Z = 2.5:1)	78
 Ph, CHO	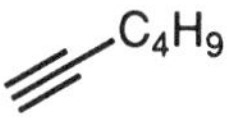 C_4H_9	1.5	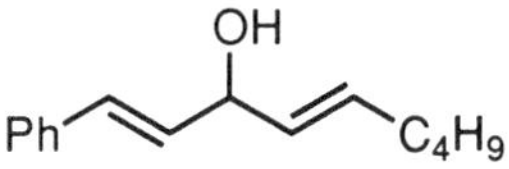 OH, Ph, C_4H_9	94
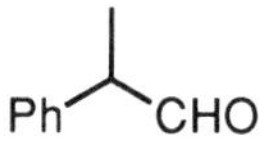 Ph, CHO	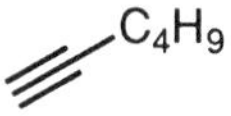 C_4H_9	1.5	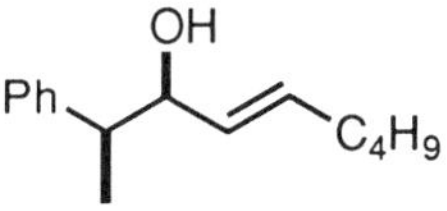 OH, Ph, C_4H_9 (syn:anti = 85:15)	92
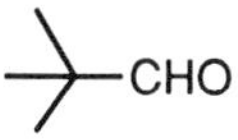 CHO	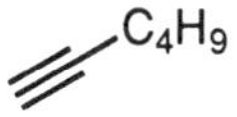 C_4H_9	2	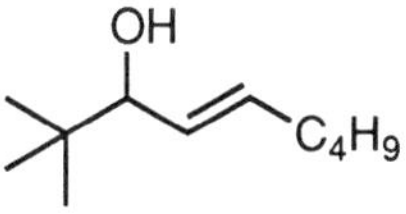 OH, C_4H_9	54
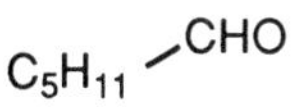 C_5H_{11}, CHO	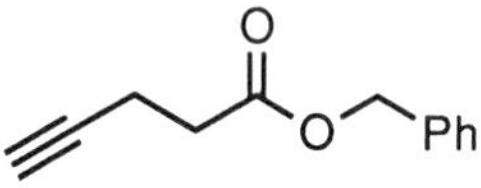 O, O, Ph	2	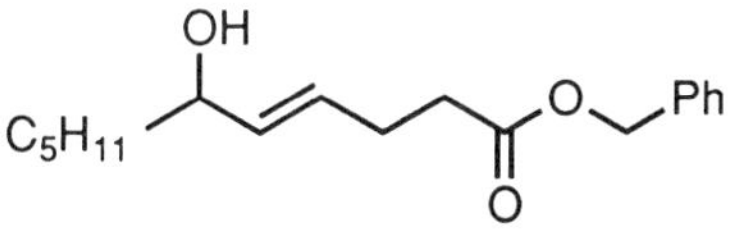 OH, C_5H_{11}, O, O, Ph	72[a]

[a] On a 5-g scale, this product was isolated in 68% yield.

PREPARATION OF 3-BROMOPROPIOLIC ESTERS: METHYL AND tert-BUTYL 3-BROMOPROPIOLATES

(2-Propynoic acid, 3-bromo-, methyl and 1,1-dimethylethyl esters)

$$\mathrm{H{-}C{\equiv}C{-}CO_2R} \;+\; \text{N-bromosuccinimide} \xrightarrow[\mathrm{Me_2CO}]{\mathrm{AgNO_3}} \mathrm{Br{-}C{\equiv}C{-}CO_2R}$$

R = Me, tert-Bu

Submitted by J. Leroy.[1]

Checked by Paul N. Devine and Ichiro Shinkai.

1. Procedure

Caution! Propiolates and their bromo derivatives are lachrymators and must be handled under an efficient hood. Distillation of bromopropiolates should be carried out behind a safety shield (Note 1).

Methyl 3-bromopropiolate. A 250-mL, one-necked, round-bottomed flask equipped with a magnetic stirring bar is charged with 100 mL of acetone (Note 2) and 4.0 g (47.6 mmol) of methyl propiolate (Note 3). To the stirred solution at room temperature is added 0.8 g (4.7 mmol) of silver nitrate. After 5 min, 9.8 g of N-bromosuccinimide (55 mmol) is added at once. The homogeneous mixture becomes cloudy and a grayish precipitate develops. Stirring is continued for 2 hr (Note 4). The solids are filtered through a pad of Celite, which is rinsed with acetone (30-50 mL). After careful rotary evaporation of the acetone at ≈20°C under 20 mm, the oily residue is bulb-to-bulb distilled at room temperature under reduced pressure (≈0.1 mm), affording methyl 3-bromopropiolate as a colorless liquid (7.0-7.5 g, 42.9-46.0 mmol, 90-97%) solidifying in the refrigerator (mp ≈20°C) (Note 5).

tert-Butyl 3-bromopropiolate. tert-Butyl propiolate (Note 6) (5.42 g, 42.8 mmol) dissolved in 150 mL of acetone is treated as above with 0.8 g of silver nitrate (4.7 mmol) and, after 5 min of stirring at room temperature, with 9.8 g (55 mmol) of N-bromosuccinimide introduced at once into the suspension. Stirring is continued for 90 min. The solids are filtered through a pad of Celite, rinsed with acetone (30-50 mL) and the filtrate is concentrated at 20-25°C (20 mm) to give a white pasty solid. Water (80 mL) is added and the mixture is extracted with ether (3 x 80 mL). The ethereal layer is dried over anhydrous magnesium sulfate and the solvent is evaporated, leaving a semi-solid residue which is bulb-to-bulb distilled in an oil-bath at 45-50°C (≈0.1 mm) to give tert-butyl 3-bromopropiolate as a white semi-solid (8.51 g, 41.5 mmol, 97%) (Note 7).

2. Notes

1. Explosions during distillation of certain bromoalkynes have been reported.[2] Although methyl 3-bromopropiolate was not specifically cited, precautionary measures are recommended. The tert-butyl ester is a new compound[3] and must be handled like the methyl ester.

2. Acetone may be redistilled before use to remove the eventual autocondensation product, 4-hydroxy-4-methyl-2-pentanone (diacetone alcohol, bp 166°C). The submitter used, as received, fresh 99% pure acetone from Prolabo.

3. Methyl propiolate 99%, N-bromosuccinimide 99%, and silver nitrate were obtained from Janssen Chimica and used as received.

4. One hour of stirring is usually sufficient for completion of the reaction, but this time can be exceeded; longer reaction times improve the succinimide particle size, which aids the subsequent filtration. The product seems to adhere to the succinimide.

5. Depending on the conditions of the preliminary evaporation this product is contaminated with up to 5% of residual acetone. Nevertheless, it can be considered as pure enough for certain uses. Complete removal of acetone may be obtained by distillation in a small Vigreux column, although a yellowing of the distillate is observed: bp 86-88°C/100 mm (lit.[2] 88°C/100 mm and lit.[4] 40-45°C/5 mm). The spectral and analytical properties of methyl 3-bromopropiolate are as follows: ^{1}H NMR (200 MHz, $CDCl_3$) δ: 3.73 (s, CH_3); ^{13}C NMR (50 MHz, $CDCl_3$) δ: 52.8 (C-3), 52.9 (CH_3), 72.4 (C-2), 152.8 (C-1). Anal. Calcd for $C_4H_3BrO_2$ (undistilled sample): C, 29.48; H, 1.86; Br, 49.03. Found: C, 29.11; H, 1.92; Br 48.79.

6. tert-Butyl propiolate was prepared from propiolic acid and isobutene in the presence of sulfuric acid.[5] It is now commercially available from Fluka Chemical Corp. and Aldrich Chemical Company, Inc.

7. Although pure enough to be used as obtained, tert-butyl 3-bromopropiolate may be distilled in a Vigreux column at 75-77°C (15 mm) to give a colorless oil crystallizing as plates: mp 25-27°C. The spectral and analytical properties of tert-butyl 3-bromopropiolate are as follows: ^{1}H NMR (200 MHz, $CDCl_3$) δ: 1.46 (s, CH_3); ^{13}C NMR (50 MHz, $CDCl_3$) δ: 27.8 (CH_3), 50.0 (C-3), 73.9 (C-2), 84.0 ($\underline{C}Me_3$), 151.3 (C-1). Anal. Calcd for $C_7H_9BrO_2$ (undistilled sample): C, 41.00; H, 4.42; Br, 38.48. Found: C, 40.48, H, 4.40; Br, 38.10.

Waste Disposal Information

All toxic materials were disposed of in accordance with "Prudent Practices in the Laboratory"; National Academic Press; Washington, DC, 1996.

3. Discussion

Methyl 3-bromopropiolate has been prepared by esterification of 3-bromopropiolic acid with methanol and sulfuric acid for 6 days (75% yield),[4] the starting bromo acid being prepared by bromination of propiolic acid with aqueous potassium hypobromite.[6] This reaction is particularly delicate to control, giving erratic results. Moreover, direct bromination of methyl propiolate with sodium hypobromite[2] could not be reproduced.

Bromination of 1-alkynes with N-bromosuccinimide in the presence of catalytic amounts of silver nitrate, was used first for the bromination of 17-ethynyl steroids.[7] Similarly, N-iodosuccinimide led to 17-iodoethynyl steroids. Iodination of propiolates in this way has not been studied. A recent method of preparation of 1-iodoalk-1-ynes under phase-transfer conditions involves molecular iodine and copper(I) iodide as catalyst, in the presence of potassium or sodium carbonate as a base. Ethyl 3-iodopropiolate was prepared by this route in 80% yield.[8]

The present procedure provides ready access to 3-bromopropiolic esters, the methyl ester requiring adapted work up, because of its low boiling point. Less volatile esters, like tert-butyl, can be conveniently isolated by a standard aqueous-extraction work up.

Methyl 3-bromopropiolate has been used in Diels-Alder reactions either as a methoxycarbonyl ketene equivalent[9] or for the synthesis of functionalized naphthalenes.[10]

1. Département de Chimie, Ecole Normale Supérieure, 24, rue Lhomond, F-75231 Paris Cedex 05, France.

2. Chodkiewicz, W. *Ann. chim.* **1957**, *2*, 819.

3. Leroy, J. *Synth. Commun.* **1992**, *22*, 567.

4. Heaton, C. D.; Noller, C. R. *J. Am. Chem. Soc.* **1949**, *71*, 2948.

5. Sondheimer, F.; Stjernstrom, N.; Rosenthal, D. *J. Org. Chem.* **1959**, *24*, 1280.

6. Straus, F.; Kollek, L.; Heyn, W. *Ber.* **1930**, *63B*, 1868.

7. Hofmeister, H.; Annen, K.; Laurent, H.; Wiechert, R. *Angew. Chem., Int. Ed. Engl.* **1984**, *23*, 727.

8. Jeffery, T. *J. Chem. Soc., Chem. Commun.* **1988**, 909.

9. (a) Chamberlain, P.; Rooney, A. E. *Tetrahedron Lett.* **1979**, 383; (b) Leroy, J. *Tetrahedron Lett.* **1992**, *33*, 2969.

10. Andersen, N. G.; Maddaford, S. P.; Keay, B. A. *J. Org. Chem.* **1996**, *61*, 2885.

Appendix

Chemical Abstracts Nomenclature (Collective Index Number); (Registry Number)

Methyl 3-bromopropiolate: Propiolic acid, bromo-, methyl ester (8); 2-Propynoic acid, 3-bromo-, methyl ester (9); (23680-40-2)

tert-Butyl 3-bromopropiolate: 2-Propynoic acid, 3-bromo-, 1,1-dimethylethyl ester (13) (140907-23-9)

Acetone (8); 2-Propanone (9); (67-64-1)

Methyl propiolate: Propiolic acid, methyl ester (8); 2-Propynoic acid, methyl ester (9); (922-67-8)

N-Bromosuccinimide: Succinimide, N-bromo- (8); 2,5-Pyrrolidinedione, 1-bromo- (9); (128-08-5)

tert-Butyl propiolate: Propiolic acid, tert-butyl ester (8); 2-Propynoic acid, 1,1-dimethylethyl ester (9); (13831-03-3)

MESITYLENESULFONYLHYDRAZINE, AND (1α,2α,6β)-2,6-DIMETHYLCYCLOHEXANECARBONITRILE AND (1α,2β,6α)-2,6-DIMETHYLCYCLOHEXANECARBONITRILE AS A RACEMIC MIXTURE

(Benzenesulfonic acid, 2,4,6-trimethyl-, hydrazide)

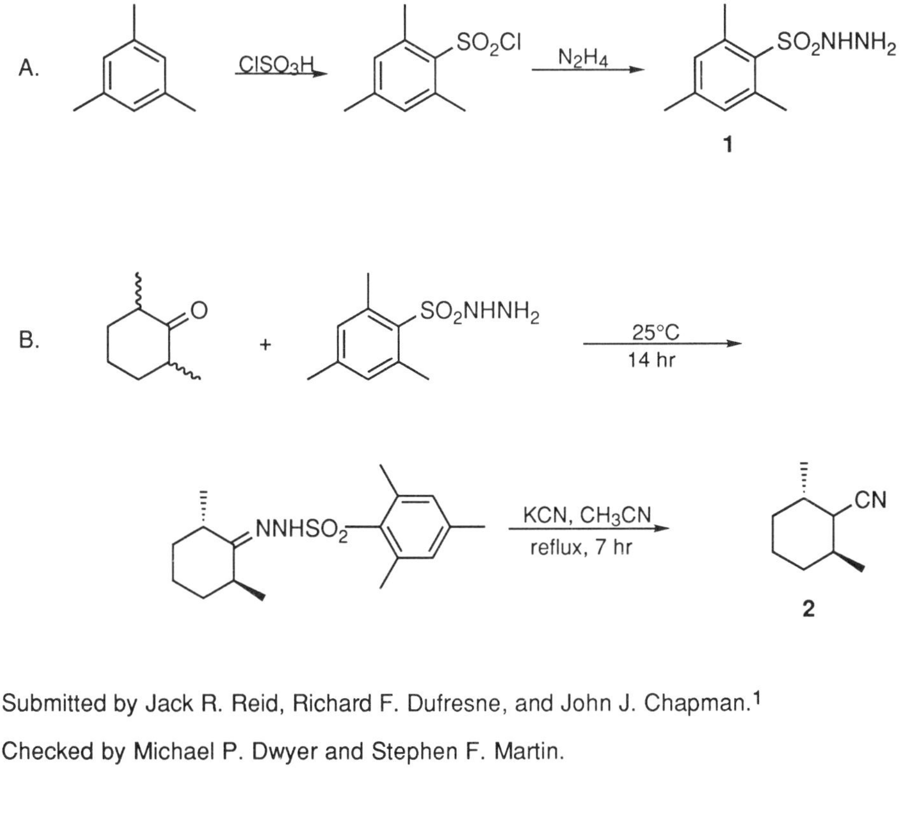

A. $ClSO_3H$ → SO_2Cl → N_2H_4 → SO_2NHNH_2

1

B. O + SO_2NHNH_2 → 25°C, 14 hr

$NNHSO_2$ → KCN, CH_3CN, reflux, 7 hr → CN

2

Submitted by Jack R. Reid, Richard F. Dufresne, and John J. Chapman.[1]

Checked by Michael P. Dwyer and Stephen F. Martin.

1. Procedure

Caution! Mesitylene is an irritant, potassium cyanide is highly toxic, chlorosulfonic acid is corrosive, hydrazine monohydrate is a toxic, cancer-suspect agent, and dichloromethane is an irritant and should be handled in a well-ventilated hood.

A. Mesitylenesulfonylhydrazine (**1**). A 500-mL, three-necked, round-bottomed flask is assembled as shown in Figure 1 (Note 1). The equipment consists of a 250-mL constant pressure addition funnel, a thermometer, and an air-cooled Friedrichs condenser attached to a gas inlet-outlet adapter. A 250-mL vacuum flask is equipped with a gas delivery tube that is inserted through a single-holed rubber stopper and positioned 1 cm above the bottom of an aqueous 20% sodium hydroxide solution. This assembly is used as a gaseous hydrogen chloride (HCl) trap (Note 2). The flask is charged with mesitylene (Note 3) (76.0 g, 0.630 mol) and its contents cooled to between -10°C and 0°C in a wet ice-acetone bath while stirring (Note 4). When the contents of the flask and HCl trap are between -10°C and 0°C, a 6-mL portion of the total quantity of the chlorosulfonic acid (Note 3) (161.0 g, 92.1 mL, 1.39 mol) is added to initiate the reaction. After HCl gas evolution begins, the remaining chlorosulfonic acid is added at a rate such that the reaction temperature remains below 60°C (Note 5). After the chlorosulfonic acid is added, the reaction mixture is heated to 60°C to disperse and dissolve any precipitated salts (Note 6) and then allowed to cool to room temperature. The reaction mixture is poured into 250 mL of ice water with stirring, and the crude crystals are recovered by suction filtration. The crystals are washed with generous portions of ice-cold water. The crude product is taken up in 125 mL of dichloromethane and a 50-mL upper aqueous layer is removed. The aqueous layer is extracted with one 50-mL portion of methylene chloride. The combined organic fractions are dried over sodium sulfate (Note 7) with vigorous stirring, filtered and then concentrated on a rotary evaporator. Crude mesitylenesulfonyl chloride (109.7 g, 80%) is recovered as an oil that solidifies into off-white crystals, mp 54-56°C (Note 8). If used immediately, no additional purification is necessary (Note 9).

Using the equipment in Figure 1, but without the HCl gas trap attached, mesitylenesulfonyl chloride (109.7 g, 0.503 mmol, Note 7) is dissolved in 175 mL of dry tetrahydrofuran (THF) and cooled to between -10°C and 0°C in a wet ice-acetone bath. Hydrazine monohydrate (63.0 g, 61.0 mL, 1.26 mol) (Note 3) is dissolved in 30 mL of ice-cold water and added so that the temperature remains below 25°C (Note 10). After the addition is completed, the mixture is stirred for an additional 45 min at room temperature and poured into 250 mL of ice water with stirring. Crude mesitylenesulfonylhydrazine is recovered by suction filtration and then dissolved in 500 mL of dichloromethane. A 150- to 175-mL portion of water separates. The water layer is removed, and the organic layer is washed with one 50-mL portion of ice-cold water, dried over sodium sulfate, filtered, and the solvent is removed on a rotary evaporator. The crystals are triturated in 250 mL of cold hexane, and the fluffy white product (95.4 g, 89%, Note 11) is recovered by suction filtration (Notes 12 and 13).

B. (1α,2α,6β)-2,6-Dimethylcyclohexanecarbonitrile and (1α,2β,6α)-2,6-Dimethylcyclohexanecarbonitrile (**2**). The apparatus depicted in Figure 1 lacking the HCl trap and with a glass stopper in place of the thermometer is used for this procedure. The 1-L reaction flask is charged with mesitylenesulfonylhydrazine (56.0 g, 0.262 mol) (Note 1) and 175 mL of acetonitrile. The mixture is stirred at room temperature until solution is attained (Note 14). 2,6-Dimethylcyclohexanone (Note 3) (32.2 g, 0.255 mol) is added in one portion and the reactants are stirred an additional 10 min. A 10-drop portion of concentrated sulfuric acid is added and the mixture is stirred at room temperature for 12 to 18 hr to facilitate complete hydrazone formation (Note 15). Water is circulated through the reflux condenser that is already attached to the flask, and potassium cyanide (27.2 g, 0.418 mol) is added (Note 3). The reaction mixture is gradually heated to reflux over 2 hr and then gently refluxed for 10 to 12 hr (Note 16). The contents of the flask are cooled to room temperature and water (125 mL), 20% aqueous sodium hydroxide solution (25 mL), hexane (175 mL), and ether (75 mL) are added in succession with vigorous mixing. The mixture is stirred for 15 min, and the lower aqueous layer is separated (Note 17). The aqueous layer is extracted with a 7:3 hexane-ether mixture (2 x 75 mL). The combined organic layers are washed with two 25-mL portions of aqueous 10% sodium hydroxide solution and dried over sodium sulfate. The organic layer is filtered and concentrated on a rotary evaporator, and the crude yellow residual oil is distilled at 100-102°C at 39 mm to give 25.5 g (73%) of colorless **2** (Notes 18 and 19).

2. Notes

1. The submitters ran the reaction on a scale four times that described. The submitters also note that the reaction may be successfully carried out in a 2-L Erlenmeyer flask equipped with a thermometer to monitor the internal reaction temperature and a large magnetic stirring assembly. The gas delivery tube and rubber stopper, used in Figure 1, is placed in the mouth of the flask, and the HCl gas is forced into a base trap or disposed of through an aspirator.

2. The HCl trap is charged with 150 mL of aqueous 20% sodium hydroxide solution and cooled in an ice-water or wet ice-acetone bath.

3. Mesitylene (1,3,5-trimethylbenzene), chlorosulfonic acid, hydrazine monohydrate, 2,6-dimethylcyclohexanone and potassium cyanide were purchased from Aldrich Chemical Company, Inc., and used without further purification.

4. A 250-g quantity of wet ice and 250 mL of acetone give a bath temperature between -15°C to -20°C. The bath generally does not need to be recharged.

5. After half of the chlorosulfonic acid has been added, the exothermic reaction and gas evolution decrease considerably. If the temperature is allowed to rise above 60°C, the yield decreases and the final product is slightly discolored.

6. Briefly warming the reaction mixture solubilizes all the mesitylenesulfonic acid and drives the reaction to completion. Mesitylenesulfonic acid is less soluble than mesitylenesulfonyl chloride. Therefore, after everything dissolves, conversion of the acid to the acid chloride is complete. The product is partially hydrolyzed if it is poured into water while it is still hot.

7. Methylene chloride partitions water from the crude product in this step and greatly reduces the drying time. However, the mixture is stirred rapidly to bring the water that naturally partitions to the top of the methylene chloride layer in contact with the drying agent that is on the bottom of the flask.

8. The spectral properties of the product are as follows: ^{1}H NMR (300 MHz, $CDCl_3$) δ: 2.35 (s, 3 H), 2.73 (s, 6 H), 7.03 (s, 2 H); ^{13}C NMR (75 MHz, $CDCl_3$) δ: 21.2, 22.9, 132.3, 139.5, 140.2, 145.3; mp 54-56°C.

9. If mesitylenesulfonyl chloride is prepared without delay, the damp crystals may be used in the next step without further drying. If the product is to be stored for long periods of time, it is necessary to remove the last traces of water. This is also a convenient place to stop if the reaction will not be completed in a single day.

10. Temperature control and efficient stirring are more important here than in the preparation of mesitylenesulfonyl chloride. Better yields are obtained when the temperature is kept below 25°C. Gas evolution accompanied by some decomposition occurs at elevated temperatures.

11. The spectral and physical properties are as follows: ^{1}H NMR (300 MHz, $CDCl_3$) δ: 2.31 (s, 3 H), 2.65 (s, 6 H), 3.61 (br s, 2 H), 5.75 (br s, 1 H), 6.99 (s, 2 H); ^{13}C NMR (75 MHz, $CDCl_3$) δ: 21.0, 22.9, 130.0, 132.1, 140.7, 143.3; mp 114-116°C (lit.[5] mp 115-116°C).

12. Additional product separates from the filtrate upon standing. Excessive work-up water results in product loss. The crude product can hold considerable volatile material. The weight of the product is usually greater than theoretical prior to thorough air or vacuum drying. Mesitylenesulfonylhydrazine must be completely dry before it is used in a reaction.

13. Mesitylenesulfonylhydrazine should be stored in a brown bottle in an area protected from light. Refrigeration of the product, though not required, tends to increase the shelf life.

14. The checkers noted that gentle warming of the solution with a heat gun greatly facilitated the dissolution of mesitylenesulfonylhydrazine.

15. A catalytic amount of sulfuric acid greatly facilitates formation of the hydrazone which precipitates from solution as it is formed. The checkers found that the reaction mixture must be stirred at room temperature for 14-18 hr to ensure complete formation of the hydrazone. It is advisable to monitor the reaction by TLC. The submitters noted that formation of the hydrazone can be reduced to several hours when the reaction mixture is heated at 40°C.

16. Unsolvated potassium cyanide and hydrazone dissolve as the reactants are converted to the thermodynamically favored trans-nitrile. Shortly after all of the solid dissolves, a second solid precipitates from solution. This solid is presumed to be potassium mesitylenesulfinate and is usually accompanied by some foaming. An oversize flask used with efficient stirring keeps the reaction mixture from foaming into the condenser.

17. Commercial Clorox, with an activity of 5.25%, is used to destroy excess potassium cyanide prior to disposing of the aqueous mixture.

18. A rotary evaporator water bath setting is kept at room temperature during the concentration of the product.

19. The spectral properties of the product are as follows: IR ($CHCl_3$) cm^{-1}: 2933, 2236, 1459, 1384, 1230; ^1H NMR (300 MHz, $CDCl_3$) δ: 0.99 (d, 3 H, J = 7.1), 1.02 (d, 3 H, J = 7.1), 1.37-1.48 (m, 4 H), 1.67-1.72 (m, 1 H), 1.86-1.93 (m, 1 H), 2.00-2.03 (m, 1 H), 2.29-2.33 (m, 1 H); ^{13}C NMR (75 MHz, $CDCl_3$) δ: 16.5, 19.2, 19.3, 28.9, 30.0, 30.5, 31.2, 40.4, 121.0.

Waste Disposal Information

All toxic materials were disposed of in accordance with "Prudent Practices in the Laboratory"; National Academic Press; Washington, DC, 1996. See Note 17.

3. Discussion

This is a simplified procedure that uses much less chlorosulfonic acid than previous methods,[2,3] while obtaining a better yield of mesitylenesulfonylhydrazine[4-6] via mesitylenesulfonyl chloride. Mesitylenesulfonylhydrazine is used for preparing medium-size cycloalkanones by the Eschenmoser fragmentation reaction,[7] diazo compounds,[8] regiospecific alkylations,[9] and hydrogenations.[5] The hydrolysis of mesitylenesulfonyl chloride and the decomposition of mesitylenesulfonylhydrazine is rather slow at low temperatures and short contact times; therefore, the preparations are conveniently worked-up in ice water. However, if damp crystals of mesitylenesulfonylhydrazine are stored at room temperature for extended periods, an odor of SO_2 becomes evident. Therefore, an alternative procedure for working up very dry mesitylenesulfonyl chloride and mesitylenesulfonylhydrazine was developed. This is especially useful if the reaction will not be completed in one session or if the material will be stored for an extended period. The hydrazine prepared in the first step is used to prepare the racemic product ($1\alpha,2\alpha,6\beta$)- and ($1\alpha,2\beta,6\alpha$)-2,6-

dimethylcyclohexanecarbonitrile via the corresponding hydrazone in good to excellent yield. This procedure represents a simple, stereoselective, large-scale, one-pot conversion of a moderately hindered ketone to the next higher nitrile analog. The crude product contains small amounts of olefin and starting ketone that can be removed by distillation. Ketones with α,α'-alkyl substituents may be used as diastereomeric mixtures, since they equilibrate to one pair of enantiomers during hydrazone formation, and this stereoselectivity is preserved during the cyanide anion reaction. A minimum of the toxic reagent potassium cyanide is used and evolution of hydrogen cyanide is avoided. The use of oxidizing agents such as bromine and strong bases such as sodium methoxide in methanol are avoided, making this method more tolerant of other substituents on the ketone such as olefins.[10-14] The reagent 2,4,6-triisopropylbenzenesulfonylhydrazine is less effective for converting moderately hindered ketones to nitriles, more difficult to prepare, and more expensive to purchase. The nitrile products are useful intermediates in the synthesis of acids by saponification, aldehydes by reduction with diisobutylaluminum hydride (DIBAL), and ketones by reaction with Grignard reagents.[15-17] The scope of this procedure is indicated by the modestly hindered nitriles shown in Table.

1. Lorillard Research Center, 420 English Street, Greensboro, NC 27420. The authors would like to thank Ms. Gertrude Rashada Um'Rani for her technical assistance.
2. Huntress, E. H.; Carten, F. H. *J. Am. Chem. Soc.* **1940**, *62*, 511.
3. Huntress, E. H.; Autenrieth, J. S. *J. Am. Chem. Soc.* **1941**, *63*, 3446.
4. Grammaticakis, P. *C.R. Acad. Sci., Ser. C* **1967**, *264*, 2067.
5. Cusack, N. J.; Reese, C. B.; Risius, A. C.; Roozpeikar, B. *Tetrahedron* **1976**, *32*, 2157.
6. Kascheres, C.; Van Fossen Bravo, R. *Org. Mass Spectrom.* **1979**, *14*, 293.
7. Reese, C. B.; Sanders, H. P. *Synthesis* **1981**, 276.
8. Dudman, C. C.; Reese, C. B. *Synthesis* **1982**, 419.
9. Lipton, M. F.; Shapiro, R. H. *J. Org. Chem.* **1978**, *43*, 1409.
10. Orere, D. M.; Reese, C. B. *J. Chem. Soc., Chem. Commun.* **1977**, 280.
11. Johnson, F.; Duquette, L. G. *J. Chem. Soc., Chem. Commun.* **1969**, 1448.
12. Jacobs, T. L.; Fenton, D. M.; Reed, R. *J. Org. Chem.* **1962**, *27*, 87.
13. Culbertson, J. B.; Butterfield, D.; Kolewe, O.; Shaw, R. *J. Org. Chem.* **1962**, *27*, 729.
14. Wender, P. A.; Eissenstat, M. A.; Sapuppo, N.; Ziegler, F. E. *Org. Synth., Coll. Vol. VI* **1988**, 334.
15. Trofimenko, S. *J. Org. Chem.* **1964**, *29*, 3046.
16. Miller, A. E. G.; Biss, J. W.; Schwartzman, L. H. *J. Org. Chem.* **1959**, *24*, 627.
17. Weiberth, F. J.; Hall, S. S. *J. Org. Chem.* **1987**, *52*, 3901.

Appendix

Chemical Abstracts Nomenclature (Collective Index Number); (Registry Number)

Mesitylenesulfonylhydrazine: 2-Mesitylenesulfonic acid, hydrazide (8);
Benzenesulfonic acid, 2,4,6-trimethyl-, hydrazide (9); (16182-15-3)

Mesitylene (8); Benzene, 1,3,5-trimethyl- (9); (108-67-8)

Chlorosulfonic acid: HIGHLY TOXIC: (8,9); (7790-94-5)

Mesitylenesulfonyl chloride: 2-Mesitylenesulfonyl chloride (8);
Benzenesulfonyl chloride, 2,4,6-trimethyl- (9); (773-64-8)

Hydrazine: HIGHLY TOXIC. CANCER SUSPECT AGENT: (8,9); (302-01-2)

Acetonitrile: TOXIC: (8,9); (75-05-8)

2,6-Dimethylcyclohexanone: Cyclohexanone, 2,6-dimethyl- (8,9); (2816-57-1)

Potassium cyanide: HIGHLY TOXIC: (8,9); (151-50-8)

TABLE

KETONES TO NITRILES VIA MESITYLENESULFONYLHYDRAZINE

Entry	Ketone	Nitrile	Yield,[a] %	Solvent[c]	BP (°C/mm)
1	4-tert-Butylcyclohexanone	4-tert-Butylcyclohexanecarbonitrile	73.0-76.7	B	140-142/30
2	4,4-Dimethylcyclohexanone	4,4-Dimethylcyclohexanecarbonitrile	73.8	B	95-97/30
3	4-Isopropylcyclohexanone	4-Isopropylcyclohexanecarbonitrile	73.0-83.0	B	130-131/35
4	Pinacolone	2,3,3-Trimethylbutanenitrile	60.3	A	54-57/33
5	3,4,4-Trimethylcyclohexanone	3,4,4-Trimethylcyclohexanecarbonitrile	80.5	B	127-130/30
6	3,3,5-Trimethylcyclohexanone	3,3,5-Trimethylcyclohexanecarbonitrile	72.6	A	110-118/26
7	3-tert-Butylcyclopentanone	3-tert-Butylcyclopentanecarbonitrile	78.7	A	117-121/28
8	2,6-Dimethylcyclohexanone	2,6-Dimethylcyclohexanecarbonitrile	73.5[b]	A	97-99/35

[a]Satisfactory analytical data, ± 0.3% C, H, were obtained for all of the products.

[b]The product is 90+% the racemic, and between 1% and 5% the meso nitriles.

[c]Solvent or solvent system used in the reaction: A = acetonitrile; B = 7:1 acetonitrile: 2-methoxyethanol.

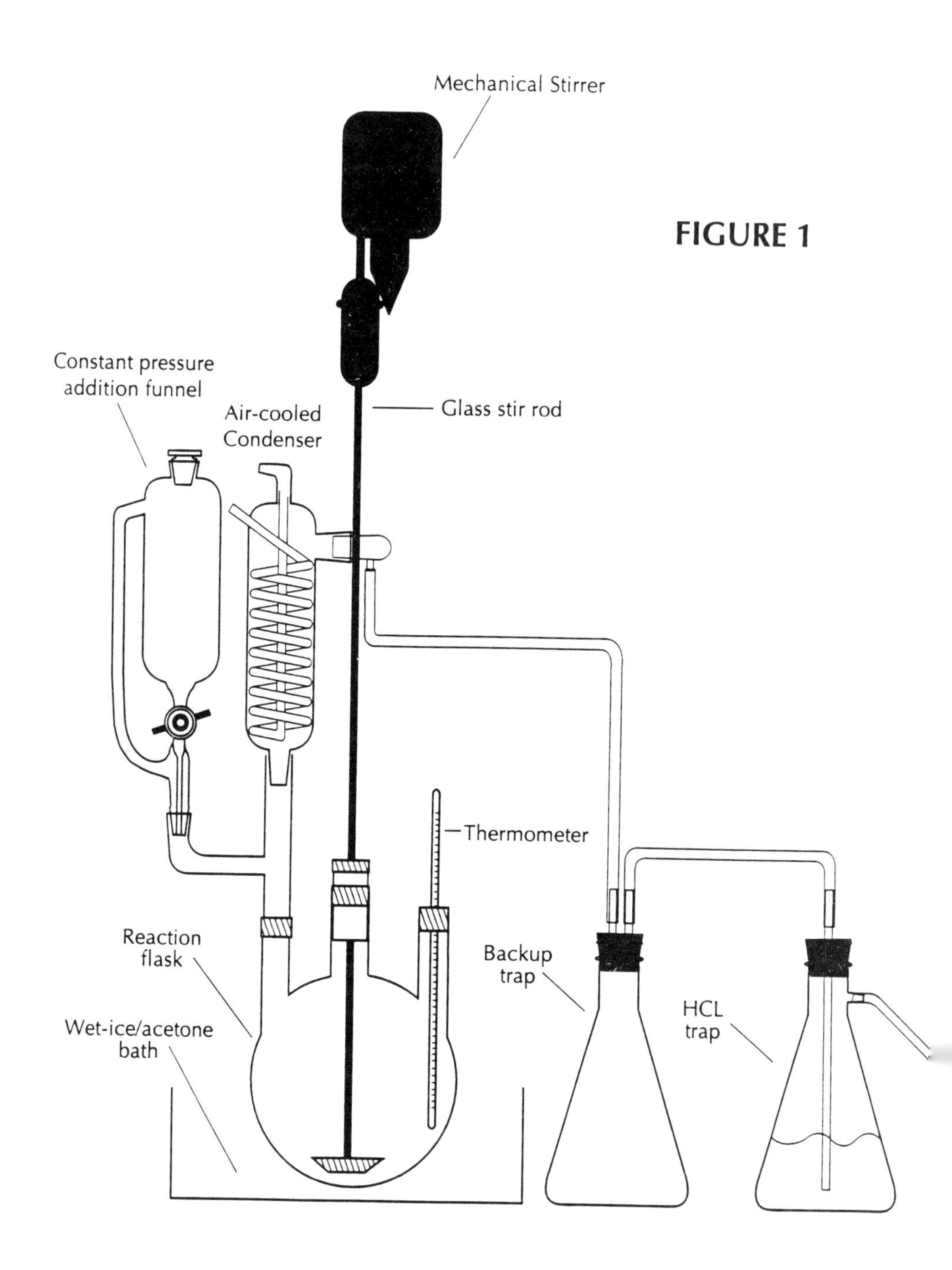
Mechanical Stirrer
FIGURE 1
Constant pressure addition funnel
Air-cooled Condenser
Glass stir rod
Thermometer
Reaction flask
Wet-ice/acetone bath
Backup trap
HCL trap

RHODIUM-CATALYZED HETEROCYCLOADDITION OF A DIAZOMALONATE AND A NITRILE: SYNTHESIS OF 4-CARBOMETHOXY-5-METHOXY-2-PHENYL-1,3-OXAZOLE

(4-Oxazolecarboxylic acid, 5-methoxy-2-phenyl-, methyl ester)

A. $CH_3CONH-C_6H_4-SO_2N_3$ → ($CH_3O_2CCH_2CO_2CH_3$; Et_3N + CH_3CN) → $CH_3O_2C-C(=N_2)-CO_2CH_3$

B. $CH_3O_2C-C(=N_2)-CO_2CH_3$ → (PhCN / $Rh_2(OAc)_4$ (cat.); $CHCl_3$, reflux) → 4-CO_2CH_3-5-OCH_3-2-Ph-oxazole

Submitted by Joshua S. Tullis and Paul Helquist.[1]
Checked by Anne K. Courtney and Stephen F. Martin.

1. Procedure

CAUTION! This experiment should be conducted behind an auxiliary safety shield in a closed fume hood because of the explosive nature of diazo and azide compounds.

A. Dimethyl diazomalonate. Into a 500-mL, round-bottomed flask is placed a large magnetic stirring bar (Note 1), 4-acetamidobenzenesulfonyl azide (72.3 mmol, 17.36 g, Note 2), acetonitrile (250 mL, Note 3), and triethylamine (76.5 mmol, 7.74 g, Note 4). The reaction mixture is cooled to 0°C, and dimethyl malonate (72.0 mmol, 9.51 g, Note 4) is slowly added over 3 - 5 min by a syringe. The reaction mixture is kept at 0°C for 10 min then stirred at room temperature for 17 hr. A thick white precipitate that forms during the reaction is removed by filtration through a Büchner filter. The filter cake is washed with a 1:1 mixture of petroleum ether and ethyl ether (200 mL, Note 5). The resulting filtrates are concentrated by rotary evaporation under reduced pressure giving a yellow oil with some visible precipitate of sulfonamide. Repeated trituration and filtration of this oil with 1:1 petroleum ether and ethyl ether (6 x 50 mL, Note 5) to remove the solid 4-acetamidobenzenesulfonamide, followed by concentration of all filtrates by rotary evaporation under reduced pressure, gives 10.92 g of the crude diazo product as a yellow oil (96%). The crude product is purified by distillation to provide 8.26 g (73%) of dimethyl diazomalonate as a bright yellow oil (Note 6).

B. 4-Carbomethoxy-5-methoxy-2-phenyl-1,3-oxazole. In an oven-dried, 100-mL, round-bottomed flask is placed a small magnetic stirring bar (Note 7), rhodium(II) acetate dimer, $Rh_2(OAc)_4$, (0.230 mmol, 0.102 g, Note 4), benzonitrile (19.5 mmol, 2.011 g, Note 8), and chloroform ($CHCl_3$, 5 mL, Note 9). The flask is placed in an oil bath and fitted with a 30-cm reflux condenser. On top of the condenser is placed a three-way stopcock having horizontal and vertical tubulations. A source of nitrogen (N_2) is connected to the system via the horizontal tubulation of the stopcock, and a septum is placed over the vertical tubulation. The entire system is flushed several times with N_2, and the solution is heated to reflux (68°C) during which time it becomes deep purple.

Into a dry, 50-mL, conical vial is placed dimethyl diazomalonate (29.4 mmol, 4.65 g) and 25 mL of chloroform (Note 9). The solution is taken up into a syringe (Note 10) fitted with a 30-cm needle. The needle is inserted about half way into the septum on the stopcock so that the needle tip extends down into the condenser. The syringe is placed on a syringe pump (Note 11) which is set to add the solution at a rate of 1.17 mL/hr over a period of about 25 hr. After the addition is complete, the syringe is again filled with 2 mL of $CHCl_3$, which is added to the system via the septum. The needle is removed, and the stopcock plug is rotated so that the septum is closed off from the system while the N_2 inlet remains open. The solution is allowed to continue stirring at reflux for another 20 hr. It is then concentrated by rotary evaporation under reduced pressure to yield a brownish-purple oil containing crude product. The crude oxazole is purified using flash column chromatography (silica gel, 40% ethyl acetate in hexane, Notes 12, 13, 14) to yield 2.73-2.96 g (60-65%) of 4-carbomethoxy-5-methoxy-2-phenyl-1,3-oxazole as white needles (Note 15).

2. Notes

1. An egg-shaped stirring bar ca. 2.5 cm in length is recommended for stirring the suspension that is produced during the reaction.

2. 4-Acetamidobenzenesulfonyl azide is obtained from Aldrich Chemical Company, Inc., and used without further purification. In addition, 4-acetamidobenzenesulfonyl azide can be synthesized from the precursor, 4-acetamidobenzenesulfonyl chloride.[2]

3. The checkers used standard reagent grade acetonitrile from EM Science.

4. Triethylamine, dimethyl malonate, and rhodium(II) acetate dimer were purchased from Aldrich Chemical Company, Inc., and used without further purification.

5. Standard reagent grade ether and petroleum ether are used during workups.

6. Crude dimethyl diazomalonate is distilled under reduced pressure using a vacuum pump and a short-path distillation apparatus equipped with a Vigreux fractionating column measuring 8 - 10 cm. The apparatus is wrapped in glass wool and aluminum foil to minimize heat loss. Pure dimethyl diazomalonate[3] is a bright yellow oil: bp 60°C, ca. 0.65 mm. (lit.[3c] bp 45°C, 0.2 mm); ^{1}H NMR (500 MHz, $CDCl_3$) δ: 3.79 (s, 6 H, $COOC\underline{H}_3$; lit.[3c] δ 3.76); ^{13}C NMR (125 MHz, $CDCl_3$) δ: 52.4, 65.6,

161.3; IR (neat) cm^{-1}: 3006 (m), 2958 (s), 2849 (w), 2138 (s), 1760 (s), 1738 (s), and 1696 (s); (lit.[3] IR).

7. The specific size of the stirring bar is not important, but the stirring rate should be adjusted so that the $Rh_2(OAc)_4$ catalyst remains suspended in solution. It should not be resting on the bottom of the round-bottomed flask, nor should it be splashed onto the sides of the flask.

8. Benzonitrile is available from Aldrich Chemical Company, Inc. and is distilled at reduced pressure (0.5 mm, 75°C) using a short-path distillation apparatus.

9. Spectroscopic grade chloroform is obtained from J. T. Baker, Inc., (99+% purity) and used without further purification.

10. The syringe used is a 50-mL, gas-tight syringe (Model #1050) available from Hamilton Company. It is very important that a gas-tight syringe be used during the addition because of the tendency of chloroform to leak with loss of starting material from other types of syringes.

11. The syringe pump used is available from Sage Instruments, Model Number 341B. *NOTE*: The slow addition of the dimethyl diazomalonate is of key importance to the success of the reaction. When scaling this reaction up or down, the rate of addition should be altered so that the addition time remains constant at ca. 25 hr. For example, in the reaction described here, there is ca. 30 mL of solution containing dimethyl diazomalonate, and it is added over 25 hr at a rate of 1.17 mL/hr. If the reaction were at 20% scale, the rate of addition would be 0.23 mL/hr.

12. On one occasion, crude product was a solid and was dissolved in a minimum amount of eluent and added to the top of the silica column.

13. To purify the product on this scale, a 35-mm wide column with a 500-mL glass bulb on top and a glass frit at the base is used. It is packed with 25-30 cm silica gel 60 (230-400 mesh, available from EM Science) and eluted with solvent to remove

any air bubbles. The purification is monitored by TLC (silica gel, 40% ethyl acetate in hexanes; R_f = 0.35).

14. Usually, the purification leads to several fractions containing a mixture of the product and the starting material, dimethyl diazomalonate, (R_f = 0.43). In this event, an alternative to performing another column chromatography has been developed based on trituration according to the following procedure. All fractions that contain the mixture are combined in a round-bottomed flask, and the solvent is removed by rotary evaporation under reduced pressure. The mixture should appear to be a yellow oil mixed with the crystalline oxazole. With an ice bath, the mixture is cooled along with a separate flask containing 10-15 mL ether. Using a pipette, 2-3 mL of the cold ether is transferred to the mixture. The mixture is swirled, and the solvent is removed with a pipette. This procedure is repeated several times until only white crystals remain. It is important to keep both flasks cool throughout the procedure so that loss of the product is minimized.

15. The product, 4-carbomethoxy-5-methoxy-2-phenyl-1,3-oxazole, is obtained as colorless needles: mp 92-95°C (lit.[4] mp 98-99°C); ^1H NMR (500 MHz, $CDCl_3$) δ: 3.86 (s, 3 H, $COOC\underline{H}_3$), 4.20 (s, 3 H, $C{=}COC\underline{H}_3$), 7.36-7.39 (m, 3 H, Ar-$\underline{H}$), 7.89-7.93 (m, 2 H, Ar-$\underline{H}$); ^{13}C NMR[4] (125 MHz, $CDCl_3$) δ: 51.7 ($COO\underline{C}H_3$), 59.7 ($C{=}CO\underline{C}H_3$), 107.4 (C-4), 125.8 (C-3', C-5'), 126.2 (C-1'), 128.6 (C-2', C-6'), 130.3 (C-4'), 150.8 (C-5), 161.7, 161.8 ($\underline{C}OOCH_3$, C-2); IR ($CDCl_3$) cm^{-1}: 1713 ($\underline{COO}CH_3$), 1619 ($\underline{C{=}C}OCH_3$); mass spectrum[4] (CI, argon) m/e (rel intensity) 233 (M^+, 27), 173 (11), 146 (15), 105 (100), 77 (24). Anal Calcd for $C_{12}H_{11}NO_4$: C, 61.80; H, 4.75. Found: C, 61.53; H, 4.75.

Waste Disposal Information

All toxic materials were disposed of in accordance with "Prudent Practices in the Laboratory"; National Academic Press; Washington, DC, 1996.

3. Discussion

Oxazoles are well-investigated compounds. The occurrence, uses, and synthesis of oxazole derivatives have been the subjects of extensive reviews.[5] The heterocyclic oxazole unit is seen with various substitution patterns in a large number of naturally occurring compounds. Furthermore, oxazoles serve as synthetic intermediates leading to many other systems.[5,6,7] In this context, oxazoles have seen, for example, numerous applications as "2-azadiene" components in 4+2 cycloadditions with several types of dienophiles. Further transformations of the products then lead to a number of other nitrogen- or oxygen-containing heterocyclic products.[6]

Of the many specific types of oxazoles, those bearing a 4-carboxy-derived group are of considerable importance. Among these compounds are several natural products, including many which contain this moiety, or the analogous thiazole group. Two examples, from the type A streptogramin family, include virginiamycin M_1 and madumycin I.[8,9]

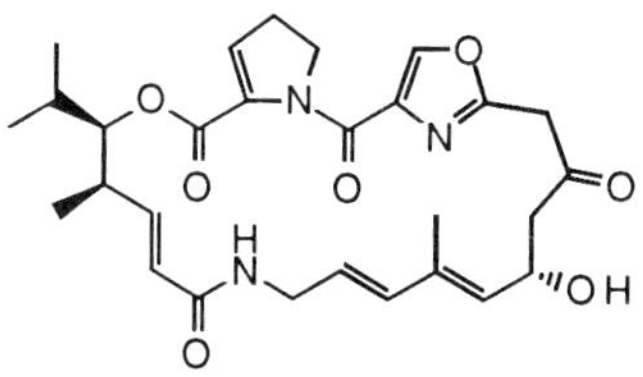

Virginiamycin M_1 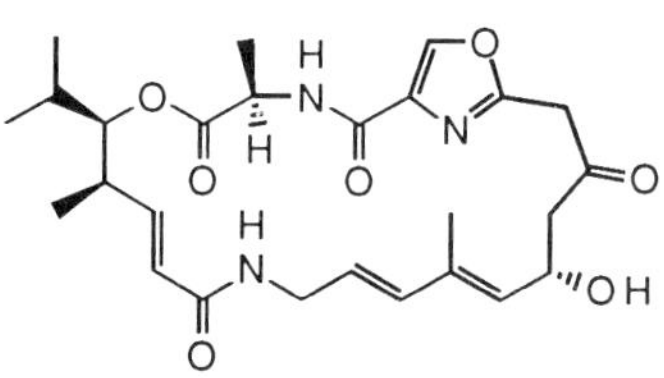 Madumycin I

Because of the many roles that these compounds play in natural products and synthetic organic chemistry, many methods have been developed for the preparation of 4-carboxy derivatives of oxazoles.[5,10] A well-established, classical method is the

Cornforth oxazole synthesis (Scheme 1). This method has been shown to be widely applicable, although it employs a complex, multi-step pathway.[11]

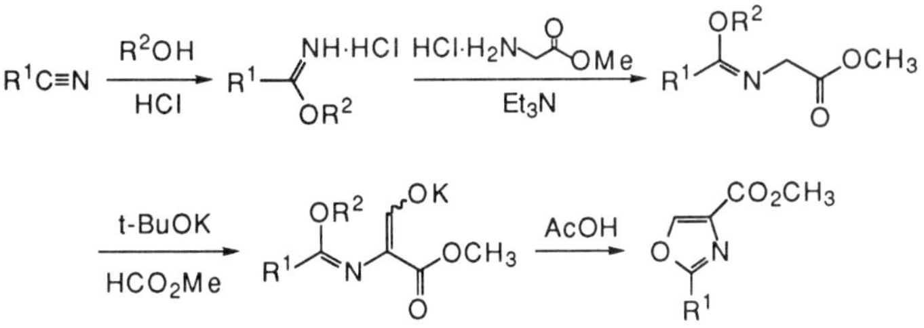

Scheme 1

In consideration of conceivable strategies for the more direct construction of these derivatives, nitriles can be regarded as simple starting materials with which the 3+2 cycloaddition of acylcarbenes would, in a formal sense, provide the desired oxazoles. Oxazoles, in fact, have previously been obtained by the reaction of diazocarbonyl compounds with nitriles through the use of boron trifluoride etherate as a Lewis acid promoter. Other methods for attaining oxazoles involve thermal, photochemical, or metal-catalyzed conditions.[12] Several recent studies have indicated that many types of rhodium-catalyzed reactions of diazocarbonyl compounds proceed via formation of electrophilic rhodium carbene complexes as key intermediates rather than free carbenes or other types of reactive intermediates.[13] If this postulate holds for the reactions described here, then the mechanism outlined in Scheme 2 may be proposed, in which the carbene complex **3** and the adduct **4** are formed as intermediates.[14]

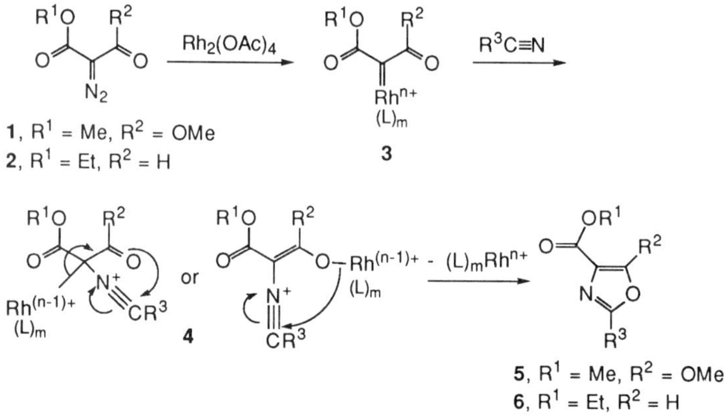

Scheme 2

The particular reaction described in Scheme 2 using dimethyl diazomalonate produces oxazoles **5** that bear a methoxy group at C-5. If desired, this substituent may be removed in some cases by reductive cleavage using $LiB(Et)_3H$ to give the 5-unsubstituted oxazoles **6**.[3,15] Alternatively, the 5-unsubstituted derivatives **6** may be obtained directly through the use of diazo formylacetate (**2**) in place of dimethyl diazomalonate (**1**).[3,15] Some additional, representative examples of the use of **1** and **2** are shown below in the Table.

In conclusion, the rhodium-catalyzed reaction of diazo dicarbonyl compounds with nitriles provides direct access to useful classes of functionalized oxazoles. This direct, one-step pathway provides a significant improvement over past preparations.

TABLE

ADDITIONAL EXAMPLES OF OXAZOLE FORMATION USING DIMETHYL DIAZOMALONATE **1** AND DIAZO FORMYLACETATE **2**[a]

Nitrile	Diazo Compound	Product	Yield
3-ClC_6H_4CN	**1**	H_3CO, O, N, CH_3O_2C, Cl	96
4-$CH_3C_6H_4CN$	**1**	H_3CO, O, N, CH_3O_2C, CH_3	93
Ph-CN	**2**	H, O, N, EtO_2C	45
$BrCH_2CN$	**2**	H, O, N, CH_2Br, EtO_2C	65

[a]Results taken from references 4 and 15.

1. Department of Chemistry and Biochemistry, University of Notre Dame, Notre Dame, IN 46556.
2. Baum, J. S.; Shook, D. A.; Davies, H. M. L.; Smith, H. D. *Synth. Commun.* **1987**, *17,* 1709-1716.
3. (a) Wulfman, D. S.; Yousefian, S.; White, J. M. *Synth. Commun.* **1988**, *18*, 2349-2352; (b) Wulfman, D. S.; Roberts, J.; Henderson, D. K.; Romine, J. C.; McDaniel, R. S., Jr.; Carver, L.; Beistel, D. W. *Can. J. Chem.* **1984**, *62*, 554-560; (c) Ciganek, E. *J. Org. Chem.* **1965**, *30*, 4366-4367.
4. Connell, R. D. Ph.D. Dissertation, University of Notre Dame, **1989**.
5. For a review on the occurrence and the chemistry of oxazoles, see: "Oxazoles"; Turchi, I. J., Ed.; Wiley: New York, 1986, and other reviews cited therein.

6. Wasserman, H. H.; McCarthy, K. E.; Prowse, K. S. *Chem. Rev.* **1986**, *86*, 845-856.

7. (a) Jacobi, P. A., "Bis-Heteroannulation: Total Synthesis of Furanoterpenes, Butenolides, Lactones and Related Materials", In *Adv. in Heterocycl. Nat. Prod. Synth.* **1992**, *2*, 215-298; (b) Subramanyam, C.; Noguchi, M.; Weinreb, S. M. *J. Org. Chem.* **1989**, *54*, 5580-5585.

8. (a) Cocito, C. *Antibiotics (N.Y.)* **1983**, *6*, 296-332; (b) Di Giambattista, M.; Chinali, G.; Cocito, C. *J. Antimicrob. Chemother.* **1989**, *24*, 485-507; (c) Paris, J. M.; Barrière, J. C.; Smith, C.; Bost, P. E. *Recent Prog. Chem. Synth. Antibiot.*; Lukacs, G.; Ohno, M., Eds.; Springer-Verlag: Berlin, 1990; Vol. 1, pp 183-248.

9. For other approaches to the oxazole unit of this antibiotic family see: (a) Meyers, A. I.; Lawson, J. P.; Walker, D. G.; Linderman, R. J. *J. Org. Chem.* **1986**, *51*, 5111-5123; (b) Nagao, Y.; Yamada, S.; Fujita, E. *Tetrahedron Lett.* **1983**, *24*, 2287-2290, 2291-2294; (c) Wood, R. D.; Ganem, B. *Tetrahedron Lett.* **1983**, *24*, 4391-4392. (d) Yokoyama, M.; Menjo, Y.; Ubukata, M.; Irie, M.; Watanabe, M.; Togo, H. *Bull. Chem. Soc. Jpn.* **1994**, *67*, 2219-2226. See also: (e) Lipshutz, B. H.; Hungate, R. W. *J. Org. Chem.* **1981**, *46*, 1410-1413.

10. For some additional, more recent preparations of 1,3-oxazole derivatives, see (a) Dow, R. L. *J. Org. Chem.* **1990**, *55*, 386-388; (b) Zhao, Z.; Scarlato, G. R.; Armstrong, R. W. *Tetrahedron Lett.* **1991**, *32*, 1609-1612; (c) Smith, A. B., III; Salvatore, B. A.; Hull, K. G.; Duan, J. J.-W. *Tetrahedron Lett.* **1991**, *32*, 4859-4862; (d) Yokoyama, M.; Sujino, K.; Irie, M.; Togo, H. *Tetrahedron Lett.* **1991**, *32*, 7269-7272; (e) Yokoyama, M.; Irie, M.; Sujino, K.; Kagemoto, T.; Togo, H.; Funabashi, M. *J. Chem. Soc., Perkin Trans. 1* **1992**, 2127-2134; (f) Evans, D. A.; Gage, J. R.; Leighton, J. L.; Kim, A. S. *J. Org. Chem.* **1992**, *57*, 1961-1963; (g) Cunico, R. F.; Kuan, C. P. *J. Org. Chem.* **1992**, *57*, 3331-3336; (h) Yokokawa, F.; Hamada, Y.; Shioiri, T. *SynLett* **1992**, 153-155; (i) Williams, E. L. *Tetrahedron*

Lett. **1992**, *33*, 1033-1036; (j) Doyle, K. J.; Moody, C. J. *Tetrahedron Lett.* **1992**, *33*, 7769-7770; (k) Eissenstat, M. A.; Weaver, J. D., III *J. Org. Chem.* **1993**, *58*, 3387-3390; (l) Kawase, M.; Miyamae, H.; Narita, M.; Kurihara, T. *Tetrahedron Lett.* **1993**, *34*, 859-862; (m) Fukuyama, T.; Xu, L. *J. Am. Chem. Soc.* **1993**, *115*, 8449-8450.

11. Cornforth, J. W.; Cornforth, R. H. *J. Chem. Soc.* **1947**, 96-102.
12. For related oxazole syntheses, see: (a) Buu, N. T.; Edward, J. T. *Can. J. Chem.* **1972**, *50*, 3730-3737; (b) Moniotte, Ph. G.; Hubert, A. J.; Teyssié, Ph. *J. Organomet. Chem.* **1975**, *88*, 115-120; (c) Ibata, T.; Sato, R. *Chem. Lett.* **1978**, 1129-1130; (d) Ibata, T.; Sato, R. *Bull. Chem. Soc. Jpn.* **1979**, *52*, 3597-3600; (e) Doyle, M. P.; Buhro, W. E.; Davidson, J. G.; Elliott, R. C.; Hoekstra, J. W.; Oppenhuizen, M. *J. Org. Chem.* **1980**, *45*, 3657-3664; (f) Mashraqui, S. H.; Keehn, P. M. *J. Am. Chem. Soc.* **1982**, *104*, 4461-4465; (g) Nagayoshi, K.; Sato, T. *Chem. Lett.* **1983**, 1355-1356; (h) Ibata, T.; Isogami, Y. *Bull. Chem. Soc. Jpn.* **1989**, *62*, 618-620; (i) Vedejs, E.; Piotrowski, D. W. *J. Org. Chem.* **1993**, *58*, 1341-1348.
13. (a) Doyle, M. P. *Chem. Rev.* **1986**, *86*, 919-939; (b) Adams, J.; Spero, D. M. *Tetrahedron* **1991**, *47*, 1765-1808.
14. (a) Huisgen, R. In "1,3-Dipolar Cycloaddition Chemistry"; Padwa, A., Ed.; Wiley: New York, 1984; Vol. 1, Chap. 1, pp 1-176; (b) "Advances in Cycloaddition"; Curran, D. P., Ed.; JAI Press: Greenwich, CT, 1988; Vol. 1; (c) Padwa, A.; Hornbuckle, S. F. *Chem. Rev.* **1991**, *91*, 263-309.
15. Connell, R. D.; Tebbe, M.; Gangloff, A. R.; Helquist, P.; Åkermark, B. *Tetrahedron* **1993**, 49, 5445-5459.

Acknowledgment. We wish to express our appreciation to the National Institutes of Health and the National Science Foundation for providing generous financial support for this work.

Appendix
Chemical Abstracts Nomenclature (Collective Index Number); (Registry Number)

4-Carbomethoxy-5-methoxy-2-phenyl-1,3-oxazole: 4-Oxazolecarboxylic acid, 5-methoxy-2-phenyl-, methyl ester (9); (53872-19-8)

Dimethyl diazomalonate: Malonic acid, diazo-, dimethyl ester (8); Propanedioic acid, diazo-, dimethyl ester (9); (6773-29-1)

p-Acetamidobenzenesulfonyl azide: Sulfanilyl azide, N-acetyl- (8); Benzenesulfonyl azide, 4-(acetylamino)- (9); (2158-14-7)

Acetonitrile: TOXIC. (8,9); (75-05-8)

Triethylamine (8); Ethanamine, N,N-diethyl- (9); (121-44-8)

Dimethyl malonate: Malonic acid, dimethyl ester (8); Propanedioic acid, dimethyl ester (9); (108-59-8)

Rhodium(II) acetate dimer: Rhodium, tetrakis(acetato-O,O')di-, (Rh-Rh) (9); (46847-37-4)

Benzonitrile (8,9); (100-47-0)

Chloroform: HIGHLY TOXIC. CANCER SUSPECT AGENT. (8); Methane, trichloro- (9); (67-66-3)

SYNTHESIS OF 5,7-DIMETHOXY-3-METHYLINDAZOLE FROM 3,5-DIMETHOXYACETOPHENONE IN A TWO-STEP PROCEDURE

(1H-Indazole, 5,7-dimethoxy-3-methyl-)

A. $Cl_3CCH_2O_2CN{=}NCO_2CH_2CCl_3$, CF_3SO_3H, CH_2Cl_2

B. Zn dust, AcOH

Submitted by Nicolas Boudreault and Yves Leblanc.[1]

Checked by John A. McCauley and Amos B. Smith, III.

1. Procedure

A. 1-(2-Acetyl-4,6-dimethoxyphenyl)-1,2-hydrazinedicarboxylic acid bis(2,2,2-trichloroethyl) ester. A 2-L, round-bottomed flask, equipped with a magnetic stirrer, thermometer, and an ice bath is charged with 3,5-dimethoxyacetophenone (20.00 g, 111.0 mmol) (Note 1) and dichloromethane (CH_2Cl_2, 555 mL) (Note 1). At 0-5°C, under nitrogen, trifluoromethanesulfonic acid (980 μL, 11.1 mmol) (Note 1) is added in one portion. To the resulting yellow solution, bis(2,2,2-trichloroethyl) azodicarboxylate

(BTCEAD) (50.70 g, 133.3 mmol) (Note 2) is gradually introduced over 15 min in order to maintain an internal temperature between 0-5°C.

The orange-yellow solution is stirred for 5.5 hr at room temperature until complete reaction has taken place. The reaction mixture is poured into saturated aqueous sodium bicarbonate ($NaHCO_3$) solution (100 mL) at 0°C. The organic phase is separated and the aqueous phase is reextracted with CH_2Cl_2 (30 mL). The combined organic layers are washed with water (H_2O) (2 x 50 mL), dried over anhydrous sodium sulfate, filtered, concentrated and placed on a vacuum pump to afford a pale yellow foam. Ethyl acetate (85 mL) is added and the crude mixture is heated under reflux until complete dissolution occurs. At reflux hexane (85 mL) is added and the solution is allowed to stand at room temperature. After a period of 15 hr the flask containing a mass of solid is cooled in an ice bath for 1.5 hr. The solid is collected by filtration and washed with a cold ethyl acetate-hexane mixture (1:4) (50 mL). The hydrazide is dried under high vacuum at 45°C for 3 days (Note 3) to provide 52.0 g (84%) of a white solid sufficiently pure for the next step (Note 4). From the mother liquor, an additional 3.8 g (6.0%) of material is recovered by flash chromatography (20% ethyl acetate in hexane) followed by recrystallization from ethyl acetate-hexane (1:1 mixture, 20 mL).

B. 5,7-Dimethoxy-3-methylindazole. A 2-L, round-bottomed flask, equipped with a magnetic stirrer and a thermometer is charged with the hydrazide (54.0 g, 96.3 mmol) and acetic acid (480 mL). To the suspension at 16°C is added zinc dust (35.6 g, 544 mmol) (Note 5) portionwise over 5 min. After a period of 10 min, the cooling bath is removed to allow the internal temperature to reach 35°C. The mixture is stirred for about 3 hr until no hydrazide is detected by TLC (30% ethyl acetate in hexane, R_f = 0.3). The mixture is filtered over Celite and washed carefully with CH_2Cl_2 (15 x 40 mL). After removal of the solvents under reduced pressure, the brown mixture is dissolved in CH_2Cl_2 (800 mL) and saturated aqueous $NaHCO_3$ solution (600 mL).

Additional solid $NaHCO_3$ is then added until pH 8 is attained. The resulting milky mixture is filtered over Celite, washed carefully with CH_2Cl_2 (10 x 100 mL), and the organic phase separated. The water phase is reextracted with CH_2Cl_2 (2 x 100 mL) and the combined organic layers are dried over sodium sulfate. After evaporation under reduced pressure, the resulting green residue is dissolved in hot CH_2Cl_2 (80 mL) and purified by flash chromatography (200 g of silica gel) (5% CH_2Cl_2/30% ethyl acetate/hexane). The indazole (14.02 g) is obtained as a yellowish solid. Crude product is recrystallized twice from ethyl acetate (100 mL) to provide 9.6 g of pure indazole (52%) as white crystals (Note 6). An additional 2.8 g (15%) of pure material is recovered from the mother liquor after recrystallization.

2. Notes

1. 3,5-Dimethoxyacetophenone 97%, anhydrous CH_2Cl_2, and trifluoromethanesulfonic acid were purchased from Aldrich Chemical Company, Inc., and used as received.

2. Bis(2,2,2-trichloroethyl) azodicarboxylate was purchased from Aldrich Chemical Company, Inc. It can also be prepared as described in reference 2.

3. The ground hydrazide was dried in order to remove trapped ethyl acetate.

4. An analytical sample was obtained by recrystallization from ethyl acetate to afford a white solid, containing 0.5 equiv of ethyl acetate, mp 137-139°C. The hydrazide exhibits the following spectral properties. IR (KBr) cm^{-1}: 3240, 1765, 1735, 1595; ^1H NMR (400 MHz, acetone-d_6, 325 K) δ: 2.65 (s, 3 H); 3.89 (s, 3 H); 3.90 (s, 3 H); 4.80 (s, 2 H); 4.90 (bs, 2 H); 6.76 (d, 1 H); 6.81 (d, 1 H); 8.91 (bs, 1 H); ^{13}C NMR (100 MHz, acetone-d_6, 325 K) δ: 56.19, 56.81, 75.44, 76.47, 96.00, 96.35, 102.39, 105.61, 121.48, 140.99, 154.57, 155.34, 158.03, 161.84, 200.40. Anal. Calcd for $C_{16}H_{16}Cl_6N_2O_7 \cdot 1/2\ C_4H_8O_2$: C, 35.70; H, 3.30; N, 4.62. Found C, 35.83; H, 3.14; N,

4.54. High-resolution mass spectrum, m/z calcd for $C_{16}H_{17}Cl_6N_2O_7$ $(M+H)^+$ 558.9167. Found 558.9166.

5. Zinc dust was purchased from Anachemia. Excess zinc was used to ensure completion of the reaction. The reactivity of the zinc dust should be verified on a small scale first. Lower reactivity has been observed with another substrate using zinc dust from Aldrich Chemical Company, Inc.

6. 5,7-Dimethoxy-3-methylindazole has the following physical properties. mp 158°C. IR (KBr) cm^{-1}: 3315, 1600; ^1H NMR (400 MHz, acetone-d_6) δ: 2.44 (s, 3 H), 3.82 (s, 3 H), 3.94 (s, 3 H), 6.45 (d, 1 H), 6.65 (d, 1 H); ^{13}C NMR (100 MHz, $CDCl_3$) δ: 12.02, 55.47, 55.74, 90.84, 98.48, 123.41, 129.41, 142.64, 145.67, 155.26. Anal Calcd for $C_{10}H_{12}N_2O_2$: C, 62.49; H, 6.29; N, 14.57. Found C, 62.54; H, 6.32; N, 14.57. High-resolution mass spectrum, m/e calcd for $C_{10}H_{13}N_2O_2$ $(M+H)^+$ 193.0977. Found 193.0976.

Waste Disposal Information

All toxic materials were disposed of in accordance with "Prudent Practices in the Laboratory"; National Academic Press; Washington, DC, 1996.

3. Discussion

The electrophilic, aromatic substitution of electron-rich arenes with the electron-deficient azodicarboxylate BTCEAD is a powerful method for the introduction, in a single operation, of a masked hydrazine or amino group.[3,4] Several activators can be used for this amination reaction; $ZnCl_2$,[4] $BF_3 \cdot Et_2O$,[4] $LiClO_4$,[3] and here, trifluoromethanesulfonic acid.[5] Trifluoromethanesulfonic acid dramatically increases the rate of the amination reaction and makes possible the use of this methodology with poorly reactive substrates.

The present, straightforward, two-step synthesis of 5,6-dimethoxyindazole from 3,5-dimethoxyacetophenone illustrates the usefulness of this amination reaction. With standard chemistry the introduction of a hydrazine group into the acetophenone molecule would have required four steps: 1) nitration, 2) reduction of the nitro group to the aniline, 3) diazotization and 4) reduction of the diazonium compound to the hydrazine.

A plausible mechanism for the formation of the indazole is illustrated below. Based on our previous observations that aryltrichlorohydrazides are readily converted to anilines, via the aryl hydrazines, it is unlikely that formation of the indazole ring **5** arises from the free hydrazine intermediate **2** since the corresponding aniline **6** was not detected at all in this case. This would suggest that the terminal trichloro ester group is selectively removed to give **3**, followed by cyclization to **4** and cleavage of the second ester unit. Intermediate **4** was isolated and characterized during a probe reaction for which an equimolar amount of zinc dust was used. However, this intermediate does not accumulate to more than 1%. It appears that cleavage of the second ester group is relatively rapid after cyclization. Traces of compound **4** have also been detected during preparation of the hydrazide **1**, presumably from acid hydrolysis.

PATHWAY A

CH_3O O CH_3 $NNHCO_2CH_2CCl_3$ CH_3O $CO_2CH_2CCl_3$

I

Zinc dust

CH_3O O CH_3 $NHNH_2$ CH_3O

II

CH_3O CH_3 N N H CH_3O

V

CH_3O O CH_3 NH_2 CH_3O

VI

PATHWAY B

I Zinc dust

CH_3O O CH_3 NNH_2 CH_3O $CO_2CH_2CCl_3$

III

CH_3O CH_3 N N CH_3O $CO_2CH_2CCl_3$

IV

V

1. Merck Frosst Centre for Therapeutic Research, P.O. Box 1005, Pointe Claire-Dorval, Québec, Canada H9R 4P8.
2. a) Mackay, D.; Pilger, C. W.; Wong, L. L. *J. Org. Chem.* **1973**, *38*, 2043; b) Little, R. D.; Venegas, M. G. *Org. Synth.* **1983**, *61*, 17. Also *Org. Synth. Coll. Vol. VII* **1990**, 56.
3. Mitchell, H.; Leblanc, Y. *J. Org. Chem.* **1994**, *59*, 682.
4. Zaltsgendler, I.; Leblanc, Y.; Bernstein, M. A. *Tetrahedron Lett.* **1993**, *34*, 2441.
5. For other examples in which trifluoromethanesulfonic acid and trifluoroacetic acid are used see: Leblanc, Y.; Boudreault, N. *J. Org. Chem.* **1995**, *60*, 4268.

Appendix

Chemical Abstracts Nomenclature (Collective Index Number); (Registry Number)

5,7-Dimethoxy-3-methylindazole: 1 H-Indazole, 5,7-dimethoxy-3-methyl- (13); (154876-15-0)

3,5-Dimethoxyacetophenone: Ethanone, 1-(3,5-dimethoxyphenyl)- (9); (39151-19-4)

1-(2-Acetyl-4,6-dimethoxyphenyl)-1,2-hydrazinedicarboxylic acid bis(2,2,2-trichloroethyl) ester: 1,2-Hydrazinedicarboxylic acid, 1-(2-acetyl-4,6-dimethoxyphenyl)-, bis(2,2,2-trichloroethyl) ester (13); (154876-13-8)

Trifluoromethanesulfonic acid: Methanesulfonic acid, trifluoro- (8,9); (1493-13-6)

Bis(2,2,2-trichloroethyl) azodicarboxylate: Diazenedicarboxylic acid, bis(2,2,2-trichloroethyl) ester (9); (38857-88-4)

Zinc (8,9); (7440-66-6)

REGIOSELECTIVE SYNTHESIS OF 3-SUBSTITUTED INDOLES: 3-ETHYLINDOLE

(1H-Indole, 3-ethyl-)

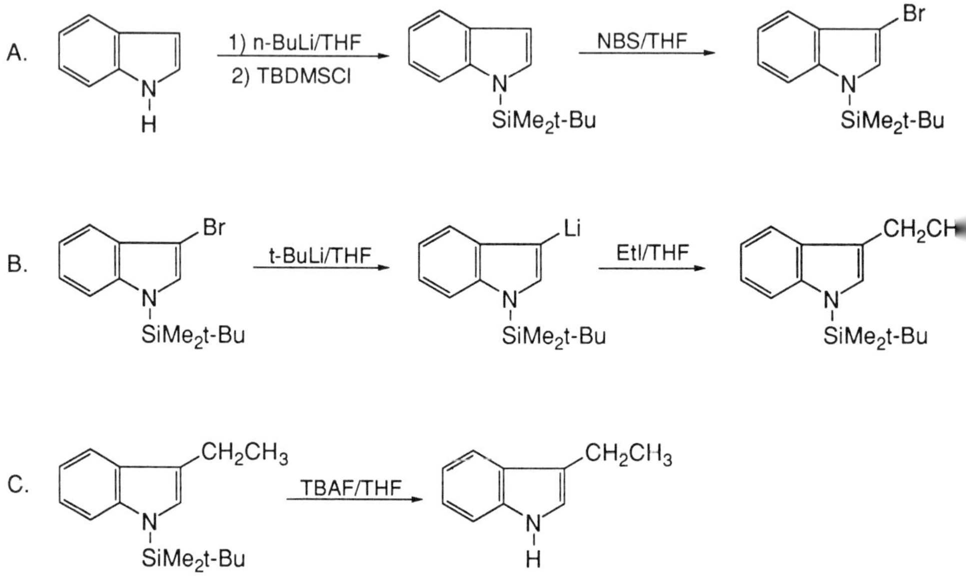

A. 1) n-BuLi/THF, 2) TBDMSCl; NBS/THF

N-H; N-$SiMe_2$t-Bu; Br, N-$SiMe_2$t-Bu

B. t-BuLi/THF; EtI/THF

Br, N-$SiMe_2$t-Bu; Li, N-$SiMe_2$t-Bu; CH_2CH_3, N-$SiMe_2$t-Bu

C. TBAF/THF

CH_2CH_3, N-$SiMe_2$t-Bu; CH_2CH_3, N-H

Submitted by Mercedes Amat, Sabine Hadida, Swargam Sathyanarayana, and Joan Bosch.[1]

Checked by Ji Liu, Chris H. Senanayake, and Ichiro Shinkai.

1. Procedure

Caution: tert-Butyllithium is extremely pyrophoric.

A. 3-Bromo-1-(tert-butyldimethylsilyl)indole. An oven-dried, 500-mL, three-necked, round-bottomed flask, equipped with a magnetic stirring bar, 100-mL pressure-equalizing addition funnel, and an argon inlet and outlet, is charged with indole (8.0 g, 0.068 mol) (Note 1) and tetrahydrofuran (200 mL) (Note 2). The solution is stirred and cooled to -78°C with a dry ice/acetone bath, and a solution of butyllithium in hexane (47 mL of a 1.6 M solution, 0.075 mol) (Note 3) is added dropwise via cannula. The mixture is warmed to -10°C, stirred for 15 min, and cooled to -50°C. A solution of tert-butyldimethylsilyl chloride (11.6 g, 0.077 mol) (Note 3) in tetrahydrofuran (60 mL) is added dropwise to this mixture. The temperature is raised to 0°C and after 3 hr the reaction mixture is cooled to -78°C. Freshly crystallized N-bromosuccinimide (12.18 g, 0.0684 mol) (Note 4) is added via a solid-addition funnel and the resulting mixture is stirred in the dark at -78°C for 2 hr and allowed to warm to room temperature. Hexane (100 mL) and pyridine (1 mL) are added and the resulting suspension is filtered through a Celite pad. The filtrate is evaporated under reduced pressure. The crude residue is purified (Note 5) by flash chromatography on silica gel (Note 6) (350 g, 30 cm x 6 cm) (100% hexane) to give 17.8 g (84%) of 3-bromo-1-(tert-butyldimethylsilyl)indole as a colorless solid (Note 7).

B. 1-(tert-Butyldimethylsilyl)-3-ethylindole. An oven-dried, 500-mL, three-necked, round-bottomed flask, equipped with a magnetic stirring bar, 50-mL pressure-equalizing addition funnel, and an argon inlet and outlet, is charged with 3-bromo-1-(tert-butyldimethylsilyl)indole (10 g, 0.032 mol) and tetrahydrofuran (100 mL). The mixture is stirred and cooled to -78°C with a dry ice/acetone bath. A solution of tert-butyllithium (Note 8) (41.7 mL of a 1.7 M solution in pentane, 0.071 mol) is transferred slowly to the above mixture from a graduated tube via a stainless steel cannula under positive argon pressure. The reaction mixture becomes yellow. Stirring is continued at -78°C for 10 min. A solution of ethyl iodide (5.2 mL, 0.065 mol) (Note 9) in tetrahydrofuran (20 mL) is added dropwise over 15 min to the resulting 1-(tert-butyldimethylsilyl)-3-lithioindole. The reaction mixture becomes colorless, and after 15 min it is allowed to reach room temperature, poured into a cold saturated sodium carbonate solution (200 mL), and extracted with methylene chloride (3 x 100 mL). The combined organic layers are washed with water (100 mL), dried over sodium sulfate, and evaporated under reduced pressure to give 8.0 g (96%) of 1-(tert-butyldimethylsilyl)-3-ethylindole as a light pink oil (Note 10).

C. 3-Ethylindole. An oven-dried, 250-mL, round-bottomed flask, equipped with a magnetic stirring bar, rubber septum, and an argon inlet and outlet, is charged with 1-(tert-butyldimethylsilyl)-3-ethylindole (8 g, 0.031 mol) and tetrahydrofuran (100 mL). The mixture is stirred and a 1 M solution of tetrabutylammonium fluoride (TBAF) in tetrahydrofuran (31 mL, 0.031 mol) (Note 11) is added. After the solution is stirred for 10 min at room temperature, it is poured into a saturated solution of sodium carbonate (200 mL) and extracted with dichloromethane (3 x 100 mL). The combined organic layers are washed with water (100 mL), dried over sodium sulfate, and evaporated under reduced pressure. The residue is subjected to flash chromatography on silica gel (175 g, 32 cm x 4.5 cm) (25% dichloromethane-hexane, v/v) to give 4.1 g (92%) of 3-ethylindole as colorless plates (Note 12).

2. Notes

1. Indole was obtained from Fluka Chemie AG, and was crystallized from hexane and dried over phosphorus pentoxide (P_2O_5) before use.

2. Tetrahydrofuran was distilled from sodium benzophenone ketyl immediately before use.

3. Butyllithium and tert-butyldimethylsilyl chloride were obtained from Fluka Chemie AG and used as received.

4. N-Bromosuccinimide was obtained from Fluka Chemie AG, and crystallized from water and dried over P_2O_5 before use.

5. Chromatographic purification of the reaction mixture must be effected as soon as possible after workup in order to separate traces of contaminating

3-bromoindole, which promotes rapid decomposition. Pure 3-bromo-1-(tert-butyldimethylsilyl)indole can be stored under argon at 4°C without appreciable decomposition.

6. Silica gel (35-70 mesh) was used as received.

7. The spectral properties for 3-bromo-1-(tert-butyldimethylsilyl)indole are as follows: ^{1}H NMR (300 MHz, $CDCl_3$) δ: 0.60 (s, 6 H), 0.93 (s, 9 H), 7.17 (s, 1 H), 7.20 (m, 2 H), 7.48 (m, 1 H), 7.54 (m, 1 H); ^{13}C NMR (75 MHz, $CDCl_3$) δ: -4.0 (CH_3Si), 19.3 [$\underline{C}(CH_3)_3$], 26.2 [$C(\underline{C}H_3)_3$], 93.6 (C-3), 114.0 (C-7), 119.1 (C-4), 120.5 (C-5), 122.5 (C-6), 129.6 (C-2), 129.8 (C-3a), 140.2 (C-7a). Anal Calcd for $C_{14}H_{20}BrNSi$: C, 54.18; H, 6.50; Br, 25.75; N, 4.51. Found: C, 54.21; H, 6.60; Br, 25.52; N, 4.62. Attempts to crystallize 3-bromo-1-(tert-butyldimethylsilyl)indole were unsuccessful.

8. tert-Butyllithium (1.7 M solution in pentane) was obtained from Aldrich Chemical Company, Inc., and used as received.

9. Ethyl iodide was obtained from Fluka Chemie AG and distilled prior to use.

10. NMR spectrum shows the presence of less than 3% of 1-(tert-butyldimethylsilyl)indole. Crude 1-(tert-butyldimethylsilyl)-3-ethylindole can be purified by flash column chromatography on silica gel (35-70 mesh) (350 g, 30 cm x 6 cm) (100% hexane) to give pure 1-(tert-butyldimethylsilyl)-3-ethylindole as a colorless oil in about 90% yield. The spectral properties are as follows: ^{1}H NMR (300 MHz, $CDCl_3$) δ: 0.57 (s, 6 H), 0.92 (s, 9 H), 1.32 (t, 3 H, J = 7.5), 2.76 (g, 2 H, J = 7.5), 6.92 (s, 1 H), 7.12 (m, 2 H), 7.47 (m, 1 H), 7.58 (m, 1 H); ^{13}C NMR (75 MHz, $CDCl_3$) δ: -3.9 (CH_3Si), 14.5 ($\underline{C}H_3CH_2$), 18.4 ($\underline{C}H_2CH_3$), 19.5 [$\underline{C}(CH_3)_3$], 26.3 [$C(\underline{C}H_3)_3$], 113.8 (C-7), 118.8 (C-4), 119.2 (C-5), 119.8 (C-3), 121.3 (C-6), 126.9 (C-2), 130.9 (C-3a), 141.6 (C-7a). Anal. Calcd for $C_{16}H_{25}NSi$: C, 74.06; H, 9.71; N, 5.40. Found: C, 74.18; H, 9.73; N, 5.46.

11. Tetrabutylammonium fluoride in tetrahydrofuran (1 M solution) was obtained from Fluka Chemie AG and used as received.

12. The spectral properties for 3-ethylindole are as follows: ^1H NMR (200 MHz, $CDCl_3$) δ: 1.37 (t, 3 H, J = 7.5), 2.82 (q, 2 H, J = 7.5), 6.98 (br s, 1 H), 7.18 (m, 2 H), 7.37 (d, 1 H, J = 7.5), 7.66 (d, 1 H, J = 7.8), 7.85 (br s, 1 H); ^{13}C NMR (50 MHz, $CDCl_3$) δ: 15.0 ($\underline{C}H_3CH_2$), 18.9 ($\underline{C}H_2CH_3$), 111.6 (C-7), 119.3 (C-3), 119.5 (C-4), 119.6 (C-5), 121.1 (C-6), 122.4 (C-2), 128.0 (C-3a), 136.9 (C-7a).

Waste Disposal Information

All toxic materials were disposed of in accordance with "Prudent Practices in the Laboratory"; National Academic Press; Washington, DC, 1996.

3. Discussion

Although there are many studies about the preparation and synthetic applications of N-protected 2-lithioindoles,[2] their isomers, the 3-lithioindoles, have received little attention. Thus, the only indole protecting group used for the preparation of simple 3-lithioindoles is the benzenesulfonyl group. 1-(Benzenesulfonyl)-3-lithioindole is prepared at -100°C by halogen-metal exchange with tert-butyllithium from the corresponding 3-iodo-[3] or 3-bromoindole.[4] At higher temperatures it rearranges to the thermodynamically more stable 2-lithio isomer. On the other hand, some 2-substituted 1-(benzenesulfonyl)-3-lithioindoles undergo ring fragmentation to give 2-aminophenylacetylene derivatives.[5] The change of the benzenesulfonyl protecting group for a trialkylsilyl group allows the preparation of 3-lithioindoles that are relatively stable species even at room temperature.

In the present procedure, indole is protected with a tert-butyldimethylsilyl group and further brominated with N-bromosuccinimide in a one-pot reaction. 3-Bromo-1-(tert-butyldimethylsilyl)indole readily undergoes halogen-lithium exchange with tert-butyllithium at -78°C. Subsequent reaction with an electrophile is exemplified by the reaction with ethyl iodide to give 1-(tert-butyldimethylsilyl)-3-ethylindole. Other electrophiles such as alkyl or allyl halides, ethylene oxide, acylating reagents, carbon dioxide, aromatic aldehydes, and trimethyltin chloride have also been employed successfully in a 500 mg scale (see Table).[6] The reactions regioselectively lead to 3-substituted indoles. 2-Substituted indoles were not detected, indicating that 1-(tert-butyldimethylsilyl)-3-lithioindole (**1**) does not undergo rearrangement to the 2-lithio isomer. Products arising from a ring fragmentation were not detected either. When the reactions of **1** with ethyl iodide and methyl iodide were carried out at room temperature, the yields of the respective 3-alkylindoles were similar to those obtained when operating at -78°C. Finally, the 1-tert-butyldimethylsilyl protecting group can be readily removed by treatment with TBAF under mild conditions.

The triisopropylsilyl group gave comparable satisfactory results.[7]

As a further synthetic application, 3-lithioindole **1** was converted to 1-(tert-butyldimethylsilyl)-3-indolylzinc chloride, which has been successfully employed in the heteroarylation of the indole 3-position by a palladium(0)-catalyzed cross-coupling reaction.[8]

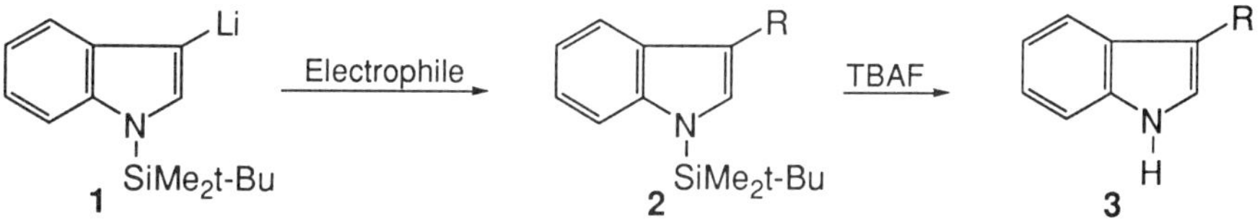

1 (N-SiMe$_2$t-Bu, 3-Li) —Electrophile→ **2** (N-SiMe$_2$t-Bu, 3-R) —TBAF→ **3** (N-H, 3-R)

TABLE

SYNTHESIS OF 3-SUBSTITUTED INDOLES

Entry	Electrophile	R	**2** Yield(%)	**3** Yield(%)[a]
a	MeI	Me	90	85
b	BuBr	Bu	64	61
c	$(CH_2CH_2)O$	CH_2CH_2OH	63	62
d	$Me_2C{=}CHCH_2Br$[b]	$CH_2CH{=}CMe_2$		69[c]
e	$HCONMe_2$[d]	CHO	e	94
f	C_6H_5COCl[d]	COC_6H_5	e	84
g	$C_6H_5CO_2CH_3$[d]	COC_6H_5	69	58
h	$ClCO_2CH_3$[d]	CO_2CH_3	84	80
i	CO_2	CO_2H	94	70[f]
j	C_6H_5CHO	$CHOHC_6H_5$	67	g
k	$4\text{-}CHO\text{-}C_5H_4N$	$4\text{-}CHOH\text{-}C_5H_4N$	55	g
l	$ClSnMe_3$	$SnMe_3$	94	h

[a]Overall yield after purification by column chromatography. [b]The 3-lithioindole **1** was converted into a cuprate by addition of 1 equiv of $CuBr{\cdot}SMe_2$. [c]An 85:15 mixture of **3d** and the isomer in which R is $CMe_2CH{=}CH_2$, respectively, is obtained. [d]The best yields were obtained by reverse addition of the lithium derivative **1** to a THF solution of the electrophile at -78°C. [e]The corresponding 3-acyl derivatives undergo partial desilylation during the work-up. [f]This desilylation was best effected by using CsF instead of TBAF. [g]These carbinols were obtained as pink oils, which partially decomposed during purification by column chromatography. [h]Attempts to deprotect the tin derivative **2l** afforded only indole.

1. Laboratory of Organic Chemistry, Faculty of Pharmacy, University of Barcelona, 08028 Barcelona, Spain.
2. Sundberg, R. J.; Russell, H. F. *J. Org. Chem.* **1973**, *38*, 3324; Gharpure, M.; Stoller, A.; Bellamy, F.; Firnau, G.; Snieckus, V. *Synthesis* **1991**, 1079, and references cited therein.
3. Saulnier, M. G.; Gribble, G. W. *J. Org. Chem.* **1982**, *47*, 757.
4. Gribble, G. W.; Barden, T. C. *J. Org. Chem.* **1985**, *50*, 5900.
5. Gribble, G. W.; Saulnier, M. G. *J. Org. Chem.* **1983**, *48*, 607.
6. Amat, M.; Hadida, S.; Sathyanarayana, S.; Bosch, J. *J. Org. Chem.* **1994**, *59*, 10.
7. Amat, M.; Hadida, S.; Sathyanarayana, S.; Bosch, J., unpublished results.
8. Amat, M.; Hadida, S.; Bosch, J. *Tetrahedron Lett.* **1994**, *35*, 793; Amat, M.; Sathyanarayana, S.; Hadida, S.; Bosch, J. *Tetrahedron Lett.* **1994**, *35*, 7123.

Appendix

Chemical Abstracts Nomenclature (Collective Index Number); (Registry Number)

3-Ethylindole: Indole, 3-ethyl- (8); 1H-Indole, 3-ethyl- (9); (1484-19-1)

3-Bromo-1-(tert-butyldimethylsilyl)indole: 1H-Indole, 3-bromo-1-[(1,1-dimethylethyl)-dimethylsilyl]- (13); (153942-69-9)

Indole (8); 1H-Indole (9); (120-72-9)

Butyllithium: Lithium, butyl- (8,9); (109-72-8)

tert-Butyldimethylsilyl chloride: Silane, chloro(1,1-dimethylethyl)dimethyl- (9); (18162-48-6)

N-Bromosuccinimide: Succinimide, N-bromo- (8); 2,5-Pyrrolidinedione, 1-bromo- (9); (128-08-5)

1-(tert-Butyldimethylsilyl)-3-ethylindole: 1H-Indole, 1-[(1,1-dimethylethyl)dimethylsilyl]-3-ethyl- (13); (153942-71-3)

tert-Butyllithium: Lithium, tert-butyl- (8); Lithium, (1,1-dimethylethyl)- (9); (594-19-4)

Ethyl iodide: Ethane, iodo- (8,9); (75-03-6)

Tetrabutylammonium fluoride: Ammonium, tetrabutyl-, fluoride (8); 1-Butanaminium, N,N,N-tributyl-, fluoride (9); (429-41-4)

3-MORPHOLINO-2-PHENYLTHIOACRYLIC ACID MORPHOLIDE AND 5-(4-BROMOBENZOYL-2-(4-MORPHOLINO)-3-PHENYLTHIOPHENE

(Morpholine, 4-[3-(4-morpholinyl)-2-phenyl-1-thioxo-2-propenyl]-,) and (Thiophene, 5-(4-bromobenzoyl)-2-(4-morpholino)-3-phenyl-)

A. Ph(C=O)CH$_3$ —[S / HN(morpholine)]→ PhCH$_2$C(=S)N(morpholine)

B. PhCH$_2$C(=S)N(morpholine) —[HC(OEt)$_3$ / HN(morpholine)]→ (morpholino)CH=C(Ph)C(=S)N(morpholine)

C. (morpholino)CH=C(Ph)C(=S)N(morpholine) —[Br-C$_6$H$_4$-C(=O)CH$_2$-Br]→ 5-(4-bromobenzoyl)-2-(4-morpholino)-3-phenylthiophene

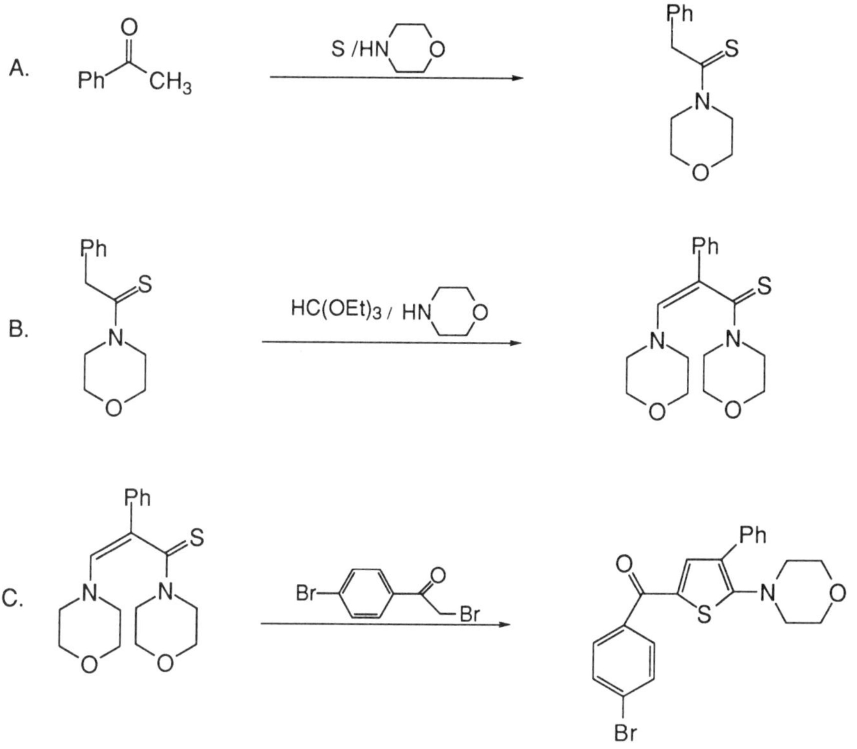

Submitted by Andreas Rolfs and Jürgen Liebscher.[1]

Checked by Andrew B. Benowitz and Amos B. Smith, III.

1. Procedure

Caution! Part A should be carried out in an efficient hood to avoid exposure to noxious vapors (hydrogen sulfide).

A. Phenylthioacetic acid morpholide. A 100-mL, round-bottomed flask is charged with 24.0 g (0.2 mol) of acetophenone, 1 g of p-toluenesulfonic acid monohydrate, 36.0 g (0.41 mol) of morpholine, and 6.4 g (0.2 mol) of sulfur (Note 1). The flask is equipped with a reflux condenser and is heated at reflux for 3 hr (Note 2). The resulting reddish-brown solution is poured into 100 mL of stirred hot methanol (55-60°C). The wall of the beaker is scratched with a glass rod for seeding. The beaker is sealed with aluminum foil and put into a refrigerator for 6 hr (Note 3). The resulting crystalline product is filtered and washed twice with ice-cold methanol (10 mL). The material (22.2 g, mp 69-79°C, almost colorless) is recrystallized by adding 25 mL of methanol and 25 mL of water and heating at reflux, followed by slow addition of methanol (about 58 mL) until complete solution is obtained. After the sides of the flask are scratched, the solution is cooled in a refrigerator for about 10 hr; 21.9 g (49.5%) (Note 4) of the pure product is obtained, mp 77.5-79°C.

B. 3-Morpholino-2-phenylthioacrylic acid morpholide. A distillation apparatus (250-mL, round-bottomed flask, distillation head, thermometer, condenser and receiver) equipped with a magnetic stirrer is charged with 22.3 g (0.1 mol) of phenylthioacetic acid morpholide, 26.1 g (0.3 mol) of morpholine and 67 mL (0.4 mol) of triethyl orthoformate (Note 5). The flask is heated in a heating bath (bath temperature ~ 160°C) for 8 hr. During this time, ethanol is distilled continuously from the reaction (about 18 mL). Further raising the temperature of the bath to 180°C gives rise to the distillation of excess morpholine and triethyl orthoformate (38 mL, bp 128-130°C) from the orange solution within 30 min (Note 6). The mixture is evaporated under reduced pressure (rotary evaporator, bath temperature ~ 80°C). The remaining crystalline material is dissolved in 30 mL of chloroform and 150 mL of hot methanol (about 60°C). The solution is allowed to cool to room temperature (Note 7) and is put into a freezer (-24°C) for 12 hr. The light yellow product is filtered and washed with methanol (30 mL) followed by diethyl ether (30 mL). After drying in the open air for 18 hr, 29.4 g (92%) (Note 8) of pure material is obtained, mp 156-156.5°C (Note 9).

C. 5-(4-Bromobenzoyl)-2-morpholino-3-phenylthiophene. A mixture of 3.18 g (0.01 mol) of 3-morpholino-2-phenylthioacrylic acid morpholide, 2.8 g (0.01 mol) of 4-bromophenacyl bromide, and 50 mL of methanol is heated to boiling. After the addition of a solution of 1.01 g (0.01 mol) of triethylamine in 10 mL of methanol to the mixture, the reaction is allowed to heat at reflux for 10 min. The reaction mixture is then cooled to 0°C in a refrigerator for 1 hr. The yellow crystalline product is collected by filtration and washed with 20 mL of cold methanol. After drying in the open air for 18 hr, 4.15 g (97%) of analytically pure product is obtained, mp 176-177°C (mp[2] 175°C). Although not necessary, the material can be recrystallized from acetonitrile.

2. Notes

1. All chemicals were obtained from the Aldrich Chemical Company, Inc. It is advisable to distill acetophenone and morpholine. The sulfur should be small particles and must not form lumps. Excess morpholine is generally recommended in Willgerodt-Kindler reactions.

2. The reflux condenser should be connected by a gas outlet and tubing directly to the hood pipe to prevent hydrogen sulfide from entering the laboratory atmosphere.

3. If crystallization fails, a small amount of the solution is evaporated, cooled in an ice/sodium chloride bath, and scratched with a glass rod. The resulting crystals are used to seed the main solution.

4. The reported yield[3] of 94% could not be achieved. The amount of sulfur can be increased to 12.4 g (0.2 mol) giving a higher yield of phenylthioacetic acid morpholide that contains some unreacted sulfur and a yellow impurity (2-morpholin-4-yl-1-phenyl-2-thioxo-ethanone) that is difficult to remove.

5. Triethyl orthoformate was distilled. Trimethyl orthoformate can not be used since no reaction occurs. Excess morpholine and triethyl orthoformate are used as solvents to achieve smooth solution. The amounts of both of these reagents can be increased without affecting the yield.

6. TLC of the remaining solution (0.25-mm Whatman precoated silica gel plate, 33% ethyl acetate in hexanes) does not show any starting phenylthioacetic acid morpholide.

7. Crystallization of the pure product begins after slight cooling.

8. Additional product can be obtained by evaporating the combined mother liquor, adding 25 mL of methanol to the remaining oil, and cooling the resulting solution to -24°C.

9. The product consists of a mixture of E/Z isomers[4] in an approximate ratio of 15:85; R_f = 0.35 and 0.25 (0.25-mm Whatman precoated silica gel plate, 33% ethyl acetate in hexanes). Anal. Calc. for $C_{17}H_{22}N_2O_2S$: C, 64.12; H, 6.96; N, 8.80; S, 10.07, Found: C, 64.38; H, 6.99; N, 8.80; S, 10.18. IR (KBr) cm^{-1}: 1611, 1586, 1440, 1429, 1413, 1236, 1226, 1116; Z-isomer (major isomer): ^{13}C NMR (75 MHz, $CDCl_3$) δ: 124.8, 128.5 (C2',C3'-C_6H_5), 125.3 (C4'-C_6H_5), 139.3 (C1'-C_6H_5), 112.6 (C2), 137.5 (C3), 198.9 (C1); ^1H NMR (300 MHz, $CDCl_3$) δ: 6.07 (s, HC=C); E-isomer (minor isomer): ^{13}C NMR (75 MHz, $CDCl_3$) δ: 126.4 (C4'-C_6H_5), 128.1, 129.9 (C2',C3'-C_6H_5), 138.3 (C1'-C_6H_5), 115.2 (C2), 147.4 (C3), 203.3 (C1).

Waste Disposal Information

All toxic materials were disposed of in accordance with "Prudent Practices in the Laboratory"; National Academic Press; Washington, DC, 1996.

3. Discussion

3-Aminothioacrylic acid amides are versatile β-dicarbonyl derivatives that have found widespread use as C-C-C and C-C-C-S building blocks in the synthesis of heterocyclic and open chain products.[5,6] To date, 2-arylthioacrylmorpholides with mainly a 3-dimethylamino leaving group have been prepared by the Vilsmeier-Haack reaction of arylthioacetic acid amides with DMF/$POCl_3$.[7,8] This synthesis is not convenient since the resulting 3-dimethylaminothioacrylic acid morpholides must be isolated as hydroperchlorates, the yields are occasionally low, and in a number of cases (especially if other amino groups beside the morpholino or dimethylamino are found in the products, or if the aryl group is ortho-substituted) no stable crystalline material is obtained. With the present procedure,[9] a very reliable synthesis of 3-morpholinothioacrylic acid amides with a wide variety of 2-aryl substituents has been elaborated. Instead of triethyl orthoformate/morpholine, the non-commercially available trismorpholinomethane[10] can be used,[9] also enabling the synthesis of the 2-unsubstituted 3-morpholinothioacrylic acid morpholide. In addition to the well known Willgerodt-Kindler reaction,[11] the starting thioacetamides can also be obtained by other methods such as reaction of styrenes with sulfur and amines,[12] and the reaction of arylthioacetic acid thiol ester with amines.[13] Furthermore the thioamide amino group of the synthesized 3-morpholinothioacrylic acid amides can also be monosubstituted. Through the use of pyrrolidine, 3-pyrrolidinothioacrylic acid amides can be obtained. Generally the procedure is very simple and safe. Scale-up can be effected without incident and very pure products are obtained. Since the 3-morpholino substituent usually acts as a leaving group in further synthetic applications, the same reactivity is found as in the well investigated 2-aryl-3-dimethylaminothioacrylic acid morpholides.

The mechanism of the formation of 3-aminothioacrylic amides from arylthioacetic acid amides and ortho ester amine is likely to resemble similar aminomethinylation reactions of other CH-acidic substrates.[14]

1. Institut für Chemie, Humboldt-Universität Berlin, Hessische Str. 1-2, D-10115 Berlin, Germany.
2. Liebscher, J.; Abegaz, B.; Areda, A. *J. Prakt. Chem.* **1983**, *325*, 168.
3. Mayer, R.; Wehl, J. *Angew. Chem.* **1964**, *76*, 861.
4. Rolfs, A.; Liebscher, J.; Jones, P. G.; Hovestreydt, E. *J. Prakt. Chem.* **1995**, *337*, 46.
5. Liebscher, J.; Abegaz, B.; Knoll, A. *Phosphorus Sulfur* **1988**, *35*, 5.
6. Rolfs, A.; Liebscher, J. *Angew. Chem. Int. Ed. Engl.* **1993**, *32*, 712.
7. Liebscher, J.; Abegaz, B. *Synthesis* **1982**, 769.
8. Liebscher, J.; Knoll, A.; Abegaz, B. *Z. Chem.* **1987**, *27*, 8.
9. Rolfs, A.; Liebscher, J. *Synthesis* **1994**, 683.
10. Swaringen, Jr., R. A.; Eaddy, J. F.; Henderson, T. R. *J. Org. Chem.* **1980**, *45*, 3986.
11. Brown, E. V. *Synthesis* **1975**, 358.
12. King, J. A.; McMillan, F. H. *J. Am. Chem. Soc.* **1946**, *68*, 2335.
13. Kornfeld, E. C. *J. Org. Chem.* **1951**, *16*, 131.
14. Wolfbeis, O. S. *Chem. Ber.* **1981**, *114*, 3471.

Appendix

Chemical Abstracts Nomenclature (Collective Index Number); (Registry Number)

3-Morpholino-2-phenylthioacrylic acid morpholide: Morpholine, 4-[3-(4-morpholinyl)-2-phenyl-1-thioxo-2-propenyl]-, (11); (86867-31-4)

Phenylthioacetic acid morpholide: Morpholine, 4-(phenylthioacetyl)- (8); Morpholine, 4-(2-phenyl-1-thioxoethyl)- (9); (949-01-9)

Acetophenone (8); Ethanone, 1-phenyl- (9); (98-86-2)

p-Toluenesulfonic acid monohydrate (8); Benzenesulfonic acid, 4-methyl-, monohydrate (9); (6192-52-5)

Morpholine (8,9); (110-91-8)

Sulfur (8,9); (7704-34-9)

Triethyl orthoformate: Orthoformic acid, triethyl ester (8); Ethane, 1,1',1"-[methylidynetris(oxy)tris- (9); (122-51-0)

2-(4-Bromobenzoyl)-5-morpholino-4-phenylthiophene: Methanone, (4-bromophenyl)[5-(4-morpholinyl)-4-phenyl-2-thienyl]- (11); (86673-62-3)

4-Bromophenacyl bromide: Acetophenone, 2,4'-dibromo- (8); Ethanone, 2-bromo-1-(4-bromophenyl)- (9); (99-73-0)

Unchecked Procedures

Accepted for checking during the period August 1, 1994 through April 1, 1996. An asterisk (*) indicates that the procedure has been subsequently checked.

Previously, *Organic Syntheses* has supplied these procedures upon request. However, because of the potential liability associated with procedures which have not been tested, we shall continue to list such procedures but requests for them should be directed to the submitters listed.

2713* Synthesis of 8,8-Dicyanoheptafulvene from Cycloheptatrienylium Tetrafluoroborate and Bromomalononitrile.
H. Takeshita, A. Mori, and K. Kubo, Institute of Advanced Material Study, 86, Kyushu University, Kasuga-koen, Kasuga, Fukuoka, 816, Japan.

2724 Ethyl Glycidate from (S)-Serine: (R)-(+)-Ethyl 2,3-Epoxypropanoate.
Y. Petit and M. Larchevêque, Laboratoire de Synthèse Organique, (URA 1381 du CNRS), E.N.S.C.P., 11, rue Pierre et Marie Curie, 75231 Paris Cedex 05, France.

2725R (1R,2R,3S,4S)-3-Amino-1,7,7-trimethylbicyclo[2.2.1]heptan-2-ol.
F. Hénin, S. Jamal-Aboulhoda, S. Letinois, and J. Muzart, URA CNRS "Réarrangements Thermiques et Photochimiques," UFR Sciences, B.P. 1039, 51687 Reims Cedex-2, France.

2726 (E)-1-[(1-Methoxy-1-propenyl)oxy]-1-(1,1-dimethylethyl)silacyclobutane: A Highly Selective syn-Aldol Reagent.
S. E. Denmark and B. D. Griedel, Roger Adams Laboratory, Department of Chemistry, University of Illinois, 600 S. Mathew Avenue, Urbana, IL 61801.

2730 Preparation of Cyanoalkynes.
F.-T. Luo and R.-T. Wang, Institute of Chemistry, Academia Sinica, Nankang, Taipei, Taiwan, ROC.

2732 Preparation of Vitamin D_2 from Ergosterol.
M. Okabe, Roche Research Center, Hoffmann-La Roche Inc., Nutley, NJ 07110.

2733* Preparation of 4-Dimethylamino-N-triphenylmethylpyridinium Chloride.
A. V. Bhatia, S. K. Chaudhary, and O. Hernandez, Chemical Development Department, D-54Z, Abbott Laboratories, North Chicago, IL 60064.

2734 R-(+)-(Hydroxymethyl)oxazolidin-2-one.
M. P. Sibi, B. J. Harris, P. A. Renhowe, J. W. Christensen, J.-L. Lu, D. Rutherford, and B. Li, Department of Chemistry, North Dakota State University, Fargo, ND 58105-5516.

2736* Asymmetric Synthesis of Diethyl (R)-(-)-(1-Amino-3-methylbutyl)phosphonate.
A. B. Smith, III, K. M. Yager, B. W. Phillips, and C. M. Taylor, Department of Chemistry, University of Pennsylvania, Philadelphia, PA 19104.

2739* (R)-Ethyl 2-azidopropionate.
A. S. Thompson, F. W. Hartner, Jr., and E. J. J. Grabowski, Department of Process Research, Merck Research Laboratories, Rahway, NJ 07065.

2740R Synthesis of Functionalized 2-Alkylsubstituted Dienes from 2-(Chloromethyl)-3-tosyl-1-propene: 2-(Phenylthiomethyl)-1,3-butadiene.
D. A. Alonso, C. Nájera, and J. M. Sansano, Departamento de Química Orgánica, Facultad de Ciencias, Universidad de Alicante, 03080 Alicante, Spain.

2742 Catalytic Asymmetric Allylation (CAA) Reactions: (S)-1-(Phenylmethoxy)-4-penten-2-ol.
G. E. Keck and D. Krishnamurthy, Department of Chemistry, University of Utah, Salt Lake City, UT 84112.

2743R* Synthesis of Chiral (E)-Crotylsilanes: [3R and 3S] (4E)-Methyl 3-(Dimethylphenyl)silyl-4-hexenoate.
R. T. Beresis, J. S. Solomon, M. G. Yang, N. F. Jain, and J. S. Panek, Department of Chemistry, Metcalf Center for Science and Engineering, Boston University, Boston, MA 02215.

2745 Cyclopropylacetylene.
E. G. Corley, A. S. Thompson, and M. Huntington, Process Research Department, Merck Research Laboratories, Division of Merck & Co., Inc., P.O. Box 2000, Rahway, NJ 07065.

2746 1-Chloro-(2S,3S)-dihydroxycyclohexa-4,6-diene.
T. Hudlicky, M. R. Stabile, D. T. Gibson, and G. M. Whited, Department of Chemistry, University of Florida, Gainesville, FL 32611.

2749 (R,R)-2,2-Dimethyl-$\alpha,\alpha,\alpha',\alpha'$-tetra(naphth-2-yl)-1,3-dioxolane-4,5-dimethanol from Dimethyl Tartrate and 2-Naphthylmagnesium Bromide.
A. K. Beck, P. Gysi, L. La Vecchia, and D. Seebach, Laboratorium für Organische Chemie, ETH-Zentrum, Universitätstrasse 16, CH-8092 Zürich, Switzerland.

2750R1 Catalytic Asymmetric Hydrogenation of Cyclic Imines.
C. A. Willoughby and S. L. Buchwald, Department of Chemistry Massachusetts Institute of Technology, Cambridge, MA 02139.

2750R2 Ethylenebis(η^5-4,5,6,7-tetrahydroindenyl)titanium Derivatives.
B. Chin, C. A. Willoughby, and S. L. Buchwald, Department of Chemistry Massachusetts Institute of Technology, Cambridge, MA 02139.

2751R* N-Vinylpyrrolidin-2-one as a 3-Aminopropyl Carbanion Equivalent in the Synthesis of Substituted 1-Pyrrolines: Preparation of 2-Phenyl-1-pyrroline.
K. L. Sorgi, C. A. Maryanoff, D. F. McComsey, and B. E. Maryanoff, The R. W. Johnson Pharmaceutical Research Institute, Welsh & McKean Roads, Spring House, PA 19477.

2752 Methyl (S)-2-Phthalimido-4-oxobutanoate.
P. Meffre, P. Durand, and F. Le Goffic, Laboratoire de Bioorganique et Biotechnologies, Associé au C.N.R.S., ENSCP, 11, rue Pierre et Marie Curie, 75231 PARIS Cédex 05, France.

2753 Resolution of 1,1'-Bi-2-naphthol.
D. Cai, D. L. Hughes, T. R. Verhoeven, and P. J. Reider, Merck Research Laboratories, P.O. Box 2000, Rahway, NJ 07065.

2754 (R)-(+)- and (S)-(-)-2,2'-Bis(diphenylphosphino)-1,1'-binaphthyl (BINAP).
D. Cai, J. F. Payack, D. R. Bender, D. L. Hughes, T. R. Verhoeven, and P. J. Reider, Merck Research Laboratories, P.O. Box 2000, Rahway, NJ 07065.

2755 Synthesis and Ru(II)-BINAP Reduction of a Ketoester Derived from Hydroxyproline.
S. A. King, J. Armstrong, and J. Keller, Merck Research Laboratories, P.O. Box 2000, Rahway, NJ 07065.

2756 Ethyl 5-Chloro-3-phenylindole-2-carboxylate.
A. Fürstner, A. Hupperts, and G. Seidel, Max-Planck-Institut für Kohlenforschung, Kaiser-Wilhelm-Platz 1, D-45470 Mülheim a. d. Ruhr, Germany.

2757 Isomerization of β-Alkynyl Allylic Alcohols to Furans Catalyzed by Silver Nitrate on Silica Gel: 2-Pentyl-3-Methyl-5-heptylfuran.
J. A. Marshall and C. A. Sehon, The University of Virginia, Department of Chemistry, McCormick Road, Charlottesville, VA 22901.

2758* Wittig Olefination of Perfluoroalkyl Carboxylic Esters: Synthesis of 1,1,1-Trifluoro-2-ethoxy-5-phenylpent-2-ene and 1-Perfluoroalkyl Epoxy Ethers: Synthesis of 1,1,1-Trifluoro-2-ethoxy-2,3-epoxy-5-phenylpentane.
J.-P. Bégué, D. Bonnet-Delpon and A. Kornilov, BIOCIS CNRS, Faculté de Pharmacie de Chatenay-Malabry, Rue J.B. Clément, 92296 Chatenay-Malabry, France.

2762R (4R,5S)-4,5-Diphenyl-3-vinyl-2-oxazolidinone.
T. Akiba, O. Tamura, and S. Terashima, Sagami Chemical Research Center, Nishi-Ohnuma, Sagamihara, Kanagawa 229, Japan.

2763R 2-Chlorophenylphosphorodichloridothioate.
V. T. Ravikumar and B. Ross, Department of Chemistry, Isis Pharmaceuticals, Carlsbad, CA 92008.

2764 Generation of 1-Propynyllithium from Z/E 1-Bromo-1-propene: Preparation of 6-Phenyl-hex-2-yn-5-en-4-ol.
D. Toussaint and J. Suffert, Laboratoire de Pharmacochimie Moléculaire (UPR 421), Centre de Neurochimie du CNRS, 5, rue Blaise Pascal 67084 Strasbourg-Cedex, France.

2765 Mono-C-Methylation of Arylacetonitriles and Methyl Arylacetates by Dimethylcarbonate: A General Method for the Synthesis of Pure 2-Arylpropionic Acids.
P. Tundo, M. Selva, and A. Bomben, Dipartimento di Scienze Ambientali dell'Università di Venezia, Calle Larga S. Marta 2137, I-30123 Venezia, Italy.

2766 A Simple and Convenient Method for the Preparation of β-Functionalized Alkynyl(phenyl)iodonium Triflates: Phenyl[p-toluenesulfonyl)ethynyl]iodonium Triflate.
J. H. Ryan, P. J. Stang, B. L. Williamson, and V. V. Zhdankin, Department of Chemistry, The University of Utah, Salt Lake City, UT 84112.

2768R Selective Protection of 1,3-Diols at the More Hindered Hydroxyl Group: 3-(Methoxymethoxy)-1-butanol.
W. F. Bailey, M. W. Carson, and L. M. J. Zarcone, Department of Chemistry, University of Connecticut, Storrs, CT 06269.

2769* Efficient Synthesis of Bromides from Carboxylic Acids Containing a Sensitive Functional Group.
D. H. R. Barton, J. MacKinnon, R. N. Perchet and C.-L. Tse, Department of Chemistry, Texas A&M University, College Station, TX 77843-3255.

2770 2-(4-Methoxyphenyl)-2-cyclohexen-1-one: Preparation of 2-Iodo-2-cyclohexen-1-one and Suzuki Coupling with 4-Methoxyphenyl-boronic Acid.
F. S. Ruel, M. P. Braun, and C. R. Johnson, Department of Chemistry, Wayne State University, Detroit, MI 48202-3489.

2771 6,7-Dihydrocyclopenta-1,3-dioxin-5(4H)-one.
K. Chen, C. S. Brook, and A. B. Smith, III, Department of Chemistry, University of Pennsylvania, Philadelphia, PA 19104.

2774 [3+4] Annulation Using (β-(Trimethylsilyl)acryloyl)silane and Lithium Enolate of α,β-Unsaturated Methyl Ketone: (1R*,6S*,7S*)-4-(tert-Butyldimethylsiloxy)-6-(trimethylsilyl)bicyclo-[5.4.0]undec-4-en-2-one.
K. Takeda, A. Nakajima, M. Takeda, and E. Yoshii, Faculty of Pharmaceutical Sciences, Toyama Medical and Pharmaceutical University, 2630 Sugitani, Toyama 930-01, Japan.

2776 Synthesis of Chiral Non-racemic Diols from (S,S)-1,2,3,4-Diepoxybutane.
M. A. Robbins, P. N. Devine, and T. Oh, Department of Chemistry, California State University, Northridge, CA 91330-8200.

2777* Accelerated Suzuki Coupling via a Ligandless Palladium Catalyst: 4-Methoxy-2'-methylbiphenyl.
F. E. Goodson, T. I. Wallow, and B. M. Novak, Department of Polymer Science and Engineering, University of Massachusetts, Amherst, MA 01003.

2778 Preparation of Optically Active N,N'-Dimethyl-1,2-diphenyl Ethylene-1,2-diamine.
A. Alexakis, I. Aujard, T. Kanger, and P. Mangeney, Laboratoire de Chimie des Organoéléments, CNRS UA 473, Université Pierre et Marie Curie, 4, Place Jussieu F-75252, Paris Cédex 05, France.

2779* Synthesis of β-Lactones by Aldolization of Ketones with Phenyl Ester Enolates: 3,3-Dimethyl-1-oxaspiro[3.5]nonan-2-one.
C. Wedler and H. Schick, Institut für Angewandte Chemie Berlin-Aldershof e.V., Rudower Chaussee 5, D-12484 Berlin, Germany.

CUMULATIVE AUTHOR INDEX

FOR VOLUMES 70, 71, 72, 73, AND 74

This index comprises the names of contributors to Volume **70**, **71**, **72**, **73**, and **74**, only. For authors to previous volumes, see either indices in Collective Volumes I through VIII or the single volume entitled *Organic Syntheses, Collective Volumes, I, II, III, IV, V, Cumulative Indices,* edited by R. L. Shriner and R. H. Shriner.

About-Jaudet, E., **73**, 152
Afarinkia, K., **73**, 231
Alami, M., **72**, 135
Alva, C. W., **73**, 174
Amat, M., **74**, 248
Archer, S., **72**, 125
Arnold, H., **70**, 111
Arnold, L. D., **70**, 1, 10

Barrett, A. G. M., **73**, 50
Barton, D. H. R., **74**, 101
Barve, P. V., **72**, 116
Baum, J. S., **70**, 93
Beak, P., **74**, 23
Becher, J., **73**, 270
Beck, A. K., **71**, 39; **72**, 62
Beifuss, U., **71**, 167
Bernardes, V., **74**, 13
Bigham, M. H., **70**, 231
Billadeau, R., **72**, 163
Bishop, R., **70**, 120
Blank, S., **72**, 62
Boeckman, Jr., R. K., **71**, 48
Boger, D. L., **70**, 79
Bollinger, F. W., **73**, 144
Bonin, M., **70**, 54
Bosch, J., **74**, 248
Boudreault, N., **74**, 241
Bowers, K. R., **73**, 116
Bradlee, M. J., **74**, 137
Bradshaw, J. S., **70**, 129
Bratz, M., **71**, 214
Braun, M., **72**, 32, 38
Breitschuh, R., **71**, 39
Breslow, R., **72**, 199; **74**, 72
Brisbois, R. G., **73**, 134
Brown, D. S., **70**, 157
Brown, J. T., **72**, 252
Bryant, J. D., **72**, 6
Buchwald, S. L., **71**, 77
Buckley, J., **73**, 159
Burgess, L. E, **73**, 215, 221

Cahiez, G., **72**, 135
Cantrell, Jr., W. R., **70**, 93
Carrasco, M., **70**, 29
Carroll, J. D., **74**, 50
Cava, M. P., **70**, 151; **73**, 270
Chan, A., **72**, 163
Chapman, J. J., **74**, 217
Chen, B.-C., **73**, 159
Chen, H. G., **70**, 195
Chen, M., **74**, 101
Chen, W., **70**, 151
Chu, K. S., **73**, 201
Clark, C., **73**, 159
Coffen, D. L., **71**, 83
Cohen, T., **72**, 173
Collignon, N., **73**, 152
Comins, D. L., **74**, 77
Confalonieri, G., **74**, 158
Corey, E. J., **71**, 22, 30
Corley, E. G., **74**, 50
Crabtree, S. R., **70**, 256
Craney, C. L., **73**, 25

Dahnke, K. R., **71**, 175, 181
Dai, H., **73**, 231
Danheiser, R. L., **71**, 133, 140; **73**, 61, 134
Davies, H. M. L., **70**, 93

Davis, F. A., **73**, 159
Deardorff, D. R., **73**, 25
Dehghani, A., **74**, 77
Dembech, P., **74**, 84
Dener, J. M., **71**, 220
Denmark, S. E., **74**, 33
deSilva, S. O., **72**, 163
Desmond, R., **74**, 50
DiBenedetto, D., **73**, 159
Dickhaut, J., **70**, 164
Dilley, G. J., **73**, 246
Dodge, J. A., **73**, 110
Dondoni, A., **72**, 21
Doyle, M. P., **73**, 13
Dragovich, P. S., **72**, 104
Dufresne, R. F., **74**, 217
Dussy, F. E., **72**, 116

Earle, M. J., **73**, 36
Echavarren, A. M., **71**, 97
Elsheimer, S., **72**, 225
Enomoto, T., **73**, 253

Fagan, P. J., **70**, 272
Flanagan, M. E., **71**, 107
Flygare, J. A., **73**, 50
Foti, C. J., **74**, 77
Foti, M. J., **72**, 225
Furuta, K., **72**, 86

Gala, D., **73**, 159
Ganem, B., **72**, 246
Gao, Q.-z., **72**, 86
Garner, P., **70**, 18
Gee, S. K., **71**, 133, 140
Gerlach, H., **71**, 48
Gibson, F. S., **70**, 101
Giese, B., **70**, 164
Gilheany, D. G., **70**, 47
Gonzalez, J., **72**, 225
Gräf, S., **72**, 32, 38
Greene, A. E., **74**, 13
Grierson, D. S., **70**, 54
Griffen, E. J., **72**, 163

Hadida, S., **74**, 248
Hamper, B. C., **70**, 246
Hansen, T. K., **73**, 270
Hartmann, P., **71**, 200
Hazen, G. G., **73**, 144
Heidelbaugh, T. M., **73**, 44
Helquist, P., **70**, 177; **74**, 137, 229
Hendrix, J. A., **71**, 97
Herzog, S., **72**, 32
Hill, J. M., **73**, 50
Horiguchi, Y., **73**, 123
Huboux, A. H., **73**, 61
Hubschwerlen, C., **72**, 1, 14
Husson, H.-P., **70**, 54
Huyer, G., **70**, 1

Imada, Y., **70**, 265
Inoue, S.-i., **72**, 74
Inoue, T., **72**, 154
Ishiyama, T., **71**, 89
Ito, Y., **73**, 94
Iwao, M., **73**, 85

Jackson, Y. A., **70**, 151
Jäger, V., **74**, 1, 130
Jamison, T. F., **71**, 220, 226
Jászberényi, J. Cs., **74**, 101
Jefford, C. W., **71**, 207
Job, K., **71**, 39; **72**, 62
Johnson, C. R., **71**, 133, 140
Johnson, M. R., **70**, 68
Jones, R. J., **70**, 29
Jørgensen, T., **73**, 270
Joseph, S. P., **74**, 77

Kabalka, G. W., **73**, 116
Kann, N., **74**, 13
Kambe, N., **72**, 154
Kamel, S., **70**, 29
Kannangara, G. S. K., **71**, 158
Kappes, D., **71**, 48
Kataoka, Y., **72**, 180; **73**, 73
Kawase, T., **73**, 240, 253
Kazala, A. P., **73**, 13
Kazlauskas, R. J., **70**, 60
Khedekar, R., **73**, 270
King, S. M., **71**, 77
Kitamura, M., **71**, 1
Kitamura, T., **70**, 215
Knapp, S., **70**, 101
Knochel, P., **70**, 195
Komada, K., **73**, 184
Konopelski, J. P., **73**, 201
Kowalski, C. J., **71**, 146
Koyama, H., **71**, 133, 140

Krakowiak, K. E., **70**, 129
Krisché, M. J., **71**, 220
Krishnamurti, R., **72**, 232
Kumar, A., **73**, 159
Kunz, T., **71**, 189
Kuraishi, T., **73**, 85
Kurata, H., **73**, 240
Kuwajima, I., **73**, 123

Lakner, F. J., **73**, 201
LaMaire, S. J., **71**, 77
Larsen, J., **72**, 265
Leblanc, Y., **74**, 241
Lee, J. H., **73**, 61
Lee, T. V., **72**, 189
Lenoir, C., **72**, 265
Lenz, G. R., **70**, 139
Leroy, J., **74**, 212
Lessor, R. A., **70**, 139
Lewis, B. M., **73**, 159
Ley, S. V., **70**, 157
Li, Y., **71**, 207
Lieberknecht, A., **74**, 1
Liebscher, J., **74**, 257
Light, J., **72**, 199
Lu, X., **72**, 112
Lubell, W. D., **71**, 220
Luh, T.-Y., **70**, 240; **74**, 187

Ma, S., **72**, 112
Maddox, J. T., **73**, 116
Magriotis, P. A., **72**, 252
Mander, L. N., **70**, 256
Marcin, L. R., **74**, 33
Marek, I., **74**, 194
Marinetti, A., **74**, 108
Marotta, E., **74**, 158
Marquais, S., **72**, 135
Maruoka, K., **72**, 95
Maruyama, K., **71**, 118, 125
Mathre D. J., **74**, 50
Matthews, D. P., **72**, 209, 216
Mattson, M. N., **70**, 177
Maw, G. N., **71**, 48
McCarthy, J. R., **72**, 209, 216
McDonald, F. E., **70**, 204
McKee, B. H., **70**, 47
Medich, J. R., **71**, 133, 140
Meister, P. G., **70**, 226
Mergelsberg, I., **73**, 159
Merino, P., **72**, 21
Meyer, C., **74**, 194
Meyers, A. I., **71**, 107;
73, 215, 221, 246
Mikami, K., **71**, 14
Miller, R. F., **73**, 61, 134
Miyai, J., **73**, 73
Miyaura, N., **71**, 89
Modi, S. P., **72**, 125
Mohan, J. J., **74**, 50
Moriwake, T., **73**, 184
Morwick, T., **74**, 169
Mozaffari, A., **70**, 68
Mudryk, B., **72**, 173
Mühlemann, C., **71**, 200
Murahashi, S.-I., **70**, 265
Murphree, S. S., **74**, 115
Murphy, C. K., **73**, 159
Murray, R. W., **74**, 91
Myers, A. G., **72**, 104
Myles, D. C., **70**, 231

Nakagawa, Y., **73**, 94
Nakai, T., **71**, 14
Nakamura, E., **73**, 123
Narisawa, S., **71**, 14
Naruta, Y., **71**, 118, 125
Negrete, G. R., **73**, 201
Negron, A., **70**, 169
Ni, Z.-J., **70**, 240; **74**, 115, 147
Nielsen, R. B., **71**, 77
Nikolaides, N., **72**. 246
Nikolic, N. A., **74**, 23
Nishigaichi, Y., **71**, 118
Nissen, J. S., **73**, 110
Normant, J.-F., **74**, 194
Nowick, J. S., **73**, 61
Noyori, R., **71**, 1; **72**, 74
Nugent, W. A., **70**, 272

O'Connor, E. J., **70**, 177
Obrecht, J.-P., **71**, 200
Oda, M., **73**, 240, 253
Oglesby, R. C., **72**, 125
Ohkuma, T., **71**, 1
Ohta, **72**, T., 74
Oi, R., **73**, 1
Okabe, M., **71**, 83; **72**, 48
Okada, T., **73**, 253
Okazoe, T., **73**, 73

Ooi, T., **72**, 95
Oshima, K., **72**, 180; **73**, 73
Overman, L. E., **70**, 111; **71**, 56, 63

Padwa, A., **74**, 115, 147
Panek, J. S., **70**, 79
Pansare, S. V., **70**, 1, 10
Paolini, J. P., **72**, 209, 216
Paquette, L. A., **70**, 226; **71**, 175, 181; **72**, 57; **73**, 36, 44; **74**, 169
Pariza, R. J., **74**, 124
Park, J. M., **70**, 18
Patel, M., **70**, 79
Patois, C., **72**, 241; **73**, 152
Pfau, M., **70**, 35
Pfiffner, A., **72**, 116
Pikul, S., **71**, 22, 30
Poggendorf, P., **74**, 130
Polin, J., **73**, 262
Porter, J. R., **72**, 189
Prakash, G. K. S., **72**, 232
Presnell, M., **73**, 110
Protopopova, M. N., **73**, 13

Rama, F., **74**, 158
Ramaiah, P., **72**, 232
Rapoport, H,. **70**, 29; **71**, 220, 226
Reddy R. E., **71**, 146
Reddy, R. T., **73**, 159
Reed, J. N., **72**, 163
Reichelt, I., **71**, 189
Reid, J. R., **74**, 217
Reissig, H.-U., **71**, 189
Reno, D. S., **74**, 124
Revial, G., **70**, 35
Ricci, A., **74**, 84
Rieke, L. I., **72**, 147
Rieke, R. D., **72**, 147
Righi, P., **74**, 158
Rishton, G. M., **71**, 56, 63
Roberts, F. E., **73**, 144
Rolfs, A., **74**, 257
Romines, K. R., **70**, 93; **71**, 133, 140
Rosini, G., **74**, 158
Royer, J., **70**, 54
Russ, W. K., **73**, 144

Saito, S., **73**, 184
Sakogawa, K., **72**, 180
Sathyanarayana, S., **74**, 248
Savignac, P., **72**, 241; **73**, 152; **74**, 108
Scherer, D., **73**, 159
Schiess, P., **72**, 116
Schipor, I., **72**, 246
Schmid, C. R., **72**, 6
Schnute, M. E., **74**, 33
Schottenberger, H., **73**, 262
Schultz, A. G., **73**, 174
Schwartz, J., **71**, 83
Seconi, G., **74**, 84
Seebach, D., **71**, 39; **72**, 62
Seman, J. J., **73**, 144
Sessler, J. L., **70**, 68
Sethi, S. P., **70**, 256
Sharp, M. J., **70**, 111
Sharpless, K. B., **70**, 47; **73**, 1
Shiota, T., **70**, 265
Shoup, T., **73**, 116
Singh, M., **74**, 91
Sivik, M. R., **70**, 226; **72**, 57
Smith, G. F., **73**, 36
Snieckus, V., **72**, 163
Soderquist, J. A., **70**, 169
Sommerfeld, Th., **72**, 62
Sonoda, N., **72**, 154
Specklin, J.-L., **72**, 1, 14
Stang, P. J., **70**, 215
Stanton, K. J., **72**, 57
Staskiewicz, S., **73**, 144
Stephenson, E. K., **70**, 151
Steuer, B., **74**, 1
Stille, J. K., **71**, 97
Streiber, J., **71**, 236
Sun, R. C., **71**, 83; **72**, 48
Suzuki, A., **71**, 89

Taddei, M., **74**, 84
Takai, K., **72**, 180; **73**, 73
Takaya, H., **72**, 74
Tamao, K., **73**, 94
Taylor, D. K., **74**, 101
Terada, M., **71**, 14
Thompson, A. S., **74**, 50
Thorarensen, A., **74**, 33
Tietze, L. F., **71**, 167, 214
Timko, J. M., **71**, 72
Tius, M. A., **71**, 158
Tokunaga, M., **71**, 1
Truong, T., **70**, 29
Tschantz, M. A., **73**, 215, 221

Tullis, J. S., **74**, 229

Utimoto, K., **72**, 180; **73**, 73

Varma, K. S., **73**, 270
Vederas, J. C., **70**, 1, 10

Wallace, E. M., **73**, 50
Wang, S., **72**, 163
Wang, Y., **71**, 207
Warmus, J. S., **73**, 246
Watson, B. T., **71**, 77
Watterson, S. H., **74**, 115, 147
Wehner, V., **74**, 1
Wender, P. A., **70**, 204
Westrum, L. J., **73**, 13
White, A. W., **70**, 204
Williams, R. M., **71**, 97
Winchester, W. R., **73**, 13
Windham, C. Q., **73**, 25
Wipf, P., **74**, 205
Witschel, M. C., **70**, 111
Wu, T.-C., **72**, 147

Yamamoto, H., **72**, 86, 95; **74**, 178
Yamashita, A., **71**, 72
Yanagisawa, A., **74**, 178
Yasue, K., **74**, 178
Yeh, M. C. P., **70**, 195
Yuan, T.-M., **74**, 187

Zhou, P., **73**, 159
Zibuck, R., **71**, 236

CUMULATIVE SUBJECT INDEX
FOR VOLUMES 70, 71, 72, 73, and 74

This index comprises subject matter for Volumes **70**, **71**, **72**, **73**, and **74** only. For subjects in previous volumes, see either the indices in Collective Volumes I through VIII or the single volume entitled *Organic Syntheses, Collective Volumes I-VIII, Cumulative Indices*, edited by J. P. Freeman.

The index lists the names of compounds in two forms. The first is the name used commonly in procedures. The second is the systematic name according to **Chemical Abstracts** nomenclature. Both are usually accompanied by registry numbers in parentheses. Also included are general terms for classes of compounds, types of reactions, special apparatus, and unfamiliar methods.

Most chemicals used in the procedure will appear in the index as written in the text. There generally will be entries for all starting materials, reagents, intermediates, important by-products, and final products. Entries in capital letters indicate compounds appearing in the title of the preparation.

Acetamidoacrylic acid, **70**, 6

p-Acetamidobenzenesulfonyl azide: Sulfanilyl azide, N-acetyl-; Benzenesulfonyl azide, 4-(acetylamino)-; (2158-14-7), **70**, 93; **73**, 16, 20, 24; **74**, 229

p-Acetamidobenzenesulfonyl chloride: Benzenesulfonyl chloride, 4-(acetylamino)-; (121-60-8), **70**, 93

Acetic acid; (64-19-7), **70**, 36, 140; **71**, 168; **74**, 24

Acetic anhydride: Acetic acid anhydride; (108-24-7), **70**, 69, 130; **73**, 25, 28, 35 **74**, 160

Acetonitrile: TOXIC. (75-05-8), **71**, 227; **73**, 15, 23, 135, 136, 143; **74**, 219, 229
anhydrous, **70**, 2

Acetophenone; (98-86-2), **74**, 258

1-ACETOXY-2-BUTYL-4-METHOXYNAPHTHALENE: 1-NAPHTHALENOL, 2-BUTYL-4-METHOXY-, ACETATE; (99107-52-5), **71**, 72

(4R)-(+)-Acetoxy-2-cyclopenten-1-one: 2-Cyclopenten-1-one, 4-(acetyloxy)-, (R)-; (59995-48-1), **73**, 36, 37, 38, 43

(±)-cis-4-Acetoxy-1-hydroxycyclopent-2-ene: 4-Cyclopentene-1,3-diol, monoacetate, cis-; (60410-18-6), **73**, 25, 27, 35

(1S,4R)-(+)-4-Acetoxy-1-hydroxycyclopent-2-ene: 4-Cyclopentene-1,3-diol, monoacetate, (1S-cis); (60410-16-4), **73**, 25, 26, 35

4-Acetoxy-3-nitrohexane: 3-Hexanol, 4-nitro-, acetate; (3750-83-2), **70**, 69

(2S,3S)-3-ACETYL-8-CARBOETHOXY-2,3-DIMETHYL-1-OXA-8-AZASPIRO[4.5]DECANE, **71**, 63

Acetyl chloride; (75-36-5), **71**, 148; **72**, 33; **73**, 185, 189, 200

Acetyl cholinesterase (from electric eel): Esterase, acetyl choline; (9000-81-1), **73**, 26, 29, 32, 33, 35

1-(2-Acetyl-4,6-dimethoxyphenyl)-1,2-hydrazinedicarboxylic acid bis (2,2,2-trichloroethyl) ester: 1,2-Hydrazinedicarboxylic acid, 1-(2-acetyl-4,6-dimethoxyphenyl)-, bis(2,2,2-trichloroethyl) ester; (154876-134-8), **74**, 241

Acetylenes, coupling reactions, **72**, 104
from aldehydes, **74**, 112
perfluoroalkyl, **70**, 246

α-ACETYLENIC ESTERS, **70**, 246

Acetylenic ethers, **74**, 13
stereoselective reduction, **74**, 19

Acetylferrocene: HIGHLY TOXIC: Ferrocene, acetyl-; (1271-55-2), **73**, 262, 264, 267, 269

O-Acetylserine, **70**, 6

Acrolein: 2-Propenal; (107-02-8), **71**, 236; **72**, 190; **73**, 2, 12

Acryloyl-2-oxazolidinone: 2-Oxazolidinone, 3-acryloyl-; (2043-21-2), **71**, 31

Activated magnesium, **73**, 52

N-Acylaziridines, **70**, 106

α-Acylmethylenephosphoranes, **70**, 251

(ACYLOXY)BORANE COMPLEXES, CHIRAL, **72**, 86

3-ACYLTETRAHYDROFURANS, Stereocontrolled preparation, **71**, 63

1-ADAMANTYL CALCIUM HALIDES, **72**, 147

1-(1-ADAMANTYL)CYCLOHEXANOL: CYCLOHEXANOL, 1-TRICYCLO[3.3.1.1^{3,7}]DEC-1-YL-; (84213-80-9), **72**, 147

L-Alanine; (56-41-7), **71**, 227

Aliquat 336: Ammonium, methyltrioctyl-, chloride; 1-Octanaminium, N-methyl-N,N-dioctyl-, chloride; (5137-55-3), **73**, 145, 147, 151, 162, 166, 168, 173

ALKENES, SYNTHESIS, **73**, 61

ALKENE TRANSFER, **74**, 205

ALKENYLCHROMIUM REAGENTS, **72**, 180

Z-Alkenyl ethers, **73**, 78

α-Alkoxyorganolithiums, **71**, 143, 144

α-Alkoxyorganostannanes, **71**, 143

α-Alkylated piperidines, **70**, 57

Alkylation-metalation of a trialkyl phosphate, **72**, 244

9-ALKYL-9-BORABICYCLO[3.3.1]NONANES, **71**, 89

Alkylzinc iodide, **70**, 201

ALKYNE-IMINIUM ION CYCLIZATIONS, **70**, 111

Alkynyl(phenyl)iodonium sulfonates, **70**, 223

ALKYNYL(PHENYL)IODONIUM TOSYLATES, **70**, 215

Allyl bromide: 1-Propene, 3-bromo-; (106-95-6), **72**, 199

ALLYLIC ALCOHOLS, **74**, 205
ASYMMETRIC HYDROGENATION OF, **72**, 74

Allylic alcohol preparation, **72**, 180

ALLYLIC DITHIOACETALS, **70**, 240

Allylic halides, coupling, **71**, 121

ALLYLIC KETALS, **71**, 63

Allylic oxidation, **72**, 61

Allylic tributyltin derivatives, synthesis, **71**, 121, 125

Allyl vinyl ethers, **73**, 79

Aluminum chloride: Aluminum trichloride; (7446-70-0), **70**, 162, 226; **74**, 3

Aluminum hydride reducing agents, **70**, 65

AMBIPHILIC DIENES, **73**, 231

Amination reaction, **74**, 245

Amino acids: α-: **70**, 6
Optically active, **73**, 194
β-: Enantiomerically pure, **73**, 201

β-Amino alcohols, **70**, 58

(1S,2S)-2-Amino-1-[((2S)-2-carbomethoxypyrrolidinyl)carbonyl]cyclohexane, **73**, 175

2-Amino-5-chloropyridine: 2-Pyridinamine, 5-chloro-; (1072-98-6), **74**, 78

AMINOCYCLOHEXANECARBOXYLIC ACID DERIVATIVES, ASYMMETRIC SYNTHESIS, **73**, 174

(3S,4S)-3-AMINO-1-(3,4-DIMETHOXYBENZYL)-4-[(R)-2,2-DIMETHYL-1,3-DIOXOLAN-4-YL]-2-AZETIDINONE, **72**, 14

2-Amino-3,3-diphenyl-1-cyclopentene-1-carbonitrile: 1-Cyclopentene-1-carbonitrile, 2-amino-3,3-diphenyl-; (3597-67-9), **74**, 33, 38

2-(2-Aminoethoxy)ethanol: Ethanol, 2-(2-aminoethoxy)-; (929-06-6), **70**, 129

Amino lactams, **70**, 106

(R)-3-AMINO-3-(p-METHOXYPHENYL)PROPIONIC ACID: BENZENEPROPANOIC ACID, β-AMINO-4-METHOXY-, (R)-; (131690-57-8), **73**, 201, 204, 213

2-Amino-2-methyl-1-propanol: 1-Propanol, 2-amino-2-methyl-; (124-68-5), **71**, 108

α-Aminonitrile, **70**, 57

(S)-3-AMINO-2-OXETANONE p-TOLUENESULFONATE: 2-OXETANONE, 3-AMINO-, (S)-, 4-METHYLBENZENESULFONATE; (112839-95-9), **70**, 10

Amipurimycin studies, **70**, 27

2-Aminopyridine: 2-Pyridinamine; (504-29-0), **74**, 77

Ammonia; (7664-41-7), **72**, 253; **73**, 87, 175, 177, 178, 183, 223

Ammonium formate: Formic acid, ammonium salt; (540-69-2), **72**, 128

Ammonium hydroxide; (1336-21-6), **73**, 87, 271, 277

Annulated lactams, **70**, 106

Antimony trichloride, **72**, 210

Arsoles, **70**, 276

Ascorbic acid, **72**, 3

L-Asparagine monohydrate: L-Asparagine; (70-47-3), **73**, 202, 205, 213

Asymmetric catalysts, **71**, 1, 14, 30; **74**, 65

ASYMMETRIC DIELS-ALDER REACTION, **72**, 86

Asymmetric dihydroxylation, **70**, 51; **73**, 8

ASYMMETRIC HYDROGENATION, **72**, 74

Asymmetric hydroxylation, **73**, 168, 170

Asymmteric reduction, **74**, 45

Asymmetric synthesis, **74**, 29, 43
OF SUBSTITUTED PIPERIDINES, **70**, 54

Azetidinones, **72**, 19

Azides, sulfonyl, **70**, 97

β-Azidoalanine, **70**, 15

Azido lactams, **70**, 106

Azidotrimethylsilane: HIGHLY TOXIC: Silane, azidotrimethyl-; (4648-54-8), **73**, 187, 191, 195, 196, 200

Azobisisobutyronitrile: Propanenitrile, 2,2'-azobis[2-methyl-; (78-67-1), **70**, 151, 167; **72**, 218

Barbier-type cross coupling, **71**, 121

BARIUM, ACTIVE, **74**, 178
ALLYLIC REAGENTS, CARBOXYLATION, **74**, 178

Barium iodide, anhydrous, **74**, 178, 180
dihydrate, **74**, 180

Barton-Zard pyrrole synthesis, **70**, 75

BEDT-TTF, **72**, 265

Benzaldehyde; (100-52-7), **74**, 2

Benzazepine, **70**, 145

Benzazepinone, **70**, 144

Benzene; (71-43-2), CANCER SUSPECT AGENT, **70**, 19, 71, 122, 231, 233; **72**, 57, 218

1,2-Benzenedimethanol; (612-14-6), **73**, 2, 5, 11

Benzenemethanamine, α-methyl-N-(2-methylcyclohexylidene)-; (76947-33-6), **70**, 35

Benzeneruthenium(II) chloride dimer: Ruthenium, bis(η^6-benzene-di-μ-chlorodichorodi-; (37366-09-9), **71**, 1; **72**, 74

2-BENZENESULFONYL CYCLIC ETHERS, **70**, 157

Benzil: Ethanedione, diphenyl-; (134-81-6), **71**, 23

BENZOANNELATION OF KETONES, **71**, 158

BENZOCYCLOBUTENONE: BICYCLO[4.2.0]OCTA-1,3,5-TRIEN-7-ONE; (3469-06-5), **72**, 116

Benzonitrile; (100-47-0), **72**, 233; **74**, 230

Benzophenone: Methanone, diphenyl-; (119-61-9), **74**, 24

p-Benzoquinone: 2,5-Cyclohexadiene-1,4-dione; (106-51-4), **73**, 253, 256, 261

Benzoyl chloride; (98-88-4), **73**, 272, 274, 277

Benzoyl peroxide: Peroxide, dibenzoyl; (94-36-0), **73**, 232, 234, 239, 240, 244

Benzyl alcohol: Benzenemethanol; (100-51-6), **72**, 87

Benzylamine; (100-46-9), **70**, 112, 131

Benzyl bromide: Toluene, α-bromo-; Benzene, (bromomethyl)-; (100-39-0), **72**, 23

Benzyl chloroformate: Carbonochloridic acid, phenylmethyl ester; (501-53-1), **70**, 29

4-BENZYL-10,19-DIETHYL-4,10,19-TRIAZA-1,7,13,16-TETRAOXA-CYCLOHENEICOSANE (TRIAZA-21-CROWN-7), **70**, 129, 132

(-)-2-O-BENZYL-L-GLYCERALDEHYDE: PROPANAL, 3-HYDROXY-2-(PHENYLMETHOXY)-, (R)-; (76227-09-3), **74**, 1

(1R)-O-Benzyl-2,3-di-O-isopropylidene-1-(2-thiazolyl)-D-glycitol: Thiazole, 2-[(2,2-dimethyl-1,3-dioxolan-4-yl)(phenylmethoxy)methyl]-, [R-(R*,R*)]-; (103795-11-5), **72**, 23

N-Benzyl-4-hexyn-1-amine: Benzenemethanamine, N-4-hexynyl-; (112069-91-7), **70**, 112

(E)-1-BENZYL-3-(1-IODOETHYLIDENE)PIPERIDINE, **70**, 111

2-O-BENZYL-3,4-ISOPROPYLIDENE-D-ERYTHROSE: 1,3-DIOXOLANE-4-ACETALDEHYDE, 2,2-DIMETHYL-α-(PHENYLMETHOXY)-, [R-(R*,R*)]-; (103795-12-6), **72**, 21

N-(Benzyloxycarbonyl)-L-methionine methyl ester: L-Methionine, (N-(phenylmethoxy)carbonyl]-, methyl ester; (56762-93-7), **70**, 29

N$^{\alpha}$-(BENZYLOXYCARBONYL)-β-(PYRAZOL-1-YL)-L-ALANINE: 1H-PYRAZOLE-PROPANOIC ACID, α-[[(PHENYLMETHOXY)CARBONYL]AMINO]-, **70**, 2

N-(BENZYLOXYCARBONYL)-L-SERINE: L-SERINE, N-[(PHENYLMETHOXY)CARBONYL]-; (1145-80-8), **70**, 1

N-(BENZYLOXYCARBONYL)-L-SERINE β-LACTONE: SERINE, N-CARBOXY-, β-LACTONE, BENZYL ESTER, L-; (26054-60-4), **70**, 1

N-(BENZYLOXYCARBONYL)-L-VINYLGLYCINE METHYL ESTER: 3-BUTENOIC ACID, 2-[[(PHENYLMETHOXY)CARBONYL]AMINO]-, METHYL ESTER, (S)-; (75266-40-9), **70**, 29

(+)-2-O-Benzyl-L-threitol: 1,2,4-Butanetriol, 3-(phenylmethoxy)-, [S-(R*,S*)]-; (124909-02-0), **74**, 2

9-Benzyl-3,9,15-triaza-6,12-dioxaheptadecane, **70**, 130

BICYCLO[3.2.0]HEPT-EN-ONES, **74**, 158

(1S-endo)-3-(Bicyclo[2.2.1]hept-5-en-2-ylcarbonyl)-2-oxazolidinone: 2-Oxazolidinone, 3-(bicyclo[2.2.1]hept-5-en-2-ylcarbonyl)-, (1S-endo)-; (109299-97-0), **71**, 30

2,2'-BI-5,6-DIHYDRO-1,3-DITHIOLO[4,5-b][1,4]DITHIINYLIDENE: BEDT-TTF; BIS(ETHYLENEDITHIO)TETRATHIAFULVALENE; 1,3-DITHIOLO[4,5-b][1,4]-DITHIIN, 2-(5,6-DIHYDRO-1,3-DITHIOLO[4,5-b][1,4]DITHIIN-2-YLIDENE)-5,6-DIHYDRO-; (66946-48-3), **72**, 265

(R)-BINAP: 2,2'-Bis(diphenylphosphino)-1,1'-binaphthyl; Phosphine, [1,1'-binaphthalene]-2,2'-diylbis(diphenyl-, (R)-; (76189-55-4), **71**, 1; **72**, 74

BINAP-Ruthenium chloride triethylamine, **72**, 82

BINAP-RUTHENIUM COMPLEXES, **71**, 1; **72**, 74

BINAP-Ruthenium iodide p-cymene, **72**, 82

BINAP-Ruthenium toluene, **72**, 82

BINAP-Ruthenium trifluoroacetate, **72**, 82

(±)-1,1'-BI-2-NAPHTHOL: [1,1'-BINAPHTHALENE]-2,2'-DIOL, (±)-; (41024-90-2), **70**, 60

(R)-(+)-1,1'-BI-2-NAPHTHOL: [1,1'-BINAPHTHALENE]-2,2'-DIOL, (R)-(+)-; (18531-94-7), **70**, 60; **71**, 15

(S)-(-)-1,1'-BI-2-NAPHTHOL: [1,1'-BINAPHTHALENE]-2,2'-DIOL, (S)-(-)-; (18531-99-2), **70**, 60

(±)-1,1'-Bi-2-naphthyl pentanoate: Pentanoic acid, [1,1'-binaphthalene]-2,2'-diyl ester, (±)-; (100465-51-8), **70**, 61

(R)-1,1'-Bi-2-naphthyl pentanoate: Pentanoic acid, [1,1'-binaphthalene]-2,2'-diyl ester, (R)-, **70**, 62

Biphenyl: 1,1'-Biphenyl; (92-52-4), **72**, 147; **74**, 178

Biphenyls, unsymmetrical, synthesis, **71**, 113

Bis(cyclopentadienyl)zirconium dichloride: Zirconocene dichloride; Zirconium, dichloro-p-cyclopentadienyl-; Zirconium, dichlorobis(h^5-2,4-cyclopentadien-1-yl)-; (1291-32-3), **71**, 77

[(R)-2,2'-Bis(diphenylphosphino)-1,1'-binaphthyl]dichlororuthenium: Ruthenium, [[1,1'-binaphthalene]-2,2'-diylbis(diphenylphosphine]-P,P']dichloro-; (115245-70-0); **71**, 1

[(R)-2,2'-Bis(diphenylphosphino)-1,1'-binaphthyl]ruthenium diacetate: Ruthenium, bis(acetato-O,O'-)[[1,1'-binaphthalene]-2,2'-diylbis[diphenylphosphine]-P,P']-, [OC-6-22-D-(R)]-; (104621-48-9), **72**, 74

[1,3-Bis(diphenylphosphino)propane]nickel(II) chloride: Nickel, dichloro[1,3-propanediylbis[diphenylphosphine]-P,P']-; (15629-92-2), **74**, 188

Bismoles, **70**, 276

1,2-Bis(2-iodoethoxy)ethane: Ethane, 1,2-bis(2-iodoethoxy)-; (36839-55-1), **70**, 132

2,3-Bis(phenylsulfinyl)-1,3-butadiene: Benzene, 1,1'-[[1,2-bis(methylene)-1,2-ethanediyl]bis(sulfinyl)]bis-; (85540-18-7), **74**, 148

2,3-BIS(PHENYLSULFONYL)-1,3-BUTADIENE: BENZENE, 1,1'-[[1,2-BIS-(METHYLENE)-1,2-ETHANEDIYL]BIS(SULFONYL)]BIS-; (85540-20-1), **74**, 147

Bis(2,2,2-trichloroethyl) azodicarboxylate: Diazenedicarboxylic acid, bis-(2,2,2-trichloroethyl) ester; (38857-88-4), **74**, 241

BIS(TRIFLUOROETHYL) (CARBOETHOXYMETHYL)PHOSPHONATE: ACETIC ACID, [BIS(2,2,2-TRIFLUOROETHOXY)PHOSPHINYL]-, ETHYL ESTER; (124755-24-4), **73**, 152, 153, 154, 156, 157

Bis(trifluoroethyl) methylphosphonate: Phosphonic acid, methyl-, bis(2,2,2-trifluoroethyl) ester; (755-95-9), **73**, 152, 153, 157

Bis(trimethylsilyl)imidate, **70**, 104

Bis(trimethylsilyl)methyllithium, **70**, 243

BIS(TRIMETHYLSILYL) PEROXIDE: SILANE, DIOXYBIS[TRIMETHYL-; (5796-98-5), **74**, 84

Bis(triphenylcyclopropenyl) ether, **74**, 74

Bis(triphenylphosphine)palladium(II) chloride: Palladium, dichlorobis(triphenylphosphine)-; (13965-03-2), **71**, 99

9-Borabicyclo[3.3.1]nonane; (280-64-8), **70**, 169; **71**, 90

9-BORABICYCLO[3.3.1]NONANE DIMER: DIBORANE (6), 1,1:2,2-DI-1,5-CYCLOOCTYLENE-; (70658-61-6) and (21205-91-4), **70**, 169

Borane-methyl sulfide complex: Borane, compd. with thiobis[methane] (1:1); (13292-87-0), **70**, 169; **74**, 35, 53, 54

Borane-tetrahydrofuran complex: Furan, tetrahydro-, compd. with borane (1:1); (14044-65-6), **70**, 131; **71**, 207; **72**, 88, 200; **73**, 117, 118, 122

Borole Diels-Alder dimers, **70**, 276

Boron trifluoride etherate: Ethyl ether, compd. with boron fluoride (BF_3) (1:1); Ethane, 1,1'-oxybis-, compd. with trifluoroborane (1:1); (109-63-7), **70**, 22, 197, 217, 240; **71**, 125, 134; **73**, 224, 230; **74**, 187

Bovine pancreas acetone powder, **70**, 61

α-BRANCHED AMINO ACIDS, **72**, 62
SYNTHESIS OF, **72**, 62

Bromine; (7726-95-6), **71**, 89; **72**, 49, 95, 104, 200, 252; **73**, 232, 234, 238; **74**, 116, 124

1-Bromoadamantane: Adamantane, 1-bromo-; Tricyclo[3; .3.1.1^{3,7}]decane, 1-bromo-; (768-90-1), **72**, 148

2-Bromoanisole: Anisole, o-bromo-; Benzene, 1-bromo-2-methoxy-; (578-57-4), **71**, 109

4-Bromoanisole: Anisole, p-bromo-; Benzene, 1-bromo-4-methoxy-; (104-92-7), **71**, 99

Bromobenzene: Benzene, bromo-; (108-86-1), **71**, 220; **72**, 32

5-(4-BROMOBENZOYL)-2-MORPHOLINO-3-PHENYLTHIOPHENE: METHANONE, (4-BROMOPHENYL)[5-(4-MORPHOLINYL)-4-PHENYL-2-THIENYL]-; (86673-62-3), **74**, 257

1-Bromobutane; (109-65-9), **70**, 218

2-Bromo-N-Boc-glycine tert-butyl ester: Acetic acid, bromo[[(1,1-dimethylethoxy)-carbonyl]amino]-, 1,1-dimethylethyl ester, (±)-; (111652-22-2), **71**, 200

3-Bromo-1-(tert-butyldimethylsilyl)indole: 1H-Indole, 3-bromo-1-[(1,1-dimethylethyl)-dimethylsilyl]-; (153942-69-9), **74**, 248

4-Bromobutyronitrile: Butanenitrile, 4-bromo-; (5332-06-9), **74**, 34

Bromocresol green spray, **70**, 5, 13

3-Bromo-1,5-cyclooctadiene: 1,5-Cyclooctadiene, 3-bromo-; (23346-40-9), **73**, 241, 243, 244

6-Bromo-1,4-cyclooctadiene: 1,4-Cyclooctadiene, 6-bromo-; (23359-89-9), **73**, 241, 243, 244

4-Bromo-2,6-di-tert-butylphenol: Phenol, 4-bromo-2,6-di-tert-butyl-; Phenol, 4-bromo-2,6-bis(1,1-dimethylethyl)-; (1139-52-2), **72**, 95

(3-Bromo-3,3-difluoropropyl)trimethylsilane: Silane, (3-bromo-3,3-difluoropropyl)trimethyl-; (134134-62-6), **72**, 226

3-Bromo-5,6-dihydro-(2H)-pyran-2-one: 2H-Pyran-2-one, 3-bromo-5,6-dihydro-; (104184-64-7), **73**, 232, 234, 238

1-Bromo-3,3-dimethoxypropane: Propane, 3-bromo-1,1-dimethoxy-; (36255-44-4), **72**, 190

2-Bromoethanol: HIGHLY TOXIC. Ethanol, 2-bromo-; (540-51-2), **73**, 94, 99, 108

2-Bromoethyl methoxymethyl ether: Ethane, 1-bromo-2-(methoxymethoxy)-; (112496-94-3), **73**, 94, 95, 99, 108

1-Bromo-3-(2-methoxyethoxy)propane: Propane, 1-bromo-3-(2-methoxyethoxy)-; (59551-75-6), **72**, 200

4-Bromophenacyl bromide: Ethanone, 2-bromo-1-(4-bromophenyl)-; (99-73-0), **74**, 259

9-BROMO-9-PHENYLFLUORENE: 9H-FLUORENE, 9-BROMO-9-PHENYL-; (55135-66-5), **71**, 220, 227, 228

1-Bromo-1-phenylthioethene; 1-Bromovinyl phenyl sulfide: Benzene, [(1-bromoethenyl) thio]-; (80485-53-6), **71**, 89; **72**, 252

2-Bromopropene: Propene, 2-bromo-; 1-Propene, 2-bromo-; (557-93-7), **71** , 64; **74**, 169

3-BROMOPROPIOLIC ESTERS, SYNTHESIS, **74**, 212

3-Bromopropionyl chloride: Propionyl chloride, 3-bromo-; Propanoyl chloride, 3-bromo-; (15486-96-1), **71**, 31

2-(3-Bromopropyl)-1H-isoindole-1,3(2H)-dione: 1H-Isoindole-1,3(2H)-dione, 2-(3-bromopropyl)-; (5460-29-7), **70**, 195

3-BROMO-(2H)-PYRAN-2-ONE: 2H-PYRAN-2-ONE, 3-BROMO-; (19978-32-6), **73**, 231, 233, 235, 236, 237, 238

2-Bromopyrrole, **70**, 155

N-Bromosuccinimide: Succinimide, N-bromo-; 2,5-Pyrrolidinedione, 1-bromo-; (128-08-5), **70**, 155; **71**, 200; **73**, 232, 239, 240, 242, 244; **74**, 212, 249

2-Bromothiazole: Thiazole, 2-bromo-; (3034-53-5), **72**, 22

Bromotrifluoromethane: Methane, bromotrifluoro-; (75-63-8), **72**, 233

Bromotris[3-(2-methoxyethoxy)propyl]stannane: 2,5,13,16-Tetraoxa-9-stannaheptadecane, 9-bromo-9-[3-(2-methoxyethoxy)propyl]-; (130691-02-0), **72**, 202

1-Bromovinyl phenyl sulfide; 1-Bromo-1-phenylthioethene: Benzene, [(1-bromoethenyl) thio]-; (80485-53-6), **71**, 89, **72**, 252

2-Butanol, (±)-; (15892-23-6), **70**, 206

2-Butanone, 4-[1-methyl-2-[(1-phenylethyl)imino]cyclohexyl]-, [S-(R*, S*)]-; (94089-44-8), **70**, 36

3-Buten-2-one; (78-94-4), **70**, 36

N-tert-BUTOXYCARBONYL-2-BROMOPYRROLE: 1H-PYRROLE-1-CARBOXYLIC ACID, 2-BROMO-, 1,1-DIMETHYLETHYL ESTER; (117657-37-1), **70**, 151

N-(tert-Butoxycarbonyl)-6,7-dimethoxy-1-methylene-1,2,3,4-tetrahydroisoquinoline: 2(1H)-Isoquinolinecarboxylic acid, 3,4-dihydro-6,7-dimethoxy-1-methylene, 1,1-dimethylethyl ester; (82044-08-4), **70**, 139

N-(tert-Butoxycarbonyl)-7,8-dimethoxy-1,3,4,5-tetrahydro-2H-3-benzazepin-2-one, **70**, 140

N-Boc-Glycine tert-butyl ester: Glycine, N-[(1,1-dimethylethoxy)carbonyl]-, 1,1-dimethylethyl ester; (111652-20-1), **71**, 200

1-(tert-BUTOXYCARBONYL)INDOLINE: 1H-INDOLE-1-CARBOXYLIC ACID, 2,3-DIHYDRO-, 1,1-DIMETHYLETHYL ESTER; (143262-10-6), **73**, 85, 86, 88, 89, 90

1-(tert-Butoxycarbonyl)-7-indolinecarboxaldehyde, **73**, 90

N-(tert-Butoxycarbonyl)pyrrolidine: 1-Pyrrolidinecarboxylic acid, 1,1-dimethylethyl ester; (86953-79-9), **74**, 23

N-tert-Butoxycarbonyl-L-serine: L-Serine, N[(1,1-dimethylethoxy)carbonyl]-; (3262-72-4), **70**, 11, 19

N-tert-BUTOXYCARBONYL-L-SERINE β-LACTONE: CARBAMIC ACID, (2-OXO-OXETANYL)-, 1,1-DIMETHYLETHYL ESTER, (S)-; (98541-64-1), **70**, 10

N-tert-BUTOXYCARBONYL-2-TRIMETHYLSILYLPYRROLE: 1H-PYRROLE-1-CARBOXYLIC ACID, 2-(TRIMETHYLSILYL)-, 1,1-DIMETHYLETHYL ESTER; (75400-57-6), **70**, 151

N-tert-Butoxy-2,5-disubstituted pyrroles, **70**, 155

tert-Butyl 2-(2-acetyl-3,4-dimethoxyphenyl)ethyl carbamate, **70**, 143

tert-Butylamine: 2-Propanamine, 2-methyl-; (75-64-9), **74**, 102

(2R,5S)-2-tert-Butyl-1-aza-3-oxabicyclo[3.3.0]octan-4-one: 1H,3H-Pyrrolo-[1,2-c]oxazol-1-one, 3-(1,1-dimethylethyl)tetrahydro-, (3R-cis)-; (81286-82-0), **72**, 63

Butyl bromide; (109-65-9), **74**, 198

tert-BUTYL 3-BROMOPROPIOLATE: 2-PROPYNOIC ACID, 3-BROMO-, 1,1-DIMETHYLETHYL ESTER; (23680-40-2), **74**, 212

tert-Butyl [1-(tert-butoxycarbonyl)-3-oxo-4-pentenyl]carbamate: 5-Hexenoic acid, 2-[[(1,1-dimethylethoxy)carbonyl]amino]-4-oxo-, 1,1-dimethylethyl ester; (117833-62-2), **71**, 201 tert-Butyl(but-3-ynyloxy)diphenylsilane: Silane, (3-butynyloxy)(1,1-dimethylethyl)-diphenyl-; (88138-68-3), **74**, 205

(S,S)-2-tert-Butyl-1-carbomethoxy-6-carboxy-2,3,5,6-tetrahydro-4(1H)-pyrimidinone: **73**, 202

2-tert-Butyl-1-carbomethoxy-2,3-dihydro-4(1H)-pyrimidinone: 1(2H)-Pyrimidinecarboxylic acid, 2-(1,1-dimethylethyl)-3,4-dihydro-4-oxo-, methyl ester, (S)- or (R)-; (S)- (131791-75-8); (R)- (131791-81-6), **73**, 201, 202

tert-Butylchlorodiphenylsilane: tert-Butyldiphenylsilyl chloride; Silane, chloro(1,1-dimethylethyl)diphenyl-; (58479-61-1), **71**, 56; **74**, 206

(4R)-(+)-tert-BUTYLDIMETHYLSILOXY-2-CYCLOPENTEN-1-ONE: 2-CYCLOPENTEN-1-ONE, 4-[[(1,1-DIMETHYLETHYL)DIMETHYLSILYL]OXY]-, (R)-; (61305-35-9), **73**, 36, 37, 40, 41, 43, 47

(4S)-(-)-tert-BUTYLDIMETHYLSILOXY-2-CYCLOPENTEN-1-ONE: 2-CYCLOPENTEN-1-ONE, 4-[[(1,1-DIMETHYLETHYL)DIMETHYLSILYL]OXY]-, (S)-; (61305-36-0), **73**, 44, 45, 46, 47, 49

(1R,4S)-(-)-4-tert-Butyldimethylsiloxy-2-cyclopentyl acetate: 2-Cyclopenten-1-ol, 4-[[(1,1-dimethylethyl)dimethylsilyl]oxy]-, acetate, (1R-cis)-; (115074-47-0), **73**, 44, 47, 49

tert-Butyldimethylsilyl chloride: CORROSIVE. TOXIC. Silane, chloro(1,1-dimethylethyl) dimethyl-; (18162-48-6), **73**, 37, 40, 43, 44, 46, 49; **74**, 249

tert-Butyldiphenylsilyl chloride: Silane, chloro(1,1-dimethylethyl)diphenyl-; (58479-61-1), **71**, 56; **74**, 206

1-(tert-Butyldimethylsilyl)-3-ethylindole: 1H-Indole, 1-[(1,1-dimethylethyl)-dimethylsilyl]-3-ethyl-; (153942-71-3), **74**, 249

1-(tert-Butyldimethylsilyl)-3-lithioindole, **74**, 249

3-(S)-[(tert-BUTYLDIPHENYLSILYL)OXY]-2-BUTANONE: 2-BUTANONE, 3-[[(1,1-DIMETHYLETHYL)DIPHENYLSILYL]OXY]-, (S)-; (135367-18-9), **71**, 56, 64

1-[(tert-BUTYLDIPHENYLSILYLOXY)]-DEC-3-EN-5-OL, **74**, 205

Butyllithium: Lithium, butyl-; (109-72-8), **70**, 152, 154, 272; **71**, 133, 140, 146, 147, 148, 220, 236; **72**, 22, 38, 63, 154, 241; **73**, 61, 63, 72, 74, 76, 84, 134, 136, 143, 153, 155, 157; **74**, 14, 33, 38, 109, 249

sec-Butyllithium: Lithium, sec-butyl; Lithium, (1-methylpropyl)-; (598-30-1), **71**, 148; **72**, 164; **73**, 86, 88, 89, 93, 96, 104, 109; **74**, 24

tert-Butyllithium: Lithium, tert-butyl-; Lithium, (1,1-dimethylethyl)-; (594-19-4), **71**, 64, 148; **73**, 215, 217, 220; **74**, 169, 248

Butylmagnesium bromide: Magnesium, butylbromo-; (693-03-8), **74**, 195

Butylmagnesium chloride: Magnesium, butylchloro-; (693-04-9), **72**, 135

(R)-2-tert-Butyl-6-(4-methoxyphenyl)-5,6-dihydro-4(1H)-pyrimidinone: 4(1H)-Pyrimidinone, 2-(1,1-dimethylethyl)-5,6-dihydro-6-(4-methoxyphenyl)-, (R)-; (131791-77-0), **73**, 203, 214

(2R,5S)-2-tert-Butyl-5-methyl-1-aza-3-oxabicyclo[3.3.0]octan-4-one: 1H,3H-Pyrrolo[1,2-c]oxazol-1-one, 3-(1,1-dimethylethyl)tetrahydro-7a-methyl-, (3R-cis)-; (86046-11-9), **72**, 63

tert-Butyl methyl ether: Ether, tert-butyl methyl; Propane, 2-methoxy-2-methyl-; (1634-04-4), **74**, 34

tert-Butyl propiolate: Propiolic acid, tert-butyl ester; 2-Propynoic acid, 1,1-dimethylethyl ester; (13831-03-3), **74**, 213

2-tert-BUTYL-1,1,3,3-TETRAMETHYLGUANIDINE: GUANIDINE, N"-(1,1-DIMETHYLETHYL)-N,N,N',N'-TETRAMETHYL-; (29166-72-1), **74**, 101

2-Butyne; (503-17-3), **70**, 272

2-Butyne-1,4-diol; (110-65-6), **74**, 148

3-Butyn-1-ol; (927-74-2), **74**, 205

CALCIUM, HIGHLY REACTIVE, **72**, 147

Calicheamycin fragment, **70**, 27

(-)-(1S,4R)-Camphanic acid: 2-Oxabicyclo[2.2.1]heptane-1-carboxylic acid, 4,7,7-trimethyl-3-oxo-, (1S)-; 2-Oxabicyclo[2.2.1]heptane-1-carboxylic acid, 4,7,7-trimethyl-3-oxo-, (-)-; (13429-83-9), **71**, 49

(-)-(1S,4R)-CAMPHANOYL CHLORIDE: 2-OXABICYCLO[2.2.1]HEPTANE-1-CARBONYL CHLORIDE, 4,7,7-TRIMETHYL-3-OXO-, (-)-; (39637-74-6), **71**, 48

(+)-(1R,3S)-Camphoric acid: Camphoric acid, cis-(+)-; 1,3-Cyclopentanedicarboxylic acid, 1,2,2-trimethyl-, (1R-cis)-; (124-83-4), **71**, 49

Camphorsulfonic acid monohydrate: Bicyclo[2.2.1]heptane-1-methanesulfonic acid, 7,7-dimethyl-2-oxo-, (±); (5872-08-2), **70**, 112

(+)-(Camphorylsulfonyl)imine: 3H-3a,6-Methano-2,1-benzisothiazole, 4,5,6,7-tetrahydro-8,8-dimethyl-, 2,2-dioxide, (3aR)-; (107869-45-4), **73**, 168, 173

(-)-(Camphorylsulfonyl)imine: 3H-3a,6-Methano-2,1-benzisothiazole, 4,5,6,7-tetrahydro-8,8-dimethyl-, 2,2-dioxide, (3aS)-; (60886-80-8), **73**, 160, 164, 172

Carbamoyllithiums, **72**, 158

5-Carbamoylpolyoxamic acid, **70**, 27

(Carbethoxymethyl)triphenylphosphonium bromide: Phosphonium, (2-ethoxy-2-oxoethyl)triphenyl-, bromide; (1530-45-6), **70**, 246

1-Carbethoxy-4-piperidone: 1-Piperidinecarboxylic acid, 4-oxo-, ethyl ester; (29976-53-2), **71**, 64

4-CARBOMETHOXY-5-METHOXY-2-PHENYL-1,3-OXAZOLE: 4-OXAZOLECARBOXYLIC ACID, 5-METHOXY-2-PHENYL-, METHYL ESTER; (53872-19-8), **74**, 229

Carbon disulfide; (75-15-0), **72**, 270; **73**, 271, 273, 276

Carbon tetrachloride: CANCER SUSPECT AGENT. Methane, tetrachloro-; (56-23-5), **71**, 200; **72**, 266; **73**, 51, 55, 59, 232, 239, 240, 244, 256

p-Carboxybenzenesulfonyl azide, **70**, 97

CARBOXYLATION, REGIO- AND STEREOSELECTIVE, **74**, 178

β,γ-Carboxylic acids, synthesis, **74**, 183

Cascade process, **74**, 174

Castro-Stephens reaction, **72**, 109

Celite filter, **70**, 122

Chemoselective intermolecular reactions, **72**, 195

Chiral auxiliary, **72**, 36, 42; **74**, 43

CHIRAL BICYCLIC LACTAMS, **73**, 221, 226, 227

Chiral borane complex, **72**, 86

Chiral dirhodium(II) carboxamide catalyst, **73**, 21

Chiral Lewis acid catalyst, **72**, 86

Chiral ligands, **72**, 60

Chiral synthons, **72**, 55

Chloroacetaldehyde dimethyl acetal: Acetaldehyde, chloro-, dimethyl acetal; Ethane, 2-chloro-1,1-dimethoxy-; (97-97-2), **71**, 175

Chloroacetyl chloride: Acetyl chloride, chloro-; (79-04-9), **72**, 266

Chlorobis(η^5-cyclopentadienyl)hydridozirconium: Bis(cyclopentadienyl)zirconium chloride hydride = Schwartz's Reagent: Zirconium, chlorodi-π-cyclopentadienylhydro-; Zirconium, chlorobis(η^5-2,4-cyclopentadien-1-yl)hydro-; (37342-97-5), **71**, 77

(Z)-4-Chloro-2-butenylammonium chloride: 2-Butenylamine, 4-chloro-, hydrochloride, (Z)-; 2-Buten-1-amine, 4-chloro-, hydrochloride, (Z)-; (7153-66-4), **73**, 247, 252

1-[(Z)-4-Chloro-2-butenyl]-1-azonia-3,5,7-triazatricyclo[3.3.1.1^{3,7}]decane chloride: 3,5,7-Triaza-1-azoniatricyclo[3.3.1.1^{3,7}]decane, 1-(4-chloro-2-butenyl)-, chloride, (Z)-; (117175-09-4), **73**, 246, 252

4-Chlorobutyronitrile: Butanenitrile, 4-chloro-; (628-20-6), **74**, 37

(-)-(1R,3R)-3-Chlorocamphoric anhydride: 3-Oxabicyclo[3.2.1]octane-2,4-dione, 1-chloro-5,8,8-trimethyl-; Oxabicyclo[3.2.1]octane-2,4-dione, 1-chloro-5,8,8-trimethyl-, (1R)- (87859-84-5), **71**, 49

3-Chloro-5,5-dimethylcyclohex-2-en-1-one: 2-Cyclohexen-1-one, 3-chloro-5,5-dimethyl-; (17530-69-7), **70**, 205

1-CHLORO-1-[(DIMETHYL)PHENYLSILYL]ALKANES, **73**, 50

1-Chloro-1-[(dimethyl)phenylsilyl]hexane: Silane, (1-chlorohexyl)dimethylphenyl-; (135987-51-8), **73**, 50, 52, 59

Chlorodimethylsilane: Silane, chlorodimethyl-; (1066-35-9), **73**, 97, 109

N-[2-(2-Chloroethoxy)ethyl]acetamide: Acetamide, 2-(2-chloroethoxy)-N-ethyl-; (36961-73-6), **70**, 130

Chloroform: HIGHLY TOXIC. CANCER SUSPECT AGENT. Methane, trichloro-; (67-66-3), **71**, 227; **72**, 209; **73**, 246, 247, 252, 272, 277

1-Chloro-3-methyl-2-butene: 2-Butene, 1-chloro-3-methyl-; (503-60-6), **71**, 118, 125

2-Chloromethyl-1,3-dithiolane: 1,3-Dithiolane, 2-(chloromethyl)-; (86147-22-0), **71**, 175

Chloromethyl methyl sulfide: Methane, chloro(methylthio)-; (2373-51-5), **70**, 177

Chloromethyltrimethylsilane: Silane, (chloromethyl)trimethyl-; (2344-80-1), **70**, 241

3-Chloro-2-oxo-1,4-dithiane: 1,4-Dithian-2-one, 3-chloro-; (88682-21-7), **72**, 266

m-Chloroperbenzoic acid or m-Chloroperoxybenzoic acid: Peroxybenzoic acid, m-chloro-; Benzocarboperoxic acid, 3-chloro-; (937-14-4), **73**, 125, 127, 130, 133, 168, 173

Chloroplatinic acid hexahydrate: Platinate(2-), hexachloro-, dihydrogen; Platinate(2-), hexachloro-, dihydrogen (OC-6-11); (16941-12-1), **73**, 100, 105, 109

N-(5-CHLORO-2-PYRIDYL)TRIFLIMIDE: METHANESULFONAMIDE, N-(5-CHLORO-2-PYRIDINYL)-1,1,1-TRIFLUORO-N-[(TRIFLUOROMETHYL)SULFONYL]-; (1455100-51-2), **74**, 77

N-Chlorosuccinimide: Succinimide, N-chloro-; 2,5-Pyrrolidinedione, 1-chloro-; (128-09-6), **72**, 266

Chlorosulfonic acid; (7790-94-5), **74**, 218

Chlorotrimethylsilane: Trimethylsilyl chloride; Silane, chlorotrimethyl-; (75-77-4), **70**, 152, 164, 196, 216; **71**, 57, 159, 190, 226, 227; **72**, 22, 191, 232; **73**, 124, 126, 132; **74**, 85

Cholesterol esterase, **70**, 61

Chromic acid, aqueous, **73**, 216, 218

Chromic acid, disodium salt, dihydrate: HIGHLY TOXIC. CANCER SUSPECT AGENT, **73**, 220

CHROMIUM CARBENE COMPLEXES, **71**, 72

Chromium carbonyl: Chromium carbonyl (OC-6-11)-; (13007-92-6), **71**, 72

Chromium(II) chloride, anhydrous: Chromium chloride; (10049-05-5), **72**, 181

Chromium hexacarbonyl: Chromium carbonyl; Chromium carbonyl (OC-6-11)-; (13007-92-6), **71**, 72

Chromium(VI) oxide: Chromium oxide; (1333-82-0), **71**, 237

(E)-Cinnamaldehyde: 2-Propenal, 3-phenyl-; (14371-10-9), **70**, 197, 240

Citric acid: 1,2,3-Propanetricarboxylic acid, 2-hydroxy-; (77-92-9), **70**, 54; **71**, 227

(R)-Citronellal: 6-Octenal, 3,7-dimethyl-, (R)-(+)-; (2385-77-5), **71**, 168

Citronellic acid: 6-Octenoic acid, 3,7-dimethyl-, (R)-(+)-; (18951-85-4), **72**, 79

β-Citronellol: 6-Octen-1-ol, 3,7-dimethyl-; (106-22-9), **72**, 80

(R)-Citronellol: 6-Octen-1-ol, 3,7-dimethyl-, (R)-; (1117-61-9), **72**, 81

(S)-CITRONELLOL: 6-OCTEN-1-OL, 3,7-DIMETHYL-, (S)-; (7540-51-4), **72**, 74

CONJUGATE ADDITION OF ORGANOMAGNESIUM REAGENTS, **72**, 135

Conjugate addition reactions, of α,β-unsaturated carbonyl compounds, **72**, 135

CONJUGATED ENYNES, **70**, 215

Contiguously substituted aromatic compounds, **72**, 168

Copper(II) acetylacetonate: Copper, bis(2,4-pentanedionato-O,O')-; (46369-53-3), **71**, 190

Copper(I) bromide-dimethyl sulfide complex: Copper, bromo[thiobis[methane]]-; (54678-23-8), **70**, 218; **73**, 52, 55, 60, 132

Copper(I) chloride: Copper chloride; (7758-89-6), **72**, 135, 225

Copper cyanide; (544-92-3), **70**, 196

Copper(I) iodide: Copper iodide; (7681-65-4), **71**, 183; **72**, 105

Copper(II) sulfate: Sulfuric acid copper (2+) salt (1:1); (7758-98-7), **71**, 208; **74**, 15

Copper(I) thiophenoxide: Benzenethiol, copper(1+) salt; (1192-40-1), **70**, 206

Copper-zinc organometallic, **70**, 196

CORTICOID SIDE CHAIN, **73**, 123

Cram addition, **73**, 58

Cross-coupling reactions, 9-alkyl-9-BBN with bromo(phenylthio)ethenes, **71**, 93

18-Crown-6: 1,4,7,10,13,16-Hexanoxacyclooctadecane; (17455-13-9), **73**, 54, 55, 60

Crown ethers, chiral, **70**, 65

Cuprous bromide-dimethyl sufide complex: Copper, bromo[thiobis[methane]]-; (54678-23-8), **73**, 52, 55, 60, 124, 126, 132

2-Cyano-5,5-diphenylcyclopentanone, **74**, 40

2-CYANO-6-PHENYLOXAZOLOPIPERIDINE: 5H-OXAZOLO[3,2-a]PYRIDINE-5-CARBONITRILE, HEXAHYDRO-3-PHENYL-, [3R-(3a,5b,8ab)]-; (88056-92-2), **70**, 54

Cyclododecanone; (830-13-7), **71**, 161

3,4-CYCLODODECENO-1-METHYLBENZENE: BENZOCYCLODODECENE, 5,6,7,8,9,10,11,12,13,14-DECAHYDRO-2-METHYL-; (81857-28-5), **71**, 158

Cycloheptanone; (502-42-1), **71**, 141

Cycloheptatriene: 1,3,5-Cycloheptatriene; (544-25-2), **71**, 182

2,4,6-CYCLOHEPTATRIEN-1-ONE (TROPONE); (539-80-0), **71**, 181

Cyclohexanecarboxaldehyde; (2043-61-0), **72**, 193

Cyclohexanone; (108-94-1), **71**, 23; **72**, 148, 234

Cyclohexene-1-carboxaldehydes, optically active, **72**, 91

2-CYCLOHEXENE-1,4-DIONE; (4505-38-8), **73**, 253, 255, 259

9-[2-(3-Cyclohexenyl)ethyl]-9-BBN: 9-Borabicyclo[3.3.1]nonane, 9-[2-(3-cyclohexen-1-yl)ethyl]-; (69503-86-2), **71**, 90

4-(3-Cyclohexenyl)-2-phenylthio-1-butene, **71**, 89

Cyclohexylporphyrin, **70**, 77

1,5-Cyclooctadiene, (Z,Z)-; (1552-12-1), **70**, 120

1,5-Cyclooctadiene; (111-78-4), **70**, 120, 169; **73**, 240, 241, 242, 243, 244

1,3,5-Cyclooctatriene; (1871-52-9), **73**, 240, 241, 243, 244

Cyclopentadiene: 1,3-Cyclopentadiene; (542-92-7), **71**, 33; **73**, 253, 256, 261

Cyclopentadienyliron dicarbonyl dimer: Iron, tetracarbonylbis (η^5-2,4-cyclopentadien-1-yl)di-, (Fe-Fe); (38117-54-3), **70**, 177

Cyclopentadienyliron dicarbonyl (dimethylsulfonium)methyl tetrafluoroborate: Iron(1+), dicarbonyl(η^5-2,4-cyclopentadien-1-yl)(dimethylsulfonium η-methylide)-, tetrafluoroborate (1-); (72120-26-4), **70**, 179

CYCLOPENTANONE ANNULATION, **74**, 137

2-Cyclopentene-1-acetamide; (72845-09-1), **70**, 101

2-Cyclopentene-1-acetic acid; (13668-61-6), **70**, 101

2-Cyclopentenone: 2-Cyclopenten-1-one; (930-30-3), **71**, 84

CYCLOPENTENONES, **74**, 115

2-(2-Cyclopentenyl)acetyl chloride, **70**, 102

Cyclopropanation, **70**, 187; **71**, 194

Cyclopropanone ethyl hemiketal: Cyclopropanol, 1-ethoxy-; (13837-45-1), **74**, 138

CYCLOPROPANONES, **74**, 137

Cyclopropenes, **70**, 97

Cyclopropenyl cation, **74**, 65

DABCO, **70**, 253; **74**, 84

DAST Pummerer reaction, **72**, 213

Decahydroquinolines, **70**, 57

Decanoyl chloride; (112-13-0), **73**, 63, 66, 72

Decylamine: 1-Decanamine; (2016-57-1), **72**, 246

DECYL DIETHYL PHOSPHATE: PHOSPHORIC ACID, DECYL DIETHYL ESTER; (20195-16-8), **72**, 246

Decyl isoamyl ether, **72**, 249

16-Dehydroprogesterone: Pregna-4,16-diene-3,20-dione; (1096-38-4), **73**, 124, 126, 132

Desilylation, **73**, 125

DETRIFLUOROACETYLATIVE DIAZO GROUP TRANSFER, **73**, 134, 138

cis-3,5-Diacetoxycyclopentene: 4-Cyclopentene-1,3-diol, diacetate, cis-; (54664-61-8), **73**, 25, 26, 28

Di-O-acetyl-3-deoxy-D-arabino-1,4-lactone: D-threo-Pentonic acid, 3-deoxy-γ-lactone, 2,5-diacetate; (79580-65-7), **72**, 50

Dialkyl benzalmalonates, **71**, 214

Dialkyl mesoxalates, **71**, 214

DIASTEREOSELECTIVE HOMOLOGATION, **72**, 21

1,4-Diazabicyclo[2.2.2]octane (DABCO); (280-57-9), **70**, 253; **74**, 84

1,4-Diazabicyclo[2.2.2]octane-hydrogen peroxide complex ($DABCO{\cdot}2H_2O_2$), **74**, 84

1,8-Diazabicylo[5.4.0]undec-7-ene: DBU; Pyrimido[1,2-a]azepine, 2,3,4,6,7,8,9,10-octahydro-; (6674-22-2), **70**, 70; **72**, 226; **73**, 161, 165, 168, 173, 252; **74**, 125

Diazo compounds, **73**, 134

Diazo ketones, **73**, 137, 138, 139

DIAZOMALONATES, HETEROCYCLOADDITION, **74**, 229

Diazomethane, **70**, 22

(E)-1-DIAZO-4-PHENYL-3-BUTEN-2-ONE: 3-BUTEN-2-ONE, 1-DIAZO-4-PHENYL-; (24265-71-2), **73**, 134, 135

Diazo transfer reaction, **70**, 97; **73**, 138, 139, 150; **74**, 229

4,5-DIBENZOYL-1,3-DITHIOLE-1-THIONE: BENZENECARBOTHIOIC ACID, S,S'-(2-THIOXO-1,3-DITHIOLE-4,5-DIYL) ESTER; (68494-08-6), **73**, 270, 272

(-)-Dibenzyl tartrate: Tartaric acid, dibenzyl ester, (+)- ; Butanedioic acid, 2,3-dihydroxy-, [R-(R*,R*)]-, bis(phenylmethyl) ester; (622-00-4), **72**, 87

Diborane-THF: Furan, tetrahydro-, compd. with borane (1:1); (14044-65-6), **70**, 131; **71**, 207; **72**, 88, 202; **73**, 122

Dibromodifluoromethane: Methane, dibromodifluoro-; (75-61-6), **72**, 225

(1,3-Dibromo-3,3-difluoropropyl)trimethylsilane: Silane, (1,3-dibromo-3,3-difluoropropyl)trimethyl-; (671-80-7), **72**, 225

3,5-Dibromo-5,6-dihydro-(2H)-pyran-2-one: 2H-Pyran-2-one, 3,5-dibromo-5,6-dihydro-, (±)-; (56207-18-2), **73**, 232, 233, 235, 238

1,3-Dibromo-5,5-dimethylhydantoin: 2,4-Imidazolidinedione, 1,3-dibromo-5,5-dimethyl-; (77-48-5), **70**, 151

1,2-Dibromoethane: Ethane, 1,2-dibromo-; (106-93-4), **70**, 196; **72**, 191, 254

Dibromomethane: Methane, dibromo-; (74-95-3), **71**, 147; **73**, 74, 76, 84

Dibromomethyllithium, **71**, 146; **73**, 73

1,1-Dibromopentane: Pentane, 1,1-dibromo-; (13320-56-4), **73**, 73, 74, 75, 84

2,3-DIBROMO-1-(PHENYLSULFONYL)-1-PROPENE: BENZENE, [(2,3-DIBROMO-1-PROPENYL)SULFONYL]-; (132604-65-0), **74**, 115

Di-tert-butyl dicarbonate: Dicarbonic acid, bis(1,1-dimethylethyl) ester; (24424-99-5), **70**, 18, 140, 152; **71**, 202; **73**, 86, 88, 93, 188, 192, 196, 198, 200; **74**, 23

4,4'-Di-tert-butylbiphenyl: Biphenyl, 4,4'-di-tert-butyl-; 1,1'-Biphenyl, 4,4'-bis(1,1-dimethylethyl)-; (1625-91-8), **72**, 173

2,6-Di-tert-butylphenol: Phenol, 2,6-di-tert-butyl-; Phenol, 2,6-bis (1,1-dimethylethyl)-; (128-39-2), **72**, 95

Dibutyl telluride: Butane, 1,1'-tellurobis-; (38788-38-4), **72**, 155

β–Dicarbonyl derivatives, **74**, 261

Dichlorobis(triphenylphosphine)nickel: Nickel, dichlorobis(triphenylphosphine)-; (14264-16-5), **70**, 241

1,4-Dichlorobutane: Butane, 1,4-dichloro-; (110-56-5), **70**, 206

cis-1,4-Dichlorobut-2-ene: 2-Butene, 1,4-dichloro-, (Z)-; (1476-11-5), **73**, 246, 248, 252

(+)-[(7,7-Dichlorocamphoryl)sulfonyl]imine: 3H-3a,6-Methano-2,1-benzisothiazole, 7,7-dichloro-4,5,6,7-tetrahydro-8,8-dimethyl-, 2,2-dioxide, (3aS)-; (127184-04-7), **73**, 161, 165, 168, 173

(+)-(2R,8aR*)-[(8,8-DICHLOROCAMPHORYL)SULFONYL]OXAZIRIDINE: 4H-4a, 7-METHANOOXAZIRINO[3,2-i][2,1]BENZISOTHIAZOLE, 8,8-DICHLORO-TETRAHYDRO-9,9-DIMETHYL-, 3,3-DIOXIDE, [2R-(2α,4aα,7α,8aR*)]-; (127184-05-8), **73**, 159, 163, 164, 167, 169, 173

(-)-(2S,8aR*)-[(8,8-Dichlorocamphoryl)sulfonyl]oxaziridine: 4H-4a,7-Methanooxazirino[3,2-i][2,1]benzisothiazole, 8,8-dichlorotetrahydro-9,9-dimethyl-, 3,3-dioxide, [2S-(2α,4aα,7α,8aR*)]-; (139628-16-3), **73**, 164, 173

Dichlorodiethoxytitanium: Titanium, dichlorodiethoxy-; (3582-00-1), **71**, 201

1,3-Dichloro-5,5-dimethylhydantoin: Hydantoin, 1,3-dichloro-5,5-dimethyl-; Imidazolidinedione, 1,3-dichloro-5,5-dimethyl-; (118-52-5), **73**, 161, 165, 168, 173

1,2-Dichloroethane: Ethane, 1,2-dichloro-; (107-06-2), **71**, 32, 39

Dichloromethane: Methane, dichloro-; (75-09-2), **71**, 168

Dichlorophenylphosphine: Phosphonous dichloride, phenyl-; (644-97-3), **70**, 273

(1α,2α,5α,6α)-2,6-Dichloro-9-thiabicyclo[3.3.1]nonane; (10502-30-4), **70**, 120

α,α-Dichlorotoluene: Benzene, (dichloromethyl)-; (98-87-3), **74**, 72

Dichromate oxidation, **72**, 61

2,6-Dicyanopiperidines, **70**, 55

DIELS-ALDER CYCLOADDITION, **73**, 231, 237
    inverse electron-demand, **70**, 79; **71**, 181

DIELS-ALDER REACTION OF AN HETEROCYCLIC AZADIENE, **70**, 79

Diels-Alder reactions, **70**, 85
    ASYMMETRIC, **72**, 86
    enantioselective catalytic, **71**, 30

Diethylamine: Ethanamine, N-ethyl-; (109-89-7), **72**, 164; **73**, 203, 206, 209, 214

2-N,N'-Diethylaminomethyl-3,4-diethylpyrrole, **70**, 74

Diethylaminosulfur trifluoride: DAST; Sulfur, (diethylaminato)trifluoro-; (38078-09-0), **72**, 210

Diethyl (2S,3R)-2-azido-3-hydroxysuccinate: Butanedioic acid, 2-azido-3-hydroxy-, diethyl ester, [S-(R*,S*)]-; (101924-62-3), **73**, 187, 188, 191, 194, 200

Diethyl azodicarboxylate: Diazenedicarboxylic acid, diethyl ester; (1972-28-7), **70**, 4, 10; **73**, 110, 111, 115 (See warning, **73**, 278)

Diethyl (-)-2,3-O-benzylidene-L-tartrate: 1,3-Dioxolane-4,5-dicarboxylic acid, 2-phenyl-, diethyl ester, [4R-(2α,4α,5β)]-; (35572-31-7), **74**, 2

Diethyl (2S,3S)-2-bromo-3-hydroxysuccinate: Butanedioic acid, 2-bromo-3-hydroxy-, diethyl ester, [S-(R*,R*)]-; (80640-14-8), **73**, 185, 186, 194, 199

DIETHYL (2S,3R)-2-(N-tert-BUTOXYCARBONYL)AMINO-3-HYDROXYSUCCINATE, **73**, 184, 188, 193, 194, 200

Diethylcarbamoyl chloride: Carbamic chloride, diethyl-; (88-10-8), CANCER SUSPECT AGENT, **72**, 154

N,N-Diethylcarbamoyllithium, **72**, 154

Diethyl chlorophosphate: HIGHLY TOXIC: Phosphorochloridic acid, diethyl ester; (814-49-3), **72**, 217, 246

Diethyl cyanophosphonate: Phosphonic acid, cyano-, diethyl ester; (2942-58-7), **72**, 80

Diethyl decylphosphoramidate: Phosphoramidic acid, decyl-, diethyl ester; (53246-96-1), **72**, 246

DIETHYL (DICHLOROMETHYL)PHOSPHONATE: PHOSPHONIC ACID, (DICHLOROMETHYL)-, DIETHYL ESTER; (3167-62-2), **74**, 108

Diethyl (2R,3R)-2,3-epoxysuccinate: 2,3-Oxiranedicarboxylic acid, diethyl ester, (2R-trans)-; (74243-85-9), **73**, 186, 187, 194, 200

Diethyl 1-formylalkylphosphonates, synthesis, **72**, 244

N,N-Diethyl-2-formyl-6-methoxybenzamide: Benzamide, N,N-diethyl-2-formyl-6-methoxy-; (70946-17-7), **72**, 164

3,4-Diethyl-5-hydroxymethylpyrrole-2-carboxylic acid, **70**, 74

N,N-DIETHYL-2-HYDROXY-4-PHENYLBUTANAMIDE: BENZENEBUTANAMIDE, N,N-DIETHYL-α-HYDROXY-; (134970-54-0), **72**, 154

N,N-Diethyl-2-methoxybenzamide: Benzamide, N,N-diethyl-2-methoxy-; (51674-10-3), **72**, 163

Diethyl oxalate: Oxalic acid, diethyl ester; Ethanedioic acid, diethyl ester; (95-92-1), **72**, 126

N,N-Diethylphenylglyoxylamide, **72**, 157

Diethyl phthalate: Phthalic acid, diethyl ester; 1,2-Benzenecarboxylic acid, diethyl ester; (84-66-2), **73**, 2, 12

DIETHYL 1-PROPYL-2-OXOETHYLPHOSPHONATE: PHOSPHONIC ACID, (1-FORMYLBUTYL)-, DIETHYL ESTER; (112292-30-5), **72**, 241

3,4-DIETHYLPYRROLE: PYRROLE, 3,4-DIETHYL-; (16200-52-5), **70**, 68

Diethyl (2R,3R)-tartrate: Tartaric acid, diethyl ester, L-(+)-; Butanedioic acid, 2,3-dihydroxy-, [R-(R*,R*)]-, diethyl ester; (87-91-2), **73**, 185, 189, 194, 199; **74**, 2

Diethyl (trichloromethyl)phosphonate: Phosphonic acid, (trichloromethyl)-, diethyl ester; (866-23-9), **74**, 109

(3,3-DIFLUOROALLYL)TRIMETHYLSILANE: SILANE, (3,3-DIFLUORO-2-PROPENYL)TRIMETHYL-; (40207-81-6), **72**, 225

Difluoromethyl group, radical addition of, **72**, 229

Dihalocarbene, **70**, 186

Dihydrocitronellol, **72**, 77, 82

2,3-Dihydrofuran: Furan, 2,3-dihydro-; (1191-99-7), **73**, 215, 217, 219, 220

(R)-2,3-Dihydro-1H-inden-1-ol: 1H-Inden-1-ol, 2,3-dihydro-, (R)-; (697-64-3), **74**, 54

Dihydroisoquinolines, **70**, 145

5,6-Dihydro-(2H)-pyran-2-one: 2H-Pyran-2-one, 5,6-dihydro-; (3393-45-1), **73**, 231, 238

1,4-DIHYDROPYRIDINE EQUIVALENT, **70**, 54

(S)-(+)-2,3-Dihydro-1H-pyrrolo[2,1-c][1,4]benzodiazepine-5,11(10H,11aH)dione: 5H-Pyrrolo[2,1-c][1,4]benzodiazepine-5,11(10H)-dione, 1,2,3,11a-tetrahydro-, L-; 1H-Pyrrolo[2,1-c]benzodiazepine-5,11(10H,11aH)-dione, 2,3-dihydro-, (S)-; (18877-34-4), **73**, 175, 177, 183

Dihydroquinidine, benzoate ester, **70**, 49

Dihydroquinidine 4-chlorobenzoate: Cinchonan-9-ol, 10,11-dihydro-6'-methoxy-, 4-chlorobenzoate (ester), (9S)-; (113162-02-0), **70**, 47

Dihydroquinidine, 2-naphthoate ester, **70**, 49

Dihydroquinidine 9-O-(9'-phenanthryl) ether: Cinchonan, 10,11-dihydro-6'-methoxy-9-(9-phenanthrenyloxy)-, (9S)-; (135042-88-5), **73**, 3, 4, 6, 7, 12

Dihydro-1,2,4,5-tetrazine-3,6-dicarboxylic acid: 1,2,4,5-Tetrazine-3,6-dicarboxylic acid, 1,2-dihydro-; (3787-09-5), **70**, 80

3-[(1S)-1,2-DIHYDROXYETHYL)]-1,5-DIHYDRO-3H-2,4-BENZODIOXEPINE: 1,2-ETHANEDIOL, 1-(1,5-DIHYDRO-2,4-BENZODIOXEPIN-3-YL)-, (S)-; (142235-22-1), **73**, 1, 3, 4, 7

Diisobutylaluminum hydride: Aluminum, hydrodiisobutyl-; Aluminum, hydrobis(2-methylpropyl)-; (1191-15-7), **70**, 20; **71**, 183; **74**, 195, 196

(+)-Diisopinocampheylborane, **73**, 116, 117

Diisopropoxytitanium(IV) dibromide: Titanium, dibromobis(1-methylethoxy)-, (T-4)-; (37943-35-4), **71**, 14

Diisopropylamine: 2-Propanamine, N-(1-methylethyl)-; (108-18-9), **71**, 133, 236; **72**, 38; **73**, 61, 63, 72, 73, 76, 84; **74**, 33, 38, 109, 138

N,N-Diisopropylethylamine: Triethylamine, 1,1'-dimethyl-; 2-Propanamine, N-ethyl-N-(1-methylethyl)-; (7087-68-5), **72**, 105

1,2:5,6-Diisopropylidene-D-mannitol; Mannitol, 1,2:5,6-di-O-isopropylidene-, D-; D-Mannitol, 1,2:5,6-bis-O-(1-methylethylidene)-; (1707-77-3), **72**, 6

(1R)-2,3-Di-O-isopropylidene-1-(2-thiazolyl)-D-glycitol: 2-Thiazolemethanol, α-(2,2-dimethyl-1,3-dioxolan-4-yl)-, [R-(R*,R*)]-; (103795-10-4), **72**, 22

Diisopropyl squarate: 3-Cyclobutene-1,2-dione, 3,4-bis(1-methylethoxy)-; (61999-62-5), **74**, 170

Diketene: 2-Oxetanone, 4-methylene-; (674-82-8), **73**, 16, 23

1,4- and 1,5-Dilithioalkanes, **70**, 211

1,4-Dilithiobutane: Lithium, μ-1,4-butanediyldi-; (2123-72-0), **70**, 205

Dimedone: 1,3-Cyclohexanedione, 5,5-dimethyl-; (126-81-8), **70**, 205

3,5-Dimethoxyacetophenone: Ethanone, 1-(3,5-dimethoxyphenyl)-; (39151-19-4), **74**, 241

2,3-Dimethoxybenzoic acid: Benzoic acid, 2,3-dimethoxy-; (1521-38-6), **71**, 108

2,3-Dimethoxybenzoyl chloride: Benzoyl chloride, 2,3-dimethoxy-; (7169-06-4), **71**, 108; **72**, 87

3,4-Dimethoxybenzylamine: Veratrylamine; Benzenemethanamine, 3,4-dimethoxy-; (5763-61-1), **72**, 15

N-[(2R,3R)-cis-1-(3,4-Dimethoxybenzyl)-2-[(S)-2,2-dimethyl-1,3-dioxolan-4-yl]-4-oxo-3-azetidinyl]phthalimide, **72**, 18

(+)-[(7,7-Dimethoxycamphoryl)sulfonyl]imine: 3H-3a,6-Methano-2,1-benzisothiazole, 4,5,6,7-tetrahydro-7,7-dimethoxy-8,8-dimethyl-, 2,2-dioxide, [3aS]-; (131863-80-4), **73**, 160, 161, 165, 168, 173

(+)-(2R,8aR*)-[(8,8-DIMETHOXYCAMPHORYL)SULFONYL]OXAZIRIDINE: 4H-4a,7-METHANOOXAZIRINO[3,2-i][2,1]BENZISOTHIAZOLE, TETRAHYDRO-8,8-DIMETHOXY-9,9-DIMETHYL-, 3,3-DIOXIDE, [2R-(2α,4aα,7α,8aR*)]-; (181863-82-5), **73**, 159, 162, 166, 169, 173

(-)-(2S,8aR*)-[(8,8-Dimethoxycamphoryl)sulfonyl]oxaziridine: 4H-4a,7-Methanooxazirino[3,2-i][2,1]benzisothiazole, tetrahydro-8,8-dimethoxy-9,9-dimethyl-, 3,3-dioxide, [2S-(2α,4aα,7α, 8aR*)]-; (132342-04-2), **73**, 163, 173

2,2'-Dimethoxy-6-(4",4"-dimethyloxazolinyl)biphenyl: Oxazole, 2-(2',6-dimethoxy[1,1'-biphenyl]-2-yl)-4,5-dihydro-4,4-dimethyl-; (57598-39-7), **71**, 109

Dimethoxyethane: Ethane, 1,2-dimethoxy-; (110-71-4), **70**, 169; **72**, 6, 203; **73**, 2, 5, 12

2,2'-DIMETHOXY-6-FORMYLBIPHENYL: [1,1'-BIPHENYL]-2-CARBOXALDEHYDE, 2',6-DIMETHOXY-; (87306-84-1), **71**, 107

Dimethoxymethane: Methane, dimethoxy-; (109-87-5), **71**, 134; **73**, 95, 99, 108

2,3-Dimethoxy-5-methyl-1,4-benzoquinone: p-Benzoquinone, 2,3-dimethoxy-5-methyl-; 2,5-Cyclohexadiene-1,4-dione, 2,3-dimethoxy-5-methyl-; (605-94-7), **71**, 125

6,7-Dimethoxy-1-methyl-3,4-dihydroisoquinoline: Isoquinoline, 3,4-dihydro-6,7-dimethoxy-1-methyl-; (4721-98-6), **70**, 139

5,7-DIMETHOXY-3-METHYLINDAZOLE: 1H-INDAZOLE, 5,7-DIMETHOXY-3-METHYL-; (154876-15-0), **74**, 241

2-(2,3-Dimethoxyphenyl)-4,4-dimethyl-2-oxazoline: Oxazole, 2-(2,3-dimethoxyphenyl)-4,5-dihydro-4,4-dimethyl-; (57598-32-0), **71**, 108

2,2-Dimethoxypropane: Acetone, dimethyl acetal; Propane, 2,2-dimethoxy-; (77-76-9), **70**, 22; **72**, 6, 51

7,8-DIMETHOXY-1,3,4,5-TETRAHYDRO-2H-3-BENZAZEPIN-2-ONE: 2H-3-BENZAZEPIN-2-ONE, 1,3,4,5-TETRAHYDRO-7,8-DIMETHOXY-; (20925-64-8), **70**, 139

4-Dimethylaminopyridine: HIGHLY TOXIC. Pyridine, 4-(dimethylamino)-; 4-Pyridinamine, N,N,-dimethyl-; (1122-58-3), **71**, 31, 72; **72**, 87; **73**, 37, 40, 43, 44, 46, 49, 187, 191, 196, 200; **74**, 206

(S)-Dimethyl aspartate hydrochloride: Aspartic acid, dimethyl ester, hydrochloride, L-; L-Aspartic acid, dimethyl ester, hydrochloride; (32213-95-9), **71**, 228

Dimethyl azodicarboxylate: Diazenedicarboxylic acid, dimethyl ester; (2446-84-6), **70**, 1 (See warning, **73**, 278)

Dimethyl benzalmalonate: Malonic acid, benzylidene-, dimethyl ester; Propanedioic acid, (phenylmethylene)-, dimethyl ester; (6626-84-2), **71**, 214

1,4-DIMETHYLBICYCLO[3.2.0]HEPT-3-EN-6-ONE: BICYCLO[3.2.0]HEPT-3-EN-6-ONE, 1,4-DIMETHYL-, cis-(±)-; (133700-21-7), **74**, 158

2,3-Dimethyl-1,3-butadiene: 1,3-Butadiene, 2,3-dimethyl-; (513-81-5), **72**, 88

($1\alpha,2\alpha,6\beta$)- and ($1\alpha,2\beta,6\alpha$)-2,6-DIMETHYLCYCLOHEXANECARBONITRILE, **74**, 217

2,6-Dimethylcyclohexanone: Cyclohexanone, 2,6-dimethyl-; (2816-57-1), **74**, 219

Dimethyl diazomalonate: Propanedioic acid, diazo-, dimethyl ester; (6773-29-1), **74**, 229

Dimethyl dihydro-1,2,4,5-tetrazine-3,6-dicarboxylate: 1,2,4,5-Tetrazine-3,6-dicarboxylic acid, 1,2-dihydro-, dimethyl ester; (3787-10-8), **70**, 80

DIMETHYLDIOXIRANE: DIOXIRANE, DIMETHYL-; (74087-85-7), **74**, 91

(E)-2,3-DIMETHYL-3-DODECENE, **73**, 61, 62

N-[(1,1-Dimethylethoxy)carbonyl]-L-serine methyl ester: L-serine, N-[(1,1-dimethylethoxy)carbonyl], methyl ester; (2766-43-0), **70**, 18

1,1-DIMETHYLETHYL (S)- OR (R)-4-FORMYL-2,2-DIMETHYL-3-OXAZOLIDINECARBOXYLATE: 3-OXAZOLIDINECARBOXYLIC ACID, 4-FORMYL-2,2-DIMETHYL, 1,1-DIMETHYLETHYL ESTER, (S)- or (R)-; (S)- (102308-32-7); (R)- (95715-87-0), **70**, 18

3-(1,1-Dimethylethyl) 4-methyl (S)-2,2-dimethyl-3,4-oxazolidinedicarboxylate: 3,4-Oxazolidinedicarboxylic acid, 2,2-dimethyl-, 3-(1,1-dimethylethyl) 4-methyl ester, (S)-; (108149-60-6), **70**, 19

N,N-Dimethylformamide: Formamide, N,N-dimethyl-; (68-12-2), CANCER SUSPECT AGENT, **71**, 99, 189; **72**, 75, 105, 127, 164, 181; **73**, 86, 89, 91, 93, 144, 146, 149, 151, 187, 191, 196, 200, 203, 241, 243, 245

2,5-DIMETHYL-2,4-HEXANEDIOL: 2,4-HEXANEDIOL, 2,5-DIMETHYL-; (3899-89-6), **72**, 173

3,6-Dimethyl-3-hydroxy-6-heptenoic acid, **74**, 159

DIMETHYL (1'R,2'R,5'R)-2-(2'-ISOPROPENYL-5'-METHYLCYCLOHEX-1'-YL)-PROPANE-1,3-DIOATE: PROPANEDIOIC ACID, [5-METHYL-2-(1-METHYLETHENYL)CYCLOHEXYL]-, DIMETHYL ESTER, [1R-($1\alpha,2\beta,5\alpha$)]-; (106431-81-6), **71**, 167

Dimethyl malonate: Malonic acid, dimethyl ester; Propanedioic acid, dimethyl ester; (108-59-8), **71**, 168; **74**, 149, 229

Dimethyl mesoxalate: Propanedioic acid, oxo-, dimethyl ester; (3298-40-6), **71**, 214

Dimethyl (E)-5-methoxycarbonyl-2-hexenedioate: 3-Butene-1,1,4-tricarboxylic acid, trimethyl ester, (E)-; (93279-60-8), **74**, 149

(E)-4,8-DIMETHYL-3,7-NONADIENOIC ACID: 3,7-NONADIENOIC ACID, 4,8-DIMETHYL-,(E)-; (459-85-8), **74**, 178

4,8-Dimethylnon-7-enol, **72**, 83

(1R,5S)-(-)-6,6-DIMETHYL-3-OXABICYCLO[3.1.0]HEXAN-2-ONE: 3-OXABICYCLO[3.1.0]HEXAN-2-ONE, 6,6-DIMETHYL-, (1R-cis)-; (71565-25-8), **73**, 13, 17, 20, 21, 23

2-(1,1-DIMETHYLPENTYL)-5-METHYLCYCLOHEXANONE: CYCLOHEXANONE, 2-(1,1-DIMETHYLPENTYL)-5-METHYL-; (109539-17-5), **72**, 135

(E)-3,3-DIMETHYL-1-PHENYL-1-BUTENE: BENZENE, (3,3-DIMETHYL-1-BUTENYL)-, (E)-; (3846-66-0), **74**, 187

DIMETHYL 4-PHENYL-1,2-DIAZINE-3,6-DICARBOXYLATE: 3,6-PYRAZINEDICARBOXYLIC ACID, 4-PHENYL-, DIMETHYL ESTER; (2166-27-0), **70**, 82

(S)-Dimethyl N-(9-phenylfluoren-9-yl)aspartate: L-Aspartic acid, N-(9-phenyl-9H-fluoren-9-yl)-, dimethyl ester; (120230-62-8), **71**, 226

DIMETHYL 3-PHENYLPYRROLE-2,5-DICARBOXYLATE: 1H-PYRROLE-2,5-DICARBOXYLIC ACID, 3-PHENYL-, DIMETHYL ESTER; (92144-12-2), **70**, 82

1-[(Dimethyl)phenylsilyl]-1-hexanol: 1-Hexanol, 1-(dimethylphenylsilyl)-; (125950-71-2), **73**, 51, 55, 59

α-(DIMETHYL)PHENYLSILYL KETONES, **73**, 50

N,N-Dimethyl-1,3-propanediamine, **72**, 19

N,N'-Dimethylpropyleneurea: 1,3-Dimethyl-3,4,5,6-tetrahydro-2(H)-pyrimidinone: DMPU: 2(1H)-Pyrimidinone, tetrahydro-1,3-dimethyl-; (7226-23-5), **74**, 17

9,9-DIMETHYLSPIRO[4.5]DECAN-7-ONE: SPIRO[4.5]DECAN-7-ONE, 9,9-DIMETHYL-; (63858-64-0), **70**, 204

Dimethyl sulfide: Methane, thiobis-; (75-18-3), **70**, 218; **71**, 214

Dimethyl sulfoxide: Methyl sulfoxide; Methane, sulfinylbis-; (67-68-5), **70**, 112, 122; **72**, 226

DIMETHYL 1,2,4,5-TETRAZINE-3,6-DICARBOXYLATE: 1,2,4,5-TETRAZINE-3,6-DICARBOXYLIC ACID DIMETHYL ESTER; (2166-14-5), **70**, 79

Dimethylzinc: Zinc, dimethyl-; (544-97-8), **74**, 206

2,6-Dinitrato-9-thiabicyclo[3.3.1]nonane 9,9-dioxide, **70**, 125

1,3-DIOLS, SYNTHESIS OF, **72**, 173

(2R,3S)-1,4-Dioxa-2,3-dimethyl-2-(1-methylethenyl)-8-carboethoxy-8-azaspiro[4.5]decane, **71**, 63

(2R,3S)- and (2S,3S)-1,4-Dioxa-2,3-dimethyl-2-(1-methylethenyl)-8-carboethoxy-8-azaspiro[4.5]decane, **71**, 63

1,4-Dioxane; (123-91-1), CANCER SUSPECT AGENT, **70**, 84, 179; **71**, 182; **73**, 263, 266, 269

DIPHENYLACETALDEHYDE: ACETALDEHYDE, DIPHENYL-; BENZENEACETALDEHYDE, α-PHENYL-; (947-91-1), **72**, 95

Diphenylacetonitrile: Benzeneacetonitrile, α-phenyl-; (86-29-3), **74**, 34, 38

Diphenylacetylene: Benzene, 1,1'-(1,2-ethynediyl)bis-; (501-65-5), **74**, 72

(R)-(-)-2,2-DIPHENYLCYCLOPENTANOL: CYCLOPENTANOL, 2,2-DIPHENYL-, (R)-; (126421-67-8), **74**, 33

2,2-Diphenylcyclopentanone: Cyclopentanone, 2,4-diphenyl-; (15324-42-2), **74**, 34

1,1-DIPHENYLCYCLOPROPANE: BENZENE, 1,1'-CYCLOPROPYLIDENEBIS-; (3282-18-6), **70**, 177

Diphenyl disulfide: Disulfide, diphenyl; (882-33-7), **74**, 124

(R,R)-1,2-DIPHENYL-1,2-ETHANEDIOL: 1,2-ETHANEDIOL, 1,2-DIPHENYL-, [R-(R*, R*)]-; (52340-78-0), **70**, 47

1,1-Diphenylethylene: Benzene, 1,1'-ethylidenebis-; (530-48-3), **70**, 179

(±)-1,2-DIPHENYLETHYLENEDIAMINE: ETHYLENEDIAMINE, 1,2-DIPHENYL-, (±)-; 1,2-ETHANEDIAMINE, 1,2-DIPHENYL-, (R*,R*)-(±)-; (16635-95-3), **71**, 23

(1R,2R)-(+)-1,2-DIPHENYLETHYLENEDIAMINE: 1,2-ETHANEDIAMINE, 1,2-DIPHENYL-, [R-(R*,R*)]-; (35132-20-8), **71**, 22

(1S,2S)-(-)-1,2-DIPHENYLETHYLENEDIAMINE: ETHYLENEDIAMINE, 1,2-DIPHENYL-, (1S,2S)-(-)-; 1,2-ETHANEDIAMINE, 1,2-DIPHENYL-, [S-(R*,R*)]-; (29841-69-8), **71**, 22, 31

(1S,2S)-1,2-Diphenylethylenediamine bistriflamide: Methanesulfonamide, N,N'-(1,2-diphenyl-1,2-ethanediyl)bis[1,1,1-trifluoro-, [S-(R*,R*)]-; (121788-77-0), **71**, 31

Diphenylethylenediamines, **71**, 27

(R)-(+)-2-(Diphenylhydroxymethyl)-N-(tert-butoxycarbonyl)pyrrolidine: 1-Pyrrolidinecarboxylic acid, 2-(hydroxydiphenylmethyl)-, 1,1-dimethylethyl ester, (R)-; (137496-68-5), **74**, 24

(R)-(+)-2-(DIPHENYLHYDROXYMETHYL)PYRROLIDINE: 2-PYRROLIDINEMETHANOL, α,α-DIPHENYL-, (R)-; (22348-32-9); **74**, 23

(S)-1,1-Diphenylprolinol or
(S)-α,α-Diphenyl-2-pyrrolidinemethanol: 2-Pyrrolidinemethanol, α,α-, (S)-; (112068-01-6), **74,** 36, 52
recovery, **74**, 36

(S)-1,1-Diphenylprolinol sulfate: 2-Pyrrolidinemethanol, α,α-diphenyl-, (S)-, sulfate (2:1)(salt); (131180-44-4), **74**, 52

Directed ortho metalation, **72**, 168

DIRHODIUM(II) TETRAKIS[METHYL 2-PYRROLIDONE-5(R)-CARBOXYLATE: Rh_2(5R-MEPY)$_4$: RHODIUM, TETRAKIS[μ-(METHYL 5-OXO-L-PROLINATO-N^1O^5)]DI-, (Rh-Rh); (132435-65-5), **73**, 13, 14, 15, 17, 22, 23

Disodium dihydro-1,2,4,5-tetrazine-3,6-dicarboxylate: 1,2,4,5-Tetrazine-3,6-dicarboxylic acid, 1,2-dihydro-, disodium salt; (96898-32-7), **70**, 80

trans-1,2-DISUBSTITUTED CYCLOHEXANES, **71**, 167

cis-, trans-2,6-Disubstituted piperidines, **70**, 57

2,6-Dithiaadamantane, **70**, 126

DITHIOACETALS, GEMINAL DIMETHYLATION, **74**, 187

9-Dithiolanobicyclo[3.2.2]nona-3,6-dien-2-one, **71**, 182

9-DITHIOLANOBICYCLO[3.2.2]NON-6-EN-2-ONE, **71**, 181

1,3-Dithiol-2-ones, reductive coupling with trialkyl phosphites, **72**, 268

1,3-Divinyltetramethyldisiloxane: Disiloxane, 1,1,3,3-tetramethyl-1,3-divinyl-; Disiloxane, 1,3-diethenyl-1,1,3,3-tetramethyl-; (2627-95-4), **73**, 100, 109

Dodecylbenzenesulfonic acid, 97%: Benzenesulfonic acid, dodecyl-; (27176-87-0), **73**, 144, 146, 151

4-Dodecylbenzenesulfonyl azide: Benzenesulfonyl azide, 4-dodecyl-; (79791-38-1), **70**, 97; **73**, 20, 135, 136, 137, 145, 149, 150, 151

4-DODECYLBENZENESULFONYL AZIDES, **73**, 144

Dodecylbenzenesulfonyl chloride: Benzenesulfonyl chloride, dodecyl-; (26248-27-1) or 4-dodecyl-; (52499-14-6), **73**, 144, 149, 151

DOUBLE HYDROXYLATION REACTION, **73**, 123, 130

Electrochemical oxidative decarboxylation, **73**, 209

ELECTROCYCLIZATION, **74**, 173
[3+2]-ANIONIC, **74**, 147
cascade, **74**, 176

Enantioenriched 3-hydroxycarboxylates, **71**, 1

ENANTIOMERICALLY PURE β-AMINO ACIDS, **73**, 201, 208, 209

Enantioselective hydrogenation, **72**, 81

ENANTIOSELECTIVE HYDROLYSIS, **73**, 25

ENANTIOSELECTIVE INTRAMOLECULAR CYCLOPROPANATION, **73**, 13, 22

Enantioselective monohydrolysis, **73**, 41

Enediynes, **72**, 109

Enol ethers, (Z)- and (E)- **74**, 13
O-alkyl, **74**, 20

Enol silanes, **72**, 189

Enzymatic catalysis, **73**, 31

Enzyme mimics, **70**, 136

Epoxide rearrangements, selective, **72**, 99

EPOXIDES, REDUCTIVE LITHIATION OF, **72**, 173
SYNTHESIS, **74**, 91

ESTER CARBONYL GROUPS, ALKYLIDENATION, **73**, 73

ESTER HOMOLOGATION, **71**, 146

1,2-Ethanedithiol; (540-63-6), **70**, 240; **71**, 175; **72**, 266; **74**, 187

Ethanolamine: Ethanol, 2-amino-; (141-43-5), **71**, 91; **72**, 225

Ethanol; (64-17-5), **71**, 148

(Z)-1-ETHOXY-1-PHENYL-1-HEXENE, **73**, 73, 76, 77

Ethyl acetate: Acetic acid, ethyl ester; (141-78-6), **71**, 57, 190, 236

Ethyl acetoacetate: Butanoic acid, 3-oxo-,ethyl ester; (141-97-9), **70**, 94

Ethyl benzoate: Benzoic acid, ethyl ester; (93-89-0), **73**, 75, 77, 84

ETHYL (-)-(R,E)-4-O-BENZYL-4,5-DIHYDROXY-2-PENTENOATE: 2-PENTENOIC ACID, 5-HYDROXY-4-(PHENYLMETHOXY)-, ETHYL ESTER, [R-(E)]-; (119770-84-2), **74**, 1

Ethyl bromoacetate: Acetic acid, bromo-, ethyl ester; (105-36-2), **70**, 248

ETHYL (2Z)-3-BROMOPROPENOATE: ACRYLIC ACID, 3-BROMO-, ETHYL ESTER, (Z)-; 2-PROPENOIC ACID, 3-BROMO-, ETHYL ESTER, (Z)-; (31930-34-4), **72**, 112

ETHYL (Z)-2-BROMO-5-(TRIMETHYLSILYL)-2-PENTEN-4-YNOATE: 2-PENTEN-4-YNOIC ACID, 2-BROMO-5-(TRIMETHYLSILYL)-, ETHYL ESTER, (Z)-; (124044-21-9), **72**, 105

Ethyl 2-(S)-[(tert-butyldiphenylsilyl)oxy]propanoate: Propanoic acid, 2-[[(1,1-dimethylethyl)diphenylsilyl]oxy]-, ethyl ester, (S)-; (102732-44-5), **71**, 56

Ethyl chloroformate: Formic acid, chloro-, ethyl ester; Carbonochloridic acid, ethyl ester; (541-41-3), **73**, 153, 155, 158

Ethyl diazoacetate: Acetic acid, diazo-, ethyl ester; (623-73-4), **70**, 80; **71**, 192

Ethyl diazoacetoacetate: Butanoic acid, 2-diazo-3-oxo-, ethyl ester; (2009-97-4), **70**, 94

ETHYL (Z)-2,3-DIBROMOPROPENOATE: ACRYLIC ACID, 2,3-DIBROMO-, ETHYL ESTER, (Z)-; 2-PROPENOIC ACID, 2,3-DIBROMO-, ETHYL ESTER, (Z)-; (26631-66-3), **72**,104

Ethyl 5-(N,N'-diethylaminomethyl)-3,4-diethylpyrrole-2-carboxylate, **70**, 74

Ethyl 3,4-diethylpyrrole-2-carboxylate: 1H-Pyrrole-2-carboxylic acid, 3,4-diethyl-, ethyl ester; (97336-41-9), **70**, 70

Ethylene: Ethene; (74-85-1), **74**, 124

Ethyl formate: Formic acid, ethyl ester; (109-94-4), **70**, 232; **71**, 161; **72**, 242

Ethyl 3-hydroxy-4-pentenoate: 4-Pentenoic acid, 3-hydroxy-, ethyl ester; (38996-01-9), **71**, 236

Ethylidene transfer reagent, **70**, 187

3-ETHYLINDOLE: 1H-INDOLE, 3-ETHYL-; (1484-19-1), **74**, 248

Ethyl iodide: Ethane, iodo-; (75-03-6), **74**, 14, 249

Ethyl (Z)-β-iodoacrylate: 2-Propenoic acid, 3-iodo-, ethyl ester, (Z)-; (3190-36-6), **74**, 194

Ethyl isocyanoacetate: Acetic acid, isocyano-, ethyl ester; (2999-46-4), **70**, 70

(S)-(-)-Ethyl lactate: Lactic acid, ethyl ester, L-; Propanoic acid, 2-hydroxy-, ethyl ester, (S)-; (687-47-8), **71**, 56

ETHYL 2-METHYL-5-PHENYL-3-FURANCARBOXYLATE: 3-FURANCARBOXYLIC ACID, 2-METHYL-5-PHENYL-, ETHYL ESTER; (29113-64-2), **70**, 93

Ethyl 1-naphthoate: 1-Naphthoic acid, ethyl ester; 1-Naphthalenecarboxylic acid, ethyl ester; (3007-97-4), **71**, 146

ETHYL 1-NAPHTHYLACETATE: 1-NAPHTHALENEACETIC ACID, ETHYL ESTER; (2122-70-5), **71**, 146

ETHYL 3-OXO-4-PENTENOATE (NAZAROV'S REAGENT): 4-PENTENOIC ACID, 3-OXO-, ETHYL ESTER; (22418-80-0), **71**, 236

Ethyl propiolate or ethyl 2-propynoate: Propiolic acid, ethyl ester; 2-Propynoic acid, ethyl ester; (623-47-2), **72**, 104, 112; **74**, 195

ETHYL 4,4,4-TRIFLUOROTETROLATE: 2-BUTYNOIC ACID, 4,4,4-TRIFLUORO-, ETHYL ESTER; (79424-03-6), **70**, 246

Ethyl 4,4,4-trifluoro-2-(triphenylphosphoranylidene)acetoacetate: Butanoic acid, 4,4,4-trifluoro-3-oxo-2-(triphenylphosphoranylidene)-, ethyl ester; (83961-56-2), **70**, 246

ETHYNYLFERROCENE: FERROCENE, ETHYNYL-; (1271-47-2), **73**, 262, 263, 264, 266, 267,268, 269

Ferric ammonium sulfate: Sulfuric acid, ammonium iron (3+) salt (2:1:1), dodecahydrate; (7783-83-7), **73**, 147, 148, 151

Ferric chloride, **70**, 180

Ferric chloride/alumina, **71**, 168

Ferric chloride, anhydrous, **72**, 253

Ferric chloride hexahydrate: Iron chloride, hexahydrate; (10025-77-1), **71**, 126

Fisher ozonizator, **71**, 215

Flash chromatography, **70**, 4

FLASH VACUUM PYROLYSIS, **72**, 116

Fluorenone: Fluoren-9-one; 9H-Fluoren-9-one; (486-25-9), **71**, 221

FLUOROALKANES, SYNTHESIS OF, **72**, 209

1-FLUOROALKENES, 2,2-DISUBSTITUTED, STEROSELECTIVE SYNTHESIS OF, **72**, 216

(Z)-[2-(FLUOROMETHYLENE)CYCLOHEXYL]BENZENE: BENZENE, [2-(FLUOROMETHYLENE)CYCLOHEXYL]-, (Z)-(±)-; (135790-02-2), **72**, 216

Fluoromethyl phenyl sulfide: Benzene, [(fluoromethyl)thio]-; (60839-94-3), **72**, 209

FLUOROMETHYL PHENYL SULFONE: SULFONE, FLUOROMETHYL PHENYL; BENZENE, [(FLUOROMETHYL)SULFONYL]-; (20808-12-2), **72**, 209, 217

(E)-[[FLUORO(2-PHENYLCYCLOHEXYLIDENE)METHYL]SULFONYL]BENZENE: BENZENE, [[FLUORO(2-PHENYLCYCLOHEXYLIDENE)METHYL]-SULFONYL]-, (E)-(±)-; (135790-01-1), **72**, 216

α-Fluoro sulfides, synthesis of, **72**, 213

Fluorotrimethylsilane: Silane, fluorotrimethyl-; (420-56-4), **72**, 234

(Fluorovinyl)stannane, **72**, 218

Fluorovinyl sulfones, **72**, 221

Formaldehyde; (50-00-0), **70**, 71, 112

Formic acid; (64-18-6), **72**, 49

Formyl anion synthon, **72**, 29

(2-Formyl-1-chlorovinyl)ferrocene: Iron, [(1-chloro-2-formylvinyl)cyclopentadienyl]cyclopentadienyl-; (12085-68-6), **73**, 262, 263, 265, 267, 269

β-Formyl esters, synthesis, **71**, 194, 195

FUNCTIONALIZED ENYNES, SYNTHESIS OF, **72**, 104

Furan, **70**, 97

FURANS, **74**, 115

(-)-Galantic acid, **70**, 27

Galloles, **70**, 276

Gas phase pyrolysis, **72**, 121

GEMINAL DIMETHYLATION, **74**, 187

Geraniol: 2,6-Octadien-1-ol, 3,7-dimethyl-, (E)-; (106-24-1), **72**, 75

Geranyl chloride: 2,6-Octadiene, 1-chloro-3,7-dimethyl-, (E)-; (5389-87-7), **74**, 179

Germoles, **70**, 276

Glutaraldehyde: Pentanedial; (111-30-8), **70**, 54

D-(R)-GLYCERALDEHYDE ACETONIDE: 1,3-DIOXOLANE-4-CARBOXALDEHYDE, 2,2-DIMETHYL-, D-; 1,3-DIOXOLANE-4-CARBOXALDEHYDE, 2,2-DIMETHYL-, (R)-; (15186-48-8), **72**, 6, 21

L-(S)-GLYCERALDEHYDE ACETONIDE: 1,3-DIOXOLANE-4-CARBOXALDEHYDE, 2,2-DIMETHYL-, L-; 1,3-DIOXOLANE-4-CARBOXALDEHYDE, 2,2-DIMETHYL-, (S)-; (22323-80-4), **72**, 1, 15

Glycerol: 1,2,3-Propanetriol; (56-81-5), **70**, 140

Glycine; (56-40-6), **70**, 30

Glycine tert-butyl ester hydrochloride: Glycine, tert-butyl ester, hydrochloride; Glycine, 1,1-dimethylethyl ester, hydrochloride; (27532-96-3), **71**, 202

Glycol cleavage, **72**, 10

GLYOXYLATE ENE REACTION, **71**, 14

Gooch tube, **70**, 226

Halogen-metal exchange, **72**, 28

Halo lactams, **70**, 106

3-Halopropenoic acids, **72**, 114

Heptanal; (111-71-7), **73**, 96, 109

2-Heptyldihydrofuran, **73**, 215, 216

(S)-(-)-5-HEPTYL-2-PYRROLIDINONE: 2-PYRROLIDINONE, 5-HEPTYL-, (S)-; (152614-98-7), **73**, 221. 226

Heterocyclic azadiene, **70**, 85

HETEROCYCLOADDITION, RHODIUM-CATALYZED, **74**, 229

Heterodienophiles, **70**, 86

Hexaethylphosphorous triamide: HEPT; Diethylamine, N,N',N"-phosphinidynetris-; Phosphorous triamide, hexaethyl-; (2283-11-6), **72**, 232

(+)-(3aS,7aR)-HEXAHYDRO-(3S,6R)-DIMETHYL-2(3H)-BENZOFURANONE: 2,3(H)-BENZOFURANONE, HEXAHYDRO-3,6-DIMETHYL-, [3R-(3α,3aβ,6β,7aα)]-; (79726-51-5), **71**, 207

(3aβ,9bβ)-1,2,3a,4,5,9b-HEXAHYDRO-9b-HYDROXY-3a-METHYL-3H-BENZ[e]INDEN-1-ONE, **74**, 137

1,1,1,3,3,3-Hexamethyldisilazane: Disilazane, 1,1,1,3,3,3-hexamethyl-; Silanamine, 1,1,1-trimethyl-N-(trimethylsilyl)-; (999-97-3), **71**, 147; **73**, 134, 136, 143, 153, 155, 157

Hexamethylenetetramine: 1,3,5,7-Tetrazatricyclo[3.3.1.1^{3,7}]decane; (100-97-0), **73**, 246, 248, 252

Hexamethylphosphoramide: HIGHLY TOXIC. CANCER SUSPECT AGENT. Phosphoric triamide, hexamethyl-; (680-31-9), **71**, 183; **73**, 73, 74, 76, 84, 123, 124, 126, 132, 217; **74**, 14

Hexanal; (66-25-1), **73**, 51, 55, 59; **74**, 206

Hexanoyl chloride; (142-61-0), **73**, 53, 60

1-Hexylethenyl triflate: Methanesulfonic acid, trifluoro-, 1-methyleneheptyl ester; (98747-02-5), **72**, 180

2-HEXYL-5-PHENYL-1-PENTEN-3-OL, **72**, 180

1-Hexyne; (693-02-7), **70**, 216; **71**, 72

4-Hexyn-1-ol; (928-93-8), **70**, 111

4-Hexyn-1-yl methanesulfonate: 4-Hexyn-1-ol, methanesulfonate; (68275-05-8), **70**, 111

1-HEXYNYL(PHENYL)IODONIUM TOSYLATE: IODINE, 1-HEXYNYL-(4-METHYLBENZENESULFONATO-O)PHENYL-; (94957-42-3), **70**, 215

Homogeraniol, **72**, 83

Hydrazine: HIGHLY TOXIC. CANCER SUSPECT AGENT; (302-01-2), **74**, 104, 218

Hydrobromic acid; (10035-10-6), **71**, 221; **73**, 185, 189, 195

Hydrocinnamaldehyde: 3-Phenylpropionaldehyde; Benzenepropanal; (104-53-0), **72**, 154, 181

Hydrogenation, **72**, 52

Hydrogen azide: HIGHLY TOXIC, **73**, 185, 191, 196

Hydrogen bromide, **72**, 190; **74**, 72

Hydrogen cyanide, **70**, 56

Hydrogen peroxide; (7722-84-1), **70**, 265; **71**, 207; **72**, 126, 207; **73**, 98, 109, 119 **74**, 84, 116, 148

Hydroiodination, **74**, 199

α-HYDROXYAMIDES, SYNTHESIS OF, **72**, 154

(R)-3-HYDROXYBUTANOIC ACID: BUTYRIC ACID, 3-HYDROXY-, (-)-; BUTANOIC ACID, 3-HYDROXY-, (R)-; (625-72-9), **71**, 39

(4R)-(+)-Hydroxy-2-cyclopenten-1-one: 2-Cyclopenten-1-one, 4-hydroxy-, (R)-; (59995-47-0), **72**, 83; **73**, 37, 39, 40, 43

(1R,4S)-(+)-4-HYDROXY-2-CYCLOPENTENYL ACETATE: 4-CYCLOPENTENE-1,3-DIOL, MONOACETATE, (1R-cis)-; (60410-16-4), **73**, 25, 26, 31, 32, 36, 38, 43, 44, 46, 47, 49

2-(1-Hydroxycyclopropyl)-2-methyl-1-tetralone, **74**, 138

N-[2-(2-Hydroxyethoxy)ethyl]acetamide: Acetamide, N-[2-(2-hydroxyethoxy)ethyl]-; (118974-46-2), **70**, 130

threo-β-Hydroxy-L-glutamic acid, **70**, 27

Hydroxy lactams, **70**, 106

HYDROXYMETHYL ANION EQUIVALENT, **71**, 133

1-(HYDROXYMETHYL)CYCLOHEPTANOL: CYCLOHEPTANEMETHANOL, 1-HYDROXY-; (74397-19-6), **71**, 140

2-(Hydroxymethylene)cyclododecanone: Cyclododecanone, 2-(hydroxymethylene)-; (949-07-5), **71**, 159

(R)-3-HYDROXY-4-METHYLPENTANOIC ACID: PENTANOIC ACID, 3-HYDROXY-4-METHYL-, (R)-; (77981-87-4), **72**, 38

(S)-3-Hydroxy-4-methylpentanoic acid, **72**, 43

1-Hydroxy-2-methylpent-1-en-3-one: 1-Penten-3-one, 1-hydroxy-2-methyl-; (50421-81-3), **70**, 232

(+)-N-[2-(1-Hydroxy-2-phenethyl)]-5-heptyl-2-pyrrolidinone: 2-Pyrrolidinone, 5-heptyl-1-(2-hydroxy-1-phenylethyl)-, [R-(R*,R*)]-; (139564-36-6), **73**, 222, 223, 229

(E)-2-(4-HYDROXY-6-PHENYL-5-HEXENYL)-1H-ISOINDOLE-1,3(2H)-DIONE, 1H-ISOINDOLE-1,3(2H)-DIONE, 2-(4-HYDROXY-6-PHENYL-5-HEXENYL)-; (121883-31-6), **70**, 195

β-Hydroxysilanes, **73**, 57, 58

(R)-(+)-2-HYDROXY-1,2,2-TRIPHENYLETHYL ACETATE: HYTRA; 1,2-ETHANEDIOL, 1,1,2-TRIPHENYL-, 2-ACETATE, (R)-; (95061-47-5), **72**, 32, 38

(S)-(-)-2-Hydroxy-1,2,2-triphenylethyl acetate, **72**, 36

Hydrozirconation, **71**, 81

(R)-HYTRA, **72**, 36

Imidazole: 1H-Imidazole; (288-32-4), **71**, 56; **73**, 25, 27, 35; **74**, 206

Imines, hydrolysis of, **70**, 36

1-Indanone: 1H-Inden-1-one, 2,3-dihydro-; (83-33-0), **74**, 54

Indole: 1H-Indole; (120-72-9), **74**, 249

INDOLE-2-ACETIC ACID METHYL ESTERS, SYNTHESIS OF, **72**, 125

INDOLES, 3-SUBSTITUTED, REGIOSELECTIVE SYNTHESIS, **74**, 248

Indoline: 1H-Indole, 2,3-dihydro-; (496-15-1), **73**, 86, 88, 93

7-INDOLINECARBOXALDEHYDE, **73**, 85, 87, 88, 90

Indolizidines, **70**, 57

Intramolecular cyclopropanation, **73**, 21, 22

INTRAMOLECULAR HYDROSILYLATION, REGIO- AND STEREOSELECTIVE, **73**, 94, 105

Intramolecular ring closure of oxo-stannanes, **72**, 195

INVERSION OF MENTHOL, **73**, 110, 113

Iodine; (7553-56-2), **70**, 103; **72**, 32, 201

(Z)-β-IODOACROLEIN: 2-PROPENAL, 3-IODO-, (Z)-; (138102-13-3), **74**, 194

(Z)- and (E)-γ-IODO ALLYLIC ALCOHOLS, **74**, 194

4-Iodoanisole: Anisole, p-iodo-; Benzene, 1-iodo-4-methoxy-; (696-62-8), **73**, 203, 214

8-exo-IODO-2-AZABICYCLO[3.3.0]OCTAN-3-ONE,: CYCLOPENTA[b]PYRROL-2(1H)-ONE, HEXAHYDRO-6-IODO-, (3aα, 6α, 6aα)-; (100556-58-9), **70**, 101

1-Iodobutane: Butane, 1-iodo-; (542-69-8), **73**, 74, 76, 84

4-Iodobutyronitrile: Butanenitrile, 4-iodo-; (6727-73-7), **74**, 37, 38

1-Iodoheptane: Heptane, 1-iodo-; (4282-40-0), **73**, 215, 217, 220

(E)-1-IODOHEPT-1-EN-3-OL: 1-HEPTEN-3-OL, 1-IODO, (E)-; (151160-08-6)), **74**, 194

(Z)-1-IODOHEPT-1-EN-3-OL: 1-HEPTEN-3-OL, 1-IODO, (Z)-; (138102-06-4),. **74**, 194

IODOLACTAMIZATION, **70**, 101

Iodomethane: Methane, iodo-; (74-88-4), **70**, 19, 113, 178; **72**, 63; **74**, 188

(3Z)-Iodopropenals, **74**, 200

2-(3-Iodopropyl)-1H-isoindole-1,3(2H)-dione: 1H-Isoindole-1,3(2H)-dione, 2-(3-iodopropyl)-; (5457-29-4), **70**, 195

Iodosobenzene: Benzene, iodosyl-; (536-80-1), **70**, 217

Iron chloride, hexahydrate; (10025-77-1), **71**, 126

Irradiation, **71**, 201

Isoamyl nitrite: Nitrous acid, isopentyl ester; Nitrous acid, 3-methylbutyl ester; (110-46-3), **72**, 247

Isobutyl alcohol: 1-Propanol, 2-methyl-; (78-83-1), **72**, 202

Isobutylene oxide: Propane, 1,2-epoxy-2-methyl-; Oxirane, 2,2-dimethyl-; (558-30-5), **72**, 174

Isobutyraldehyde: Propanal, 2-methyl-; (78-84-2), **71**, 190; **72**, 39, 174

Isooctane: Pentane, 2,2,4-trimethyl-; (540-84-1), **71**, 221

Isopentane: Butane, 2-methyl-; (78-78-4), **72**, 39

(+)-ISOPINOCAMPHEOL: BICYCLO[3.1.1]HEPTAN-3-OL, 2,6,6-TRIMETHYL-, [1S-(1α,2β,3α,5α)]-; (24041-60-9), **73**, 116, 117, 119, 122

Isoprene; 1,3-Butadiene, 2-methyl-; (78-79-5), **70**, 256

Isopropyl acetate: Acetic acid, 1-methylethyl ester; (108-21-4), **70**, 158

2,3-O-Isopropylidene-D-glyceraldehyde: 1,3-Dioxolane-4-carboxaldehyde, 2,2-dimethyl-, (D)-; 1,3-Dioxolane-4-carboxaldehyde, 2,2-dimethyl-, (R)-; (15186-48-8), **72**, 7

5,6-O-Isopropylidene-L-gulono-1,4-lactone: L-Gulonic acid, 5,6-O-(1-methylethylidene)-, γ-lactone; (94697-68-4), **72**, 16

Isopropylmagnesium chloride; (1068-55-9), **70**, 157; **74**, 109

Isopulegol: Cyclohexanol, 5-methyl-2-(1-methylethenyl)-, [1R-(1α,2β,5α)]-; (89-79-2), **71**, 207

ISOQUINOLINE ENAMIDES, **70**, 139

Jones Reagent [CrO_3/H_2SO_4], **71**, 237; **72**, 79, 80, 237

Karl Fisher titration, **74**, 55

Ketene dithioacetals, **71**, 175

Ketene intermediates, α,β-unsaturated, **74**, 164

Ketene silyl acetals, **70**, 162

α-Ketoamides, **72**, 159

Keto carbenoids, **70**, 97

β-Ketoesters, C-acylation, **70**, 256

KETONES, α,β-ETHYLENIC, CONJUGATE ADDITION TO, **72**, 135

4-Ketopipecolic acid hydrochloride: Pipecolic acid, 4-oxo-, hydrochloride; (99979-55-2), **71**, 200

4-KETOUNDECANOIC ACID: UNDECANOIC ACID, 4-OXO-; (22847-06-9), **73**, 215, 216, 220, 222, 229

Kieselgel, **70**, 13

Kieselgel 60 silica gel, Merck, **70**, 258

β-Lactone, **70**, 1

β-LACTONES, SYNTHESIS, **73**, 61, 67, 68

Lead(II) chloride: Lead chloride; (7758-95-4), **73**, 75, 77, 80, 84

Lead nitrate: Nitric acid, lead (2+) salt; (10099-74-8), **71**, 226

Lead tetraacetate: Acetic acid, lead (4+) salt; (546-67-8),
CANCER SUSPECT AGENT, **70**, 140; **72**, 57
for oxidative ring expansion of isoquinoline enamides, **70**, 139

Lewis acids catalysts, **70**, 65

Liquid ammonia, **70**, 256

LITHIATION, ORTHO-DIRECTED, **73**, 90, 91

2-Lithiofuran, **70**, 154

N-Lithiopyrrole, **70**, 154

2-Lithiothiazole, **72**, 28

2-Lithiothiophene, **70**, 154

Lithium; (7439-93-2), **70**, 164; **71**, 23; **72**, 147, 173; **74**, 178
powder, **70**, 205
wire; **73**, 51, 55, 223, 225, 230

Lithium aluminum hydride: Aluminate (1-), tetrahydro-, lithium; Aluminate (1-), tetrahydro-, lithium, (T-4)-; (16853-85-3), **70**, 55; **71**, 77; **73**, 2, 11, 20; **74**, 2, 15

Lithium biphenylide, **72**, 148; **74**, 179

Lithium bis(trimethylsilyl)amide: LiHMDS: Disilazane, 1,1,1,3,3,3-hexamethyl-, lithium salt; Silanamine, 1,1,1-trimethyl-N-(trimethylsilyl), lithium salt; **72**, 217

Lithium butanetelluroate, **72**, 156

Lithium chloride; (7447-41-8), **70**, 196; **72**, 135

Lithium 4,4'-di-tert-butylbiphenylide: 1,1'-Biphenyl, 4,4'-bis(1,1-dimethylethyl)-, radical ion (1-), lithium; (61217-61-6), **72**, 173

Lithium ethoxide: Ethyl alcohol, lithium salt; Ethanol, lithium salt; (2388-07-0), **71**, 147

Lithium hexamethyldisilazide: Disilazane, 1,1,1,3,3,3-hexamethyl-, lithium salt; Silanamine, 1,1,1-trimethyl-N-(trimethylsilyl)-, lithium salt; (4039-32-1), **71**, 147

LITHIUM β-LITHIOALKOXIDES, **72**, 173

LITHIUM-TELLURIUM EXCHANGE REACTIONS, **72**, 154

Lithium 2,2,6,6-tetramethylpiperidide: Piperidine, 2,2,6,6-tetramethyl-, lithium salt; (38227-87-1), **71**, 146

MABR catalyst , **72**, 99

Magnesium; (7439-95-4), **71**, 98; **72**, 32, 180, 201; **74**, 188, 198

Magnesium bromide etherate: Magnesium, dibromo(ethyl ether); Magnesium, dibromo[1,1'-oxybis[ethene]]-; (29858-07-9), **73**, 52, 55, 60; **74**, 137

(R)-(-)-Mandelic acid: Mandelic acid, D-(-)-; Benzeneacetic acid, α-hydroxy-, (R)-; (611-71-2), **72**, 34

Manganese(II) chloride: Manganese chloride; (7773-01-5), **72**, 135

MANGANESE-COPPER CATALYSIS, **72**, 135

Manganese dioxide, active, **73**, 45, 47

D-Mannitol; (69-65-8), **72**, 6

(-)-(1R,3R,4S,8R)-p-Menthane-3,9-diol: p-Menthane-3,9-diol, (1R,3R,4S,8R)-(-)-; Cyclohexaneethanol, 2-hydroxy-β,4-dimethyl-, [1S-[1α(S*),2β,4β]]-; (13834-07-6), **71**, 207

(1R,2S,5R)-MENTHOL: MENTHOL, (-)-; CYCLOHEXANOL, 5-METHYL-2-(1-METHYLETHYL)-, [1R-(1α,2β,5α)]-; (2216-51-5), **74**, 13
INVERSION, **73**, 110, 111

1-Menthoxyacetylene: Cyclohexane, 2-ethynyloxy-4-methyl-1-(1-methylethyl)-, [1S-(1α,2β,4β)]-; **74**, 16

(E)-1-MENTHOXY-1-BUTENE: CYCLOHEXANE, 2-(1-BUTENYLOXY)-4-METHYL-1-(1-METHYLETHYL)-, {1S-(1α,2β(E),4β)]-; (107941-63-9), **74**, 13

(Z)-1-MENTHOXY-1-BUTENE: CYCLOHEXANE, 2-(1-BUTENYLOXY)-4-METHYL-1-(1-METHYLETHYL)-, {1S-(1α,2β(Z),4β)]-; (107941-62-8), **74**, 13

1-MENTHOXY-1-BUTYNE: CYCLOHEXANE, 2-(1-BUTYNYLOXY)-4-METHYL-1-(1-METHYLETHYL)-, {1S-(1α,2β,4β)]-; (108266-28-0), **74**, 13

Mercury(II) chloride; (7487-94-7), **72**, 24

Mesitylene: Benzene, 1,3,5-trimethyl-; (108-67-8), **74**, 218

Mesitylenesulfonyl chloride: Benzenesulfonyl chloride, 2,4,6-trimethyl-; (773-64-8), **74**, 219

MESITYLENESULFONYLHYDRAZINE: BENZENESULFONIC ACID, 2,4,6-TRIMETHYL-, HYDRAZIDE; (16182-15-3), **74**, 217

Metal-ammonia reduction, **73**, 180, 227

Methacrolein: 2-Propenal, 2-methyl-; (78-85-3), **72**, 88

Methallyl chloride: 1-Propene, 3-chloro-2-methyl-; (563-47-3), **71**, 159

Methanesulfonyl azide; (1516-70-7), **70**, 97; **73**, 20, 136, 143

Methanesulfonyl chloride; (124-63-0), **70**, 111

L-Methionine; (63-68-3), **70**, 31

L-Methionine methyl ester hydrochloride: L-Methionine, methyl ester, hydrochloride; (2491-18-1), **70**, 29

4-Methoxybenzaldehyde: Benzaldehyde, 4-methoxy-; (123-11-5), **74**, 110

2-Methoxybenzoic acid: o-Anisic acid; Benzoic acid, 2-methoxy-;
(579-75-9), **72**, 163

9-Methoxy-9-borabicyclo[3.3.1]nonane: 9-Borabicyclo[3.3.1]nonane, 9-methoxy-;
(38050-71-4), **70**, 173

Methoxycarbonylmethylation, **71**, 189

2-Methoxyethanol: Ethanol, 2-methoxy-; (109-86-4), **72**, 199

3-(2-Methoxyethoxy)propene: 1-Propene, 3-(2-methoxyethoxy)-; (18854-48-3),
72, 199

1-[(Methoxymethoxy)methyl]cycloheptanol: Cycloheptanol,
1-[(methoxymethoxy)methyl]-; (115384-52-6), **71**, 140

(METHOXYMETHOXY)METHYLLITHIUM: LITHIUM,
[(METHOXYMETHOXY)METHYL]-; (115384-62-8), **71**, 140

2,3-syn-2-METHOXYMETHOXY-1,3-NONANEDIOL: 1,3-NONANEDIOL,
2-(METHOXYMETHOXY)-, (R*,R*)-(±)-; (114675-32-0), **73**, 94, 96, 98, 102, 103,
108

2-Methoxymethoxy-1-nonen-3-ol: 1-Nonen-3-ol, 2-(methoxymethoxy)-, (±)-;
(114675-31-9), **73**, 96, 100, 108

1-Methoxy-2-methyl-1-penten-3-one, **70**, 233

(E,Z)-1-METHOXY-2-METHYL-3-(TRIMETHYLSILOXY)-1,3-PENTADIENE: SILANE,
[[1-(2-METHOXY-1-METHYLETHENYL)-1-PROPENYL]OXY]TRIMETHYL-,
(Z,E)-; (72486-93-2), **70**, 231

Methoxymethyl vinyl ether: Ethene, (methoxymethoxy)-; (63975-05-3), **73**, 95, 96, 100,
104, 108

4-METHOXY-4'-NITROBIPHENYL: 1,1'-BIPHENYL, 4-METHOXY-4'-NITRO-;
(2143-90-0), **71**, 97

5-Methoxy-2-nitrophenylacetic acid: Acetic acid, (5-methoxy-2-nitrophenyl)-;
Benzeneacetic acid, 5-methoxy-2-nitro-; (20876-29-3), **72**, 126

(4-METHOXYPHENYL)ETHYNE: BENZENE, 1-ETHYNYL-4-METHOXY-; (768-60-5),
74, 108

9-Methoxy-9-phenylfluorene: 9H-Fluorene, 9-methoxy-9-phenyl-;
(56849-87-7), **71**, 228

7-METHOXYPHTHALIDE: PHTHALIDE, 7-METHOXY-; 1(3H)-ISOBENZOFURANONE,
7-METHOXY-; (28281-58-5), **72**, 163

Methyl acetoacetate: Acetoacetic acid, methyl ester; Butanoic acid, 3-oxo-,
methyl ester; (105-45-3), **72**, 127

1-(2-Methyl)allyl-2-(trimethylsiloxy)methylenecyclododecanol, **71**, 158

METHYLALUMINUM BIS(4-BROMO-2,6-DI-tert-BUTYLPHENOXIDE): ALUMINUM, BIS[4-BROMO-2,6-BIS(1,1-DIMETHYLETHYL)PHENOLATO]METHYL-; (118495-99-1), **72**, 95

Methylaluminum bis(4-bromo-2,6-diisopropylphenoxide), **72**, 100

(S)-(-)-α-Methylbenzylamine: Benzenemethanamine, α-methyl-, (S)-; (2627-86-3), **70**, 35

Methyl L-2-(benzyloxycarbonylamino)-4-(methylsulfinyl)butanoate: Butenoic acid, 4-(methylsulfinyl)-2-[[(phenylmethoxy)carbonyl]amino]-, ethyl ester, (S)-; (75266-39-3), **70**, 30

Methyl bromoacetate: Acetic acid, bromo-, methyl ester; (96-32-2), **74**, 159

Methyl 4-bromocrotonate: 2-Butenoic acid, 4-bromo-, methyl ester; (1117-71-1), **74**, 149

METHYL 3-BROMOPROPIOLATE: 2-PROPYNOIC ACID, 3-BROMO-, METHYL ESTER; (23680-40-2), **74**, 212

2-Methylbutane: Isopentane: Butane, 2-methyl-; (78-78-4), **72**, 39

3-Methyl-2-butanone: TOXIC. 2-Butanone, 3-methyl-; (563-80-4), **73**, 61, 64, 72

3-Methyl-2-buten-1-ol: 2-Buten-1-ol, 3-methyl-; (556-82-1), **73**, 15, 23

3-Methyl-2-butenyl acetoacetate, **73**, 15

3-Methyl-2-butenyl diazoacetate, **73**, 16

Methyl chloroformate: Formic acid, chloro-, methyl ester; Carbonochloridic acid, methyl ester; (79-22-1), **73**, 202, 205, 213

16a-METHYLCORTEXOLONE: PREGN-4-ENE-3,20-DIONE, 17,21-DIHYDROXY-16-METHYL-, (16α)-(±)-; (122405-63-4), **73**, 123, 124, 126

Methyl cyanoformate: Carbonocyanidic acid, methyl ester; (17640-15-2), **70**, 256

2-Methylcyclohexanone: Cyclohexanone, 2-methyl; (583-60-8), **70**, 35

3-Methyl-2-cyclohexen-1-one, **70**, 261

2-METHYL-1,3-CYCLOPENTANEDIONE: 1,3-CYCLOPENTANEDIONE, 2-METHYL-; (765-69-5), **70**, 226

Methyl diazoacetate: Acetic acid, diazo-, methyl ester; (6832-16-2), **71**, 190

Methyl 3,6-dimethyl-3-hydroxy-6-heptenoate, **74**, 158

METHYL 3,3-DIMETHYL-4-OXOBUTANOATE: BUTANOIC ACID, 3,3-DIMETHYL-4-OXO-, METHYL ESTER; (52398-45-5), **71**, 189

Methyl 2,2-dimethyl-3-(trimethylsiloxy)-1-cyclopropanecarboxylate: Cyclopropanecarboxylic acid, 2,2-dimethyl-3-[(trimethylsilyl)oxy]-, methyl ester; (77903-45-8), **71**, 190

6-METHYL-6-DODECENE, **73**, 50, 52

(E)-6-Methyl-6-dodecene: 6-Dodecene, 6-methyl, (E)-; (101146-61-6), **73**, 54, 60

(Z)-6-Methyl-6-dodecene: 6-Dodecene, 6-methyl-, (Z)-; (101165-44-0), **73**, 54, 60

2-Methylene-1,3-dithiolane: 1,2-Dithiolane, 2-methylene-; (26728-22-3), **71**, 175, 182

Methylene transfer reagent, **70**, 187

Methyl glyoxylate: Glyoxylic acid, methyl ester; Acetic acid, oxo-, methyl ester; (922-68-9), **71**, 15

5-Methyl-5-hexen-2-one: 5-Hexen-2-one, 5-methyl-; (3240-09-3), **74**, 158

N-Methylhydrazine: Hydrazine, methyl-; (60-34-4), **72**, 16

(R)-(-)-METHYL 3-HYDROXYBUTANOATE: BUTYRIC ACID, 3-HYDROXY-, METHYL ESTER, D-(-)-; BUTANOIC ACID, 3-HYDROXY-, METHYL ESTER, (R)-; (3976-69-0), **71**, 1, 39

METHYL (2R)-2-HYDROXY-4-PHENYL-4-PENTENOATE: BENZENEBUTANOIC ACID, α-HYDROXY-γ-METHYLENE, METHYL ESTER, (R)-; (119072-58-1), **71**, 14

Methyl indole-2-acetate: Indole-2-acetic acid, methyl ester; 1H-Indole-2-acetic acid, methyl ester; (21422-40-2), **72**, 132

Methyl iodide: Methane, iodo-; (74-88-4), **70**, 19, 113; **71**, 99; **72**, 24; **74**, 188

Methyllithium-lithium bromide complex: Lithium, methyl-; (917-54-4), **70**, 165; **71**, 176; **73**, 54, 60

Methylmagnesium bromide: Magnesium, bromomethyl-; (75-16-1), **71**, 31; **73**, 124, 126, 133; **74**, 138

Methylmagnesium iodide: Magnesium, iodoomethyl-; (917-64-6), **74**, 188

Methyl (R)-(-)-mandelate: Mandelic acid, methyl ester, (R)- or (R)-(-)-; Benzeneacetic acid, α-hydroxy-, methyl ester, (R)-; (20698-91-3), **72**, 33

Methyl (5R)-2-(methoxycarbonyl)-5,9-dimethyldeca-2,8-dienoate: Propanedioic acid, (3,7-dimethyl-6-octenylidene)-, dimethyl ester, (R)-; (106431-76-9), **71**, 167

Methyl 5-methoxyindole-2-acetate: Indole-2-acetic acid, 5-methoxy-, methyl ester; 1H-Indole-2-acetic acid, 5-methoxy-, methyl ester; (27798-66-9), **72**, 125

Methyl (5-methoxy-2-nitrophenylacetyl)acetoacetate: Benzenebutanoic acid, α-acetyl-, 5-methoxy-2-nitro-, β-oxo-, methyl ester; (130916-40-4), **72**, 127

Methyl 4-(5-methoxy-2-nitrophenyl)-3-oxobutyrate: Benzenebutanoic acid, 5-methoxy-2-nitro-, β-oxo-, methyl ester; (130916-41-5), **72**, 128

(1S,2S,5R)-5-METHYL-2-(1-METHYLETHYL)CYCLOHEXYL 4-NITROBENZOATE: (1S,2S,5R)-1-(4-NITROBENZOYL)-2-(1-METHYLETHYL)-5-METHYLCYCLOHEXAN-1-OL: CYCLOHEXANOL, 5-METHYL-2-(1-METHYLETHYL)-, 4-NITROBENZOATE, [1S-(1α,2α,5β)]-; (27374-00-1), **73**, 110

4-Methylmorpholine N-oxide: Morpholine, 4-methyl-, 4-oxide; (7529-22-8), **70**, 47

3-Methyl-4-nitroanisole: Anisole, 3-methyl-4-nitro-; Benzene, 4-methoxy-2-methyl-1-nitro-; (5367-32-8), **72**, 126

(R)-(-)-10-METHYL-1-OCTAL-2-ONE: 2(3H)-NAPHTHALENONE, 4,4a,5,6,7,8-HEXAHYDRO-4a-METHYL-, (R)-; (63975-59-7), **70**, 35

Methyl γ-oxoalkanoates, synthesis, **71**, 194

Methyl 3-oxobutanoate: Butanoic acid, 3-oxo-, methyl ester; (105-45-3), **71**, 2

(R)-(+)-2-Methyl-2-(3-oxobutyl)cyclohexanone: Cyclohexanone, 2-methyl-2-(3-oxobutyl)-, (R)-; (91306-30-8), **70**, 36

METHYL (1α,4aβ,8aα)-2-OXODECAHYDRO-1-NAPHTHOATE, **70**, 256

3-Methyl-2,4-pentanedione: 2,4-Pentanedione, 3-methyl-; (815-57-6), **74**, 117

Methyl 3-phenyl-5-carboxamidopyrrole-2-carboxylate, **70**, 85

(E)-2-Methyl-2-(2-phenylethenyl)-1,3-dithiolane: 1,3-Dithiolane, 2-methyl-2-(2-phenylethenyl)-, (E)-; (107389-59-3), **74**, 187

2-METHYL-4-[(PHENYLSULFONYL)METHYL]-2-CYCLOPENTEN-1-ONE: 2-CYCLOPENTEN-1-ONE, 2-METHYL-4-[(PHENYLSULFONYL)METHYL]-; (132604-66-1), **74**, 115

2-METHYL-4-[(PHENYLSULFONYL)METHYL]FURAN: FURAN, 2-METHYL-4-[(PHENYLSULFONYL)METHYL]-; (128496-98-0), **74**, 115

Methyl phenyl sulfoxide: Benzene, (methylsulfinyl)-; (1193-82-4), **72**, 209

Methylphosphonic dichloride: Phosphonic dichloride, methyl-; (676-97-1), **73**, 153, 154, 157

2-Methylpiperidine: Piperidine, 2-methyl-; (109-05-7), **70**, 265

(R)-2-Methylproline: D-Proline, 2-methyl-; (63399-77-9), **72**, 69

(S)-2-METHYLPROLINE: L-PROLINE, 2-METHYL-; (42856-71-3), **72**, 62

2-Methylpropanal: Isobutyraldehyde; Propanal, 2-methyl-; (78-84-2), **72**, 39

2-Methyl-1-propanol: Isobutyl alcohol; 1-Propanol, 2-methyl-; (78-83-1), **72**, 202

Methyl propiolate: 2-Propynoic acid, methyl ester; (922-67-8), **74**, 212

Methyl 2-pyrrolidone-5(R)-carboxylate: D-Proline, 5-oxo-, methyl ester; (64700-65-8), **73**, 14, 15, 19, 21, 22, 23

α-Methylstyrene: Styrene, α-methyl-; Benzene, (1-methylethenyl)-; (98-83-9), **71**, 15

Methyl sulfide: Methane, thiobis-; (75-18-3), **71**, 214

6-METHYL-2,3,4,5-TETRAHYDROPYRIDINE N-OXIDE: PYRIDINE, 2,3,4,5-TETRAHYDRO-6-METHYL-, 1-OXIDE; (55386-67-9), **70**, 265

2-Methyl-1-tetralone: 1(2H)-Naphthalenone, 3,4-dihydro-2-methyl-; (1590-08-5), **74**, 137
magnesium enolate, **74**, 137

Methyl trifluoromethanesulfonate: Methanesulfonic acid, trifluoro-, methyl ester; (333-27-7), **71**, 110

(Z)-16a-Methyl-20-trimethylsiloxy-4,17(20)-pregnadien-3-one: Pregna-4,17(20)-dien-3-one, 16-methyl-20-[(trimethylsilyl)oxy]-, (16α,17Z)-(±)-; (122315-01-9), **73**, 124, 132

2-Methyl-1-(trimethylsiloxy)propene: Silane, trimethyl[(2-methyl-1-propenyl)oxy]-; (6651-34-9), **71**, 189

Methyl vinyl ketone, **70**, 35

Michael-type alkylation, **70**, 40

Mitsunobu conditions, **70**, 6

MITSUNOBU INVERSION OF STERICALLY HINDERED ALCOHOLS, **73**, 110, 113

Molybdophosphoric acid, **70**, 31

Monobactam antibiotics, **72**, 19

Mono(2,6-dimethoxybenzoyl)tartaric acid: Butanedioic acid, 2-[(2,6-dimethoxybenzoyl)oxy-3-hydroxy-, [R-(R*,R*)]-; (116212-44-3), **72**, 87

Morpholine; (110-91-8), **74**, 258

3-MORPHOLINO-2-PHENYLTHIOACRYLIC ACID MORPHOLIDE: MORPHOLINE, 4-[3-(4-MORPHOLINYL)-2-PHENYL-1-THIOXO-2-PROPENYL]-; (86867-31-4), **74**, 257

Mosher amide, formation, **74**, 59

Mosher ester, **70**, 24, 48; **73**, 7, 30

NAPHTHALENEDIOLS, 2-SUBSTITUTED, SYNTHESIS, **71**, 72

1-Naphthalenesulfonyl azide, **70**, 97

(R)-1-(1-Naphthyl)ethylamine: 1-Naphthalenemethylamine, α-methyl-, (R)-(+)-; 1-Naphthalenemethanamine, α-methyl-, (R)-; (3886-70-2), **72**, 79

Nerol, **72**, 82

Nickel acetylacetonate: Nickel, bis(2,4-pentanedionato)-; Nickel, bis(2,4-pentanedionato-O,O')-, (SP-4-1)-; (3264-82-2), **71**, 84

Nickel-catalyzed coupling reactions, **70**, 243

NICKEL-CATALYZED GEMINAL DIMETHYLATION, **74**, 187

Nickel(II) chloride, anhydrous: Nickel chloride; (7718-54-9), CANCER SUSPECT AGENT, **72**, 181

Nitriles, from lower analog ketones, **74**, 224
HETEROCYCLOADDITION, **74**, 229

NITROACETALDEHYDE DIETHYL ACETAL: ACETALDEHYDE, NITRO-, DIETHYL ACETAL; (34560-16-2), **74**, 130

Nitroalkanes, **74**, 132

4-Nitrobenzoic acid: Benzoic acid, p-nitro-; Benzoic acid, 4-nitro-; (62-23-7), **73**, 110, 111, 112, , 113, 114

1-Nitrocyclohexene, **70**, 77

4-Nitro-3-hexanol: 3-Hexanol, 3-nitro-; (5342-71-2), **70**, 68

Nitromethane: Methane, nitro-; (75-52-5), **70**, 183, 184, 226; **71**, 65; **74**, 130

NITRONES, **70**, 265

4-Nitrophenol: Phenol, p-nitro-; Phenol, 4-nitro-; (100-02-7), **71**, 98

Nitrophenylacetyl chloride, **72**, 132

4-Nitrophenyl trifluoromethanesulfonate: Methanesulfonic acid, trifluoro-, p-nitrophenyl ester; Methanesulfonic acid, trifluoro-, 4-nitrophenyl ester; (17763-80-3), **71**, 98

1-Nitropropane: Propane, 1-nitro-; (108-03-2), **70**, 69

Nitrosomethylurea: Urea, N-methyl-N-nitroso-; (684-93-5), **70**, 22

Nitrous gases, **70**, 81, 84

2,3,7,8,12,13,17,18-OCTAETHYLPORPHYRIN: 21H, 23H-PORPHINE, 2,3,7,8,12,13,17,18-OCTAETHYL-; (2683-82-1), **70**, 68

Octahydrobinaphthol, **70**, 65

$\Delta^{1(9)}$-Octalone-2: 2(3H)-Naphthalenone, 4,4a,5,6,7,8-hexahydro-; (1196-55-0), **70**, 256

3-(1-OCTEN-1-YL)CYCLOPENTANONE: CYCLOPENTANONE, 3-(1-OCTENYL)-, (E)-; (64955-00-6), **71**, 83

1-Octyne; (629-05-0), **71**, 84; **72**, 180

Organic tellurium compounds, **72**, 158

Organoaluminum compounds, **72**, 98

ORGANOBIS(CUPRATES), **70**, 204

ORGANOBORANES, OXIDATION, **73**, 116, 119

ORGANOCALCIUM REAGENTS, **72**, 147

Organochromium reagents, **72**, 180

ORGANOMAGNESIUM REAGENTS, **72**, 135

Organotin compounds, **72**, 190, 199

Osmium tetroxide; (20816-12-0), **70**, 47

Oxalic acid: Ethanedioic acid; (144-62-7), **71**, 110

Oxalyl chloride; (79-37-8), **70**, 101, 121, 205

Oxazaborolidinones, **74**, 29, 65
    borane reduction catalyzed by, **74**, 47

Oxazoles, rhodium-catalyzed synthesis, **74**, 234

Oxazolidine, **70**, 57

2-Oxazolidinone; (497-25-6), **71**, 31

Oxazolines, in aromatic substitution reactions, **71**, 113

Oxidation, α,ω-diols to lactones, **71**, 211

Oxidation, heterogeneous, **71**, 207

OXIDATION OF SECONDARY AMINES, **70**, 265

(-)-[(3-Oxocamphoryl)sulfonyl]imine: 3H-3a,6-Methano-2,1-benzisothiazole-7(4H)-one, 5,6-dihydro-8,8-dimethyl-, 2,2-dioxide, (3aS)-; (119106-38-6), **73**, 160, 164, 168, 172

3-OXO CARBOXYLATES, **71**, 1

2-Oxo-5,6-dihydro-1,3-dithiolo[4,5-b][1,4]dithiin: 1,3-Dithiolo[4,5-b][1,4]dithiin-2-one, 5,6-dihydro-; (74962-29-1), **72**, 267

2-Oxo-1,4-dithiane: 1,4-Dithian-2-one; (74637-14-2), **72**, 266

2-OXO-5-METHOXYSPIRO[5.4]DECANE: SPIRO[4.5]DECAN-1-ONE, 4-METHOXY-; (108264-15-9), **72**, 189

Oxone: Peroxymonosulfuric acid, monopotassium salt, mixt. with dipotassium sulfate and potassium hydrogen sulfate; (37222-66-5), **72**, 210; **74**, 91

2-Oxo-3-(2-propoxythiocarbonylthio)-1,4-dithiane: Carbonodithioic acid, O-(1-methylethyl)-S-(3-oxo-1,4-dithian-2-yl) ester; (120627-40-9), **72**, 267

Ozone; (10028-15-6), **71**, 214

Ozonolysis, **71**, 214

10% Palladium on activated charcoal, **72**, 88, 128; **73**, 188, 192, 196, 197
on barium sulfate, **74**, 15

PALLADIUM-CATALYZED REACTIONS, **71**, 89, 97, 104

PALLADIUM/COPPER-CATALYZED COUPLING REACTIONS OF ACETYLENES, **72**, 104

Paraformaldehyde; (30525-89-4), **71**, 133

Parr apparatus, **72**, 88

Pentacarbonyl[phenyl(methoxy)chromium]carbene: Chromium, pentacarbonyl(α-methoxybenzylidene)-; Chromium, pentacarbonyl(methoxyphenylmethylene)-, (OC-6-21)-; (27436-93-7), **71**, 72

2,4-Pentanedione; (123-54-6), **74**, 117

3-Pentanone; (96-22-0), **70**, 232

Pentanoyl chloride; (638-29-9), **70**, 61

4-Pentyn-1-ol; (5390-04-5), **70**, 113

Peracetic acid: Peroxyacetic acid; Ethaneperoxoic acid; (79-21-0), **73**, 162, 163, 166, 173

Perfluoroalkyl acetylenes, **70**, 246

(5aS,9aS,11aS)-Perhydro-5H-pyrrolo[2,1-c][1,4]benzodiazepine-5,11-dione: 1H-Pyrrolo[2,1-c][1,4]benzodiazepine-5,11(5aH,11aH)-dione, octahydro-, [5aS-(5aα,9aβ,11aβ)]-; (110419-86-8), **73**, 175, 183

Periodate glycol cleavage, **72**, 4, 10

Peterson olefination reaction, **73**, 57

Phase transfer catalysis, **73**, 95, 100, 145, 147

1,10-Phenanthroline; (66-71-7), **70**, 206, 274; **72**, 202; **73**, 126

Phenylacetylene: Benzene, ethynyl-; (536-74-3), **70**, 94, 157, 218

p-Phenylbenzoyl chloride: 4-Biphenylcarbonyl chloride; [1,1'-Biphenyl]-4-carbonyl chloride; (14002-51-8), **72**, 53

trans-4-Phenyl-3-buten-2-one: 3-Buten-2-one, 4-phenyl-, (E)-; (1896-62-4), **73**, 134, 136, 143; **74**, 187

Phenylchlorocarbene, **74**, 74

2-Phenylcyclohexanone: Cyclohexanone, 2-phenyl-; (1444-65-1), **72**, 217

S-Phenyl decanoate: Decanethioic acid, S-phenyl ester; (51892-25-2), **73**, 61, 63, 72

Phenyldimethylsilyl chloride: Aldrich: Chlorodimethylphenylsilane: Silane, chlorodimethylphenyl-; (768-33-2), **73**, 51, 55, 59

5-Phenyldimethylsilyl-6-dodecanone, **73**, 53

(E)-5-PHENYLDODEC-5-EN-7-YNE: BENZENE, (1-BUTYL-1-OCTEN-3-YNYL)-, (E)-; (111525-79-2), **70**, 215

2-(2-Phenylethenyl)-1,3-dithiolane: 1,3-Dithiolane, 2-(2-phenylethenyl)-; (5616-58-0), **70**, 240

9-Phenyl-9-fluorenol: Fluoren-9-ol, 9-phenyl-; 9H-Fluoren-9-ol, 9-phenyl-; (25603-67-2), **71**, 220, 228

(S)-N-(9-Phenylfluoren-9-yl)alanine: L-Alanine, N-(9-phenyl-9H-fluoren-9-yl)-; (105519-71-9), **71**, 226

(-)-Phenylglycine, **70**, 55

(S)-(+)-2-Phenylglycine: Glycine, 2-phenyl-, L-; Benzeneacetic acid, α-amino-, (S)-; (2935-35-3), **73**, 224, 230

(S)-(+)-2-Phenylglycinol: Phenethyl alcohol, β-amino-, L-; (20989-17-7), **73**, 222, 224, 227, 229

(-)-Phenylglycinol: Benzeneethanol, β-amino-, (R)-; (56613-80-0), **70**, 54

Phenyllithium: Lithium, phenyl-; (591-51-5), **71**, 223

Phenylmagnesium bromide: Magnesium, bromophenyl-; (100-58-3), **72**, 32

Phenylmagnesium chloride; (100-59-4), **74**, 52

Phenyl methyl sulfide (thioanisole); (100-68-5), **74**, 92

(+)-3-Phenyl-5-oxo-7a-heptyl-2,3,5,6,7,7a-hexahydropyrrolo[2,1-b]oxazole: Pyrrolo[2,1-b]oxazol-5(6H)-one, tetrahydro-7a-heptyl-3-phenyl-, (3R-cis)-; (132959-41-2), **73**, 227, 229

3-Phenylpropanal: Hydrocinnamaldehyde; Benzenepropanal; (104-53-0), **72**, 154, 181

Phenylsulfenyl chloride; (931-59-9), **74**, 116, 148

1-(Phenylsulfonyl)-1,2-propadiene: Benzene, (1,3-propanedienylsulfonyl)-; (2525-42-0), **74**, 116

2-(Phenylsulfonyl)tetrahydro-2H-pyran: 2H-Pyran, tetrahydro-2-(phenylsulfonyl)-; (96754-03-9), **70**, 157

1-PHENYL-2,3,4,5-TETRAMETHYLPHOSPHOLE: 1H-PHOSPHOLE, 2,3,4,5-TETRAMETHYL-1-PHENYL-; (112549-07-2), **70**, 272

Phenylthioacetic acid morpholide: Morpholine, 4-(2-phenyl-1-thioxoethyl)-; (949-01-9), **74**, 258

PHENYLTHIOACETYLENE: SULFIDE, ETHYNYL PHENYL; BENZENE, (ETHYNYLTHIO)-; (6228-98-4), **72**, 252

Phenylthio alkynes, preparation of from phenylthioacetylene, **72**, 259

1-Phenylthio-2-bromoethane: Benzene, [(2-bromoethyl)thio]-; (4837-01-8), **74**, 124

Phenylthio-1,2-dibromoethane, **72**, 252

1-Phenyl-1-(trimethylsiloxy)ethylene: Silane, trimethyl[(1-phenylethenyl)oxy]-; (13735-81-4), **70**, 82

PHENYL VINYL SULFIDE: BENZENE, (ETHENYLTHIO)-; (1822-73-7), **71**, 89; **72**, 252; **74**, 124

Phosgene: Carbonic dichloride; (75-44-5), **74**, 51

PHOSPHO ESTERS, FROM AMINES, **72**, 246

Phosphomolybdic acid, **70**, 22

Phosphonic aldehydes, **72**, 244

Phosphonocarboxylates, **73**, 156

Phosphoric acid; (7664-38-2), **74**, 24

Phosphorus chloride or phosphorus pentachloride: Phosphorane, pentachloro-; (10026-13-8), **71**, 49

Phosphorus oxide: Phosphorus pentoxide; (1314-56-3), **71**, 161; **73**, 95, 99, 108

Phosphorus oxychloride: HIGHLY TOXIC. Phosphoryl chloride; (10025087-3), **73**, 262, 265, 267, 269

Phthalides, **72**, 168

3-(N-Phthalimido)propyl(cyano)(iodozinc)cuprate, **70**, 196

Phthaloylglycine: 2-Isoindolineacetic acid, 1,3-dioxo-; 2H-Isoindole-2-acetic acid, 1,3-dihydro-1,3-dioxo-; (4702-13-0), **72**, 15

Phthaloylglycyl chloride: Isoindolineacetyl chloride, 1,3-dioxo-; 2H-Isoindole-2-acetyl chloride, 1,3-dihydro-1,3-dioxo-; (6780-38-7), **72**, 15

(+)-α-Pinene: (1R)-(+)-α-Pinene: 2-Pinene, (1R,5R)-(+)-; Bicyclo[3.1.1]hept-2-ene, 2,6,6-trimethyl-, (1R)-; (7785-70-8), **72**, 57

(-)-α-Pinene: Bicyclo[3.1.1]hept-2-ene, 2,6,6-trimethyl-, (1S)-; (7785-26-4), **73**, 117, 118, 122

Piperidine; (110-89-4), **71**, 168

Pirkle Type 1-A column, **70**, 64

Pivalaldehyde: Propanal, 2,2-dimethyl-; (630-19-3), **72**, 63; **73**, 202, 205, 213

Platinum catalyst solution, **73**, 97, 100

Polyaza-crown compounds, N-alkyl-substituted, **70**, 135

Polyaza-crowns, **70**, 135, 136

POLY-[(R)-3-HYDROXYBUTYRIC ACID] = PHB: BUTYRIC ACID, 3-HYDROXY-, D-(-)-, POLYESTERS; BUTANOIC ACID, 3-HYDROXY-, (R)-, HOMOPOLYMER; (29435-48-1), **71**, 39

POLYQUINANES, **74**, 169

Poly(triphenylmethyl)methacrylate on silica gel (Chiralpak OT), **70**, 64

Polyvalent iodine compounds, **70**, 222

Porphyrins, synthetic, **70**, 74

Potassium, **73**, 52, 55, 175, 183

Potassium tert-butoxide; (865-47-4), **74**, 34, 38, 72

Potassium cyanide; (151-50-8), **70**, 55; **74**, 218

Potassium dihydrogen phosphate: Phosphoric acid, monopotassium salt; (7778-77-0), **71**, 182

Potassium ferricyanide: Ferrate (3-), hexacyano-, tripotassium; Ferrate (3-), hexakis(cyano-C)-, tripotassium, (OC-6-11); (13746-66-2), **73**, 3, 9, 12

Potassium fluoride; (7789-23-3), **71**, 126; **73**, 98, 102

Potassium hydride; (7693-26-7), **74**, 14

Potassium iodide-starch test paper, **70**, 267

Potassium osmate(VI) dihydrate: Osmic acid, dipotassium salt; (19718-36-3), **73**, 3, 12

Potassium permanganate: Permanganic acid, potassium salt; (7722-64-7), **70**, 55; **71**, 208

Potassium O-(2-propyl) dithiocarbonate: Carbonic acid, dithio-, O-isopropyl ester, potassium salt; Carbonodithioic acid, O-(1-methylethyl) ester, potassium salt; (140-92-1), **72**, 267

Potassium sodium tartrate: Tartaric acid, monopotassium monosodium salt, tetrahydrate, L-(+)-; Butanedioic acid, 2,3-dihydroxy-, [R-(R*,R*)]-, monopotassium monosodium salt, tetrahydrate; (6381-59-5), **74**, 196

L-Proline: (S)-(-)-Proline; (147-85-3), **72**, 63; **74**, 41, 51

(S)-Proline-N-carboxyanhydride: 1H,3H-Pyrrolo[1,2-c]oxazole-1,3-dione, tetrahydro-, (S)-; (45736-33-2), **74**, 51

Propargyl alcohol: 2-Propyn-1-ol; (107-19-7), **74**, 116

2-Propenyllithium, **74**, 169

Propionaldehyde: Propanal; (123-38-6), **70**, 69

Propionyl chloride: Propanoyl chloride; (79-03-8), **70**, 226

Protected 1,2-diamines, **73**, 197

Protecting groups, for indoles, **74**, 252

Protection, nitrogen, **71**, 200, 226

Protective group, **73**, 91

(R)-(+)-PULEGONE: p-MENTH-4(8)-EN-3-ONE, (R)-(+)-; CYCLOHEXANONE, 5-METHYL-2-(1-METHYLETHYLIDENE)-, (R)-; (89-82-7), **72**, 135

Pummerer reaction, **72**, 213

Pyrazole: 1H-Pyrazole; (288-13-1), **70**, 1

Pyridine; (110-86-1), **71**, 98; **74**, 15, 77, 249

Pyridinium chlorochromate: CANCER SUSPECT AGENT. Pyridine, chlorotrioxochromate (1-); (26299-14-9), **73**, 36, 43

Pyridinium dichromate: Dichromic acid, compd. with pyridine (1:2); Chromic acid ($H_2Cr_2O_7$), compd. with pyridine (1:2); (20039-37-6), **72**, 195

Pyridocarbazoles, **72**, 133

N-(2-PYRIDYL)TRIFLIMIDE: METHANESULFONAMIDE, 1,1,1-TRIFLORO-N-2-PYRIDINYL-N-[(TRIFLUOROMETHYL)SULFONYL]-; (145100-50-1), **74**, 77

Pyrrole; (109-97-7), **70**, 151

Pyrrolidine; (123-75-1), **74**, 23

Pyrrolidinemethanol analogs, α,α-diphenyl-, **74**, 65,
α,α-disubstituted, **74**, 30

PYRROLIDINES, **73**, 221, 226
3-hydroxy-4-substituted, **74**, 43

(R)-(+)-2-Pyrrolidone-5-carboxylic acid: D-Proline, 5-oxo-; (4042-36-8), **73**, 14, 18, 23

PYRROLIDINONES, **73**, 221, 226, 227

3-PYRROLINE; 1H-PYRROLE, 2,5-DIHYDRO-; (109-96-6), **73**, 246, 248, 249, 250

PYRROLOBENZODIAZEPINE-5,11-DIONES, **73**, 174, 180

Radex study, **73**, 167

Radical addition of dibromodifluoromethane to alkenes, **72**, 229

Radiometer pH meter, **72**, 2

Raney nickel; (7440-02-0), **72**, 50

Reductive coupling with hexaethylphosphorous triamide, **72**, 232

REDUCTIVE LITHIATION OF EPOXIDES, **72**, 173

REGIOSELECTIVE SYNTHESIS, 3-SUBSTITUTED INDOLES, **74**, 248

Rhodium(II) acetate dimer: Acetic acid, rhodium (2+) salt; (5503-41-3), **70**, 93; **73**, 15, 18, 23; **74**, 230

RHODIUM-CATALYZED HETEROCYCLOADDITION, **74**, 229

Rhodium(III) chloride hydrate: Rhodium chloride, hydrate; (20765-98-4), **73**, 18, 24

Robinson-Schopf condensation, **70**, 56

SCHWARTZ'S REAGENT: CHLOROBIS(η^5-CYCLOPENTADIENYL)-HYDRIDOZIRCONIUM; ZIRCONIUM, CHLORODI-π-CYCLOPENTADIENYLHYDRO-; ZIRCONIUM, CHLOROBIS(η^5-2,4-CYCLOPENTADIEN-1-YL)HYDRO-; BIS(CYCLOPENTADIENYL)ZIRCONIUM CHLORIDE HYDRIDE; (37342-97-5), **71**, 77, 83; **74**, 206

Selective rearrangement of epoxides, **72**, 99

Selenium dioxide: Selenium oxide; (7446-08-4), **71**, 182; **73**, 160, 164, 172

Selenophenes, **70**, 276

Self-regeneration of a stereogenic center, **72**, 68

Serine: L-Serine; (56-45-1), **70**, 6, 18

Silicon chloride, **70**, 164

Siloles, **70**, 276

Siloxycyclopropanes, **71**, 189

Siloxy dienes, **70**, 237

Silver nitrate; (7761-88-8), **74**, 212

Silyl enol ethers, **70**, 162

α-Silyl ketones, preparations, **73**, 57, 58

SILYLOLEFINATION, **70**, 240

Simmons-Smith reaction, **70**, 186

Sodium; (7440-23-5), **70**, 177; **72**, 253
shavings, **73**, 271, 273

Sodium amide; (7782-92-5), **70**, 256; **72**, 253

Sodium azide; (26628-22-8), **70**, 94; **73**, 26, 28, 35, 145, 147, 149, 151

Sodium-benzophenone ketyl, **70**, 23

Sodium borohydride: Borate(1-), tetrahydro-, sodium; (16940-66-2), **72**, 24, 165, 203, 226; **73**, 204, 214, 224, 230

Sodium borohydride reduction of α-bromosilanes, **72**, 229

Sodium cyclopentadienyldicarbonylferrate: Ferrate(1-), dicarbonyl (η^5-2,4-cyclopentadien-1-yl-, sodium; (12152-20-4), **70**, 177, 181

Sodium dichromate dihydrate: Chromic acid, disodium salt, dihydrate; (7789-12-0), HIGHLY TOXIC. CANCER SUSPECT AGENT, **72**, 58; **73**, 220

Sodium ethoxide, **73**, 186, 190, 195, 275

Sodium fluoride, **72**, 96

Sodium hydride; (7646-69-7), **70**, 231; **71**, 161; **72**, 23, 127, 161; **74**, 4, 139, 149

Sodium iodide; (7681-82-5), **70**, 112, 130, 195; **71**, 189; **74**, 37, 194

Sodium metabisulfate; (7681-57-4), **70**, 48; **71**, 209

25 wt % Sodium methoxide in methanol, **70**, 38; **74**, 117

SODIUM PERBORATE TETRAHYDRATE: PERBORIC ACID SODIUM SALT, TETRAHYDRATE; (10486-00-7), **73**, 116, 117, 118, 119, 120

Sodium periodate or sodium metaperiodate: Periodic acid, sodium salt; (7790-28-5), **70**, 30; **72**, 1, 7; **74**, 3

Sodium sulfite: Sulfurous acid, disodium salt; (7757-83-7), **72**, 96

Sodium taurocholate, **70**, 61

Sodium tetrafluoroborate: Borate (1-), tetrafluoro-, sodium; (13755-29-8), **70**, 178

Sodium thiosulfate pentahydrate solution, **73**, 98, 102

Sodium tungstate dihydrate: Tungstic acid, disodium salt, dihydrate; (10213-10-2), **70**, 265

Sonication, **71**, 118

(-)-Sparteine: 7,14-Methano-2H,6H-dipyrido[1,2-a:1',2'-e][1,5]diazocine, dodecahydro-,[7S-(7α,7aα,14α,14aβ)]-; (90-39-1), **74**, 23
recovery, **74**, 24

(-)-Sparteine sulfate pentahydrate: Sparteine, sulfate(1:1), pentahydrate; (6160-12-9), **74**, 26

D-erythro-Sphingosine, **70**, 27

Spiroacetals, **70**, 160

SPIROANNELATION, **70**, 204
OF ENOL SILANES, **72**, 189

Spirobicyclic compound synthesis, **72**, 197

2,2-Spirocyclohexane-4,5-diphenyl-2H-imidazole: 1,4-Diazaspiro[4.5]deca-1,3-diene, 2,3-diphenyl-; (5396-98-5), **71**, 22

SQUARATE ESTERS, **74**, 169

Stannoles, **70**, 276

Stannous chloride: Tin chloride; (7772-99-8), **72**, 6

Starch-iodide test, **73**, 98, 102

Staudinger reaction, **72**, 19

STEREOSELECTIVE ALDOL REACTION, **72**, 38

STEREOSELECTIVE ALKENE SYNTHESIS, **73**, 50, 57, 69

Stereospecific addition to acetylenic esters, **72**, 112

STEREOSPECIFIC COUPLING WITH VINYLCOPPER REAGENTS, **70**, 215

Stereospecific synthesis, **72**, 112

Stilbene, (E)-: (E)-1,2-Diphenylethene: Benzene, 1,1'-(1,2-ethenediyl)bis-, (E)-; (103-30-0), **70**, 47; **74**, 92

trans-STILBENE OXIDE: OXIRANE, 2,3-DIPHENYL-, trans-; (1439-07-2), **72**, 95; **74**, 91

Stiboles, **70**, 276

2-Substituted dihydropyrans, **70**, 160

7-SUBSTITUTED INDOLINES, **73**, 85, 90

2-Substituted tetrahydrofurans, **70**, 160

Succinic acid: Butanedioic acid; (110-15-6), **70**, 226

Sulfonyloxaziridines, oxidizing reagents, **73**, 168

SULFOXIDES, REACTIONS WITH DIETHYLAMINOSULFUR TRIFLUORIDE, **72**, 209

Sulfur; (7704-34-9), **74**, 258

Sulfur dichloride: Sulfur chloride; (10545-99-0), **70**, 121

Swern oxidation, in presence of sulfur moiety, **70**, 126

D-Tartaric acid: Tartaric acid, D-(-)-; Butanedioic acid, 2,3-dihydroxy-, [S-(R*,R*)]-; (147-71-7), **71**, 24

L-Tartaric acid: Tartaric acid, L-(-)-; Butanedioic acid, 2,3-dihydroxy-, [R-(R*,R*)]-; (87-69-4), **71**, 24; **72**, 87

Tellurium; (13494-80-9), **72**, 154

TERMINAL ACETYLENES, **73**, 262

Tetrabutylammonium fluoride: Ammonium, tetrabutyl-, fluoride; 1-Butanaminium, N,N,N-tributyl-, fluoride; (429-41-4), **71**, 64; **72**, 23, 234; **74**, 250

Tetrabutylammonium hydrogen sulfate: Ammonium, tetrabutyl-, sulfate (1:1); 1-Butanaminium, N,N,N-tributyl-, sulfate (1:1); (32503-27-8), **72**, 253

Tetrabutylammonium iodide: 1-Butanaminium, N,N,N-tributyl-, iodide; (311-28-4), **72**, 24

Tetrachlorosilane: Silane, tetrachloro-; (10026-04-7), **70**, 164

Tetraethylammonium bis(1,3-dithiole-2-thione-4,5-dithiol) zincate: Ethanaminium, N,N,N-triethyl-, (T-4)-bis[4,5-dimercapto-1,3-dithiole-2-thionato)-S^4,S^5]zincate (2-) (2:1); (72022-68-5), **73**, 270, 276

Tetraethylammonium bromide: Ammonium, tetraethyl-, bromide; Ethanaminium, N,N,N-trimethyl, bromide; (71-91-0), **73**, 271, 277

Tetraethyl orthotitanate: Ethyl alcohol, titanium(4+) salt; Ethanol, titanium(4+) salt; (3087-36-3), **71**, 203

Tetrahydrobenzazepine, **70**, 145

TETRAHYDRO-3-BENZAZEPIN-2-ONES, **70**, 139

4,5,6,6a-TETRAHYDRO-3a-HYDROXY-2,3-DIISOPROPOXY-4,6a-DIMETHYL-1(3H)-PENTALENONE, **74**, 169

(S)-Tetrahydro-1-methyl-3,3-diphenyl-1H,3H-pyrrolo[1,2-c][1,3,2]oxazaborole: 1H,3H-Pyrrolo[1,2-c][1,3,2]oxazaborole, tetrahydro-1-methyl-3,3-diphenyl-, (S)-; (112068-01-6), **74**, 41, 53

(S)-TETRAHYDRO-1-METHYL-3,3-DIPHENYL-1H,3H-PYRROLO[1,2-c]-[1,3,2]OXAZABOROLE-BORANE COMPLEX: BORON, TRIHYDRO-(TETRAHYDRO-1-METHYL-3,3-DIPHENYL-1H,3H-PYRROLO [1,2-c][1,3,2]OXAZABOROLE-N^7)-, [I-4-(3aS-cis)]-, **74**, 42, 50

Tetrahydro-2-(4-pentynyloxy)-2H-pyran: 2H-Pyran, tetrahydro-2-(4-pentynyloxy)-; (62992-46-5), **70**, 113

TETRAHYDRO-2-(PHENYLETHYNYL)-2H-PYRAN: 2H-PYRAN, TETRAHYDRO-2-(PHENYLETHYNYL)-; (70141-82-1), **70**, 157

Tetrahydropyranyl ethers, **70**, 160

Tetrakis(triphenylphosphine)palladium(0): Palladium, tetrakis(triphenylphosphine)-; Palladium, tetrakis(triphenylphosphine)-, (T-4)-; (14221-01-3), **71**, 90; **72**, 90, 105; **73**, 207, 214

Tetra[3-(2-methoxyethoxy)propyl]stannane: 2,5,13,16-Tetraoxa-9-stanna–heptadecane, 9,9-bis[3-(2-methoxyethoxy)propyl]-; (130691-01-9), **72**, 201

N,N,N',N'-Tetramethylethylenediamine: Ethylenediamine, N,N,N',N'-tetramethyl-; 1,2-Ethanediamine, N,N,N',N'-tetramethyl-; (110-18-9), **72**, 164; **73**, 75, 77, 81, 83, 84, 86, 88, 93

2,2,6,6-Tetramethylpiperidine: Piperidine, 2,2,6,6-tetramethyl-; (768-66-1), **71**, 146

N,N,N',N'-Tetramethylurea: Urea, tetramethyl- (8,9); (632-22-4), **74**, 102

Tetrathiafulvalene derivatives, **72**, 265

Thermal decomposition of N-nitrosoamides to esters, **72**, 249

2-Thiaadamantane, **70**, 126

(endo,endo)-9-Thiabicyclo[3.3.1]nonane-2,6-diol: 9-Thiabicyclo[3.3.1]nonane-2,6-diol, (endo,endo)-; (22333-35-3), **70**, 121

9-THIABICYCLO[3.3.1]NONANE-2,6-DIONE; (37918-35-7), **70**, 120

2-Thiabrexane, **70**, 126

Thiacycloheptane, **70**, 126

Thiacyclohexane, **70**, 126

Thiazolyl-to-formyl deblocking, **72**, 24

Thioanisole (Phenyl methyl sulfide); (100-68-5), **74**, 93

THIOL ESTERS, **73**, 61, 67

Thionyl chloride; (7719-09-7), **70**, 80, 130; **71**, 50, 108; **72**, 15, 127, 164; **73**,14, 23, 144, 146, 151

Thiophenes, **70**, 276

Thiophenol: Benzenethiol; (108-98-5), **72**, 254; **73**, 63, 72

Thorpe-Ziegler cyclization, **74**, 45

Thymine polyoxin, **70**, 27

Tin(II) chloride: Tin chloride; (7772-99-8), **72**, 6

Tin(IV) chloride: Tin chloride; Stannane, tetrachloro-; (7646-78-8), **71**, 65; **72**, 202

Tin hydride reagent in free radical chemistry, **72**, 206

TIN HYDRIDE, WATER SOLUBLE, **72**, 199

Tischenko product, **72**, 98

Titanium(IV) bromide: Titanium bromide; (7789-68-6), **71**, 14

Titanium(IV) chloride: Titanium tetrachloride; Titanium chloride; HIGHLY TOXIC. (7550-45-0), **71**, 203; **72**, 190; **73**, 75, 77, 82, 83, 84, 222, 225, 230

Titanium(IV) ethoxide: Ethyl alcohol, titanium(4+) salt; Ethanol, titanium(4+) salt; (3087-36-3), **71**, 203

Titanium(IV) isopropoxide: Isopropyl alcohol, titanium (4+) salt; 2-Propanol, titanium (4+) salt; (546-68-9), **71**, 14

Titanium-mediated additions, **71**, 200

Toluene: Benzene, methyl-; (108-88-3), **71**, 160, 221

p-Toluenesulfonic acid monohydrate: Benzenesulfonic acid, 4-methyl-, monohydrate; (6192-52-5), **70**, 13, 19, 71, 113, 217, 233; **71**, 40, 64, 160; **73**, 2, 12, 53, 60; **74**, 2, 258

p-Toluenesulfonyl azide, **70**, 97; **73**, 150

p-Toluenesulfonyl chloride; Benzenesulfonyl chloride, 4-methyl-; (98-59-9), **73**, 176, 178, 183

2-Toluic acid: o-Toluic acid; Benzoic acid, 2-methyl-; (118-90-1), **72**, 119

2-Toluic acid chloride: o-Toluoyl chloride; Benzoyl chloride, 2-methyl-; (933-88-0), **72**, 117

(1S,2S)-2-(N-Tosylamino)-1-[((2S)-2-carbomethoxypyrrolidinyl)carbonyl]cyclohexane: L-Proline, 1-[[2-[[(4-methylphenyl)sulfonyl]amino]cyclohexyl]carbonyl]-, methyl ester, (1S-trans)-; (110419-91-5), **73**, 176, 183

(1S,2S)-2-(N-Tosylamino)cyclohexanecarboxylic acid: Cyclohexanecarboxylic acid, 2-[[(4-methylphenyl)sulfonyl]amino]-, (1S-trans)-; (110456-11-6), **73**, 177, 183

Transacetalization, **74**, 132

Transition metal carbene complexes, **70**, 186

Transmetalation, alkenylzirconocenes to organozincs, **74**, 209

Tri-O-acetyl-D-xylono-1,4-lactone: D-Xylonic acid, γ-lactone, triacetate; (79580-60-2), **72**, 49

Triaza-21-crown-7, **70**, 132

(E)-Tributyl[fluoro(2-phenylcyclohexylidene)methyl]stannane: Stannane, tributyl[fluoro(2-phenylcyclohexylidene)methyl]-, (E)-(±)-; (135789-96-7), **72**, 218

TRIBUTYL[(METHOXYMETHOXY)METHYL]STANNANE: STANNANE, TRIBUTYL[(METHOXYMETHOXY)METHYL]-; (100045-83-8), **71**, 133, 140

Tributyl(4-methoxyphenyl)stannane: Stannane, tributyl(4-methoxyphenyl)-; (70744-47-7), **71**, 98

Tributyl(3-methyl-2-butenyl)tin: Stannane, tributyl(3-methyl-2-butenyl)-; (53911-92-5), **71**, 118, 125

(Tributylstannyl)methanol: Methanol, (tributylstannyl)-; (27490-33-1), **71**, 133

Tributyltin chloride: Stannane, tributylchloro-; (1461-22-9), **71**, 99, 118, 126

Tributyltin hydride: Stannane, tributyl-; (688-73-3), **70**, 166; **71**, 133; **72**, 206, 218

trans-4,7,7-TRICARBOMETHOXY-2-PHENYLSULFONYLBICYCLO[3.3.0]OCT-1-ENE, **74**, 147

Trichloroethylene: Ethene, trichloro-; (79-01-6), **74**, 14

Trichloromethyl chloroformate: Carbonochloridic acid, trichloromethyl ester; (503-38-8), **70**, 72

endo-Tricyclo[6.2.1.0^{2,7}]undeca-4,9-diene-3,6-dione: 1,4-Methanonaphthalene-5,8-dione, 1,4,4a,8a-tetrahydro-; (1200-89-1), **73**, 253, 261

endo-Tricyclo[6.2.1.0^{2,7}]undec-9-ene-3,6-dione: 1,4-Methanonaphthalene-5,8-dione, 1,4,4a,6,7,8a-hexahydro-; (21428-54-6), **73**, 254, 261

Triethylamine: Ethanamine, N,N-diethyl-; (121-44-8), **70**, 61, 94, 103, 111; **71**, 31, 32, 72, 159, 182, 189, 227; **74**, 51, 104, 116, 148, 229, 259

Triethylamine hydrofluoride: Ethanamine, N,N-diethyl-, hydrofluoride; (29585-72-6), **71**, 191, 193

Triethylamine trishydrofluoride: Ethanamine, N,N-diethyl-, trihydrofluoride; (73602-61-6), **71**, 193

Triethyl orthoformate: Ethane, 1,1',1''-[methylidynetris(oxy)]tris-; (122-51-0), **74**, 130, 258

Triethyl phosphate: Phosphoric acid, triethyl ester; (78-40-0), **72**, 241

Triethyl phosphonoacetate: Acetic acid, (diethoxyphosphinyl)-, ethyl ester; (867-13-0), **74**, 4

Triethylsilane: Silane, triethyl-; (617-86-7), **73**, 222, 225, 230

TRIFLATING AGENTS, **74**, 77

Trifluoroacetic acid: Acetic acid, trifluoro-; (76-05-1), **70**, 11, 141; **72**, 63

Trifluoroacetic anhydride: Acetic acid, trifluoro-, anhydride; (407-25-0), **70**, 246

Trifluoroethanol: Ethanol, 2,2,2-trifluoro-; (75-89-8), **73**, 153, 154, 157

2,2,2-Trifluoroethyl trifluoroacetate: Acetic acid, trifluoro-, 2,2,2-trifluoroethyl ester; (407-38-5), **73**, 135, 136, 143

Trifluoromethanesulfonic acid: Methanesulfonic acid, trifluoro-; (1493-13-6), **72**, 180; **74**, 241

Trifluoromethanesulfonic anhydride: Methanesulfonic acid, trifluoro-, anhydride; (358-23-6), **71**, 31, 98; **74**, 78

Trifluoromethide equivalent, **72**, 238

1-TRIFLUORMETHYL-1-CYCLOHEXANOL: CYCLOHEXANOL, 1-(TRIFLUOROMETHYL)-; (80768-55-4), **72**, 232

(Trifluoromethyl)trimethylsilane: Silane, trimethyl(trifluoromethyl)-; (81290-20-2), **72**, 232

(2S,4S)-2,4,5-TRIHYDROXYPENTANOIC ACID 4,5-ACETONIDE METHYL ESTER: D-erythro-PENTONIC ACID, 3-DEOXY-4,5-O-(1-METHYLETHYLIDENE)-, METHYL ESTER; (134455-80-4), **72**, 48

2,4,5-Trihydroxypentanoic acid derivatives, **72**, 55

Trimethylaluminum: Aluminum, trimethyl-; (75-24-1), **71**, 32; **72**, 96

Trimethyl borate: Boric acid, trimethyl ester; (121-43-7), **74**, 158

Trimethylboroxine: Boroxin, trimethyl-; (823-96-1), **74**, 53

2,2,6-Trimethylcyclohexanone: Cyclohexanone, 2,2,6-trimethyl-; (2408-37-9), **74**, 104

2,2,6-Trimethylcyclohexanone hydrazone, **74**, 102

(1R)-1,3,4-TRIMETHYL-3-CYCLOHEXENE-1-CARBOXALDEHYDE: 3-CYCLOHEXENE-1-CARBOXALDEHYDE, 1,3,4-TRIMETHYL-, (-)-; (130881-20-8), **72**, 86

2,2,6-TRIMETHYLCYCLOHEXEN-1-YL IODIDE, **74**, 101

(3R,7R)-3,7,11-Trimethyldodecanol, **72**, 83

Trimethyl orthoformate: Orthoformic acid, trimethyl ester; Methane, trimethoxy-; (149-73-5), **72**, 190; **73**, 2, 5, 12, 160, 173

Trimethyloxonium tetrafluoroborate: Oxonium, trimethyl-, tetrafluoroborate(1-); (420-37-1), **72**, 29

(E,E)-TRIMETHYL(4-PHENYL-1,3-BUTADIENYL)SILANE: SILANE, TRIMETHYL(4-PHENYL-1,3-BUTADIENYL)-, (E,E)-; (70960-88-2), **70**, 240

Trimethyl phosphite: Phosphorous acid, trimethyl ester; (121-45-9), **72**, 268

2-Trimethylsiloxybutadiene: Silane, trimethyl[(1-methylene-2-propenyl)oxy]-; (38053-91-7), **71**, 201

[(1-Trimethylsiloxy)methylene]cyclohexane: Silane, (cyclohexylidenemethoxy)-trimethyl-; (53282-55-6), **72**, 191

Trimethylsilylacetylene: Silane, ethynyltrimethyl-; (1066-54-2), **72**, 105

Trimethylsilyl chloride: Silane, chlorotrimethyl-; (75-77-4), **70**, 152, 164, 196, 216; **71**, 57, 159, 190, 226, 227; **72**, 22, 191, 232; **73**, 132

1-Trimethylsilyl-1-hexyne: Silane, 1-hexynyltrimethyl-; (3844-94-8), **70**, 215

(Trimethylsilyl)methylmagnesium chloride: Magnesium, chloro[(trimethylsilyl)methyl]-; (13170-43-9), **70**, 241

2-(Trimethylsilyl)thiazole, **72**, 28

2-(TRIMETHYLSILYL)THIAZOLE: THIAZOLE, 2-(TRIMETHYLSILYL)-; (79265-30-8), **72**, 21

Trimethylsilyl trifluoromethanesulfonate: Methanesulfonic acid, trifluoro-, trimethylsilyl ester; (27607-77-8), **70**, 101, 234; **72**, 190

1-Trimethylstannyl-3,3-dimethoxypropane: Stannane, (3,3-dimethoxypropyl)-trimethyl-; (102402-80-2), **72**, 190

Triphenylborane, **73**, 121

1,2,3-TRIPHENYLCYCLOPROPENIUM BROMIDE: CYCLOPROPENYLIUM, TRIPHENYL-, BROMIDE; (4919-51-5), **74**, 72

1,2,3-Triphenylcyclopropenyl tert-butyl ether, **74**, 73

(R)-(+)-1,1,2-Triphenylethanediol: 1,2-Ethanediol, 1,1,2-triphenyl-, (R)-; (95061-46-4), **72**, 33, 41

Triphenylphosphine: Phosphine, triphenyl-; (603-35-0), **70**, 1, 10, 248; **71**, 90; **73**, 51, 59, 110, 113, 115

Triphosgene: Carbonic acid, bis(trichloromethyl) ester; (32315-10-9), **74**, 101

Tripotassium phosphate: Phosphoric acid, tripotassium salt; (7778-53-2), **71**, 226

Tris[2-(2-methoxyethoxy)ethyl]amine: TDA-1: Ethanamine, 2-(2-methoxyethoxy)-N,N-bis[2-(2-methoxyethoxy)ethyl]-; (70384-51-9), **73**, 95, 100, 108

TRIS[3-(2-METHOXYETHOXY)PROPYL]STANNANE: 2,5,13,16-TETRAOXA-9-STANNAHEPTADECANE, 9-[3-(2-METHOXYETHOXY)PROPYL]-; (130691-03-1), **72**, 199

TRIS(TRIMETHYLSILYL)SILANE: Trisilane, 1,1,1,3,3,3-hexamethyl-2-(trimethylsilyl)-; (1873-77-4), **70**, 164

1,2,3-Trisubstituted benzene derivatives, **72**, 168

TROPONE; (539-80-0), **71**, 181

UBIQUINONE-1: p-BENZOQUINONE, 2,3-DIMETHOXY-5-METHYL-6-(3-METHYL-2-BUTENYL)-; 2,5-CYCLOHEXADIENE-1,4-DIONE, 2,3-DIMETHOXY-5-METHYL-6-(3-METHYL-2-BUTENYL)-; (727-81-1), **71**, 125

Ultrasonic processor, **71**, 118

Vacuum pyrolysis apparatus, **72**, 123

(1R,5R)-(+)-VERBENONE: BICYCLO[3.1.1]HEPT-3-EN-2-ONE-, 4,6,6-TRIMETHYL-, (1R-CIS)-; (18309-32-5), **72**, 57

Vilsmeier salt, **74**, 103

4-Vinyl-1-cyclohexene: Cyclohexene, 4-vinyl-; Cyclohexene, 4-ethenyl-;
(100-40-3), **71**, 90

3-Vinyl-1,5-dihydro-3H-2,4-benzodioxepine: 2,4-Benzodioxepin, 3-ethyl-1,5-dihydro-;
(142169-23-1), **73**, 2, 3, 6, 12

L-Vinylglycine hydrochloride: 3-Butenoic acid, 2-amino-, hydrochloride, (S)-;
(75266-38-5), **70**, 30
Vinyl iodides, synthesis, **74**, 105

Vinylsiloxane/platinum(0) complex, **73**, 105

Vinyl triflates,**74**, 79
from acetylenes, **72**, 180

Vinyltrimethylsilane: Silane, trimethylvinyl-; Silane, ethenyltrimethyl-; (754-05-2),
72, 225

VINYLZIRCONIUM REAGENTS, CONJUGATE ADDITION OF, **71**, 83

"Walk-in" hood, **74**, 86

Wheat germ lipase: Lipase, triacylglycerol; (9001-62-1), **73**, 37, 39, 43

Wittig-Horner-Emmons reagents, **72**, 241

D-Xylose; (58-86-6), **72**, 49

Ynolate anion, **71**, 146

Zinc; (7440-66-6), **70**, 195; **73**, 75, 77, 80, 81, 82, 83, 84; **74**, 242
activation, **74**, 161
alkenylalkyl reagents, **74**, 211

Zinc bromide; (7699-45-8), **70**, 55, 157

Zinc chloride; (7476-85-7), **73**, 271, 273, 277; **74**, 130

Zinc-copper couple, **70**, 186

Zinc dust, **70**, 85; **73**, 75, 77, 80, 254, 257

Zirconium(IV), organometallic derivatives, **74**, 209

Zirconium, bis(η^5-2,4-cyclopentadien-1-yl)(1,2,3,4-tetramethyl-1,3-butadiene-1,4-diyl)-; (84101-39-3), **70**, 272

Zirconium metallacycle, **70**, 274

Zirconocene dichloride: Bis(cyclopentadienyl)zirconium dichloride; Zirconium, dichloro-π-cyclopentadienyl-; Zirconium, dichlorobis(η^5-2,4-cyclopentadien-1-yl)-; (1291-32-3), **70**, 272; **71**, 77

Zirconocene hydrochloride (Schwartz's Reagent): Zirconium, chlorobis(η^5-2,4-cyclopentadien-1-yl)hydro-; (37342-97-5), **74**, 206